Bacterial Metagenomics

Sustainable Water and Wastewater Treatment

Edited by
Maulin P. Shah

Volume 6

Bacterial Metagenomics

In Industrial Wastewater Treatment

Edited by
Maulin P. Shah and Sumira Malik

DE GRUYTER

Editors
Dr. Maulin P. Shah
Industrial Waste Water Research Lab
Enviro Technology Limited
Opp. Champapuri Jain Mandir
A/103 Satsang Park
Ankleshwar 393002
Gujarat
India

Dr. Sumira Malik
Amity Institute of Biotechnology
Amity University Jharkhand
Ranchi 834001
Jharkhand
India

ISBN 978-3-11-079355-0
e-ISBN (PDF) 978-3-11-079363-5
e-ISBN (EPUB) 978-3-11-079368-0
ISSN 2747-4208

Library of Congress Control Number: 2024951410

Bibliographic information published by the Deutsche Nationalbibliothek
The Deutsche Nationalbibliothek lists this publication in the Deutsche Nationalbibliografie; detailed bibliographic data are available on the Internet at http://dnb.dnb.de.

Cover image: fatido/iStock/Getty Images Plus
Typesetting: Integra Software Services Pvt. Ltd.
Printing and binding: CPI books GmbH, Leck

www.degruyterbrill.com
Questions about General Product Safety Regulation:
productsafety@degruyterbrill.com

MIX
Papier | Fördert gute Waldnutzung
FSC® C083411

Contents

List of contributing authors

Nairouz H. Mahmoud
Botany and Microbiology Department
Faculty of Science
Alexandria University
Alexandria, Egypt

Nourhan A. Abozahra
Faculty of Aquatic and Fisheries Sciences
Kafrelsheikh University
Kafrelsheikh, Egypt

Bahaa Abdella
Faculty of Aquatic and Fisheries Sciences
Kafrelsheikh University
Kafrelsheikh, Egypt

Susmita Paul
Royal School of Biosciences
The Assam Royal Global University
Guwahati 781035
Assam, India
And
Microbiology Laboratory
Department of Biotechnology and Bioinformatics
North-Eastern Hill University
Shillong 793022
Meghalaya, India

Trishnamoni Gautom
Royal School of Biosciences
The Assam Royal Global University
Guwahati 781035
Assam, India

Santa Ram Joshi
Microbiology Laboratory
Department of Biotechnology and Bioinformatics
North-Eastern Hill University
Shillong 793022
Meghalaya, India

Hiren K. Patel
School of Science
P. P. Savani University
Surat 394125
Gujarat, India

Divyesh K. Vasava
R&D Department
Organic Pesticides Industries
Ankleshwar 393002
Gujarat, India

Gaurav Yadav
Natural Product Laboratory
Department of Bioengineering
Integral University
Lucknow
Uttar Pradesh, India

Surati Kumari
School of Bio Sciences and Technology
Vellore Institute of Technology
Vellore
Tamil Nadu, India

Aneesha Polisety
Department of Industrial Engineering
University of Rome Tor Vergata
Rome, Italy

Priyanka Prajapati
Centre for Advanced Studies in Botany
Department of Botany
Institute of Science
Banaras Hindu University
Varanasi
Uttar Pradesh, India

Devendra Singh
Department of Biotechnology
Motilal Nehru National Institute of Technology
Allahabad
Allahabad
Uttar Pradesh, India

Sam Joy
National Institute of Malaria Research
Indian Council of Medical Research
New Delhi, India

https://doi.org/10.1515/9783110793635-203

Iffat Zareen Ahmad
Natural Product Laboratory
Department of Bioengineering
Integral University
Lucknow
Uttar Pradesh, India

Shreya Anand
Department of Biotechnology and Bioengineering
Birla Institute of Technology
Mesra 835215
Jharkhand, India

Abhimanyu Kumar
Department of Biotechnology and Bioengineering
Birla Institute of Technology
Mesra 835215
Jharkhand, India

Vishakha Chaturvedi
Department of Biotechnology and Bioengineering
Birla Institute of Technology
Mesra 835215
Jharkhand, India

Padmini Padmanabhan
Department of Biotechnology and Bioengineering
Birla Institute of Technology
Mesra 835215
Jharkhand, India

Sougata Ghosh
Department of Physics
Faculty of Science
Kasetsart University
Bangkok, Thailand
And
Department of Microbiology
School of Science
R. K. University
Rajkot
Gujarat, India

Bishwarup Sarkar
College of Science
Northeastern University
Boston, MA, USA

Sirikanjana Thongmee
Department of Physics
Faculty of Science
Kasetsart University
Bangkok, Thailand

Raturi Manisha Hemwhati Nandan
Department of Biochemistry and Biotechnology
Sardar Bhagwan Singh University
Balawala, Dehradun 248001
Uttarakhand, India

Vijaylakshmi
Department of Biochemistry and Biotechnology
Sardar Bhagwan Singh University
Balawala, Dehradun 248001
Uttarakhand, India

Santosh Kumar Karn
Department of Biochemistry and Biotechnology
Sardar Bhagwan Singh University
Balawala, Dehradun 248001
Uttarakhand, India

Debapriya Maitra
Department of Microbiology
St. Xavier's College (Autonomous)
Kolkata
West Bengal, India

Bedaprana Roy
Department of Microbiology
St. Xavier's College (Autonomous)
Kolkata
West Bengal, India

Debdatta Das
Department of Microbiology
St. Xavier's College (Autonomous)
Kolkata
West Bengal, India

Ananya Datta
Department of Microbiology
St. Xavier's College (Autonomous)
Kolkata
West Bengal, India

Ruchira Pal
Department of Biotechnology
St. Xavier's College (Autonomous)
Kolkata
West Bengal, India

Sudeshna Shyam Choudhury
Department of Microbiology
St. Xavier's College (Autonomous)
Kolkata
West Bengal, India

Arup Kumar Mitra
Department of Microbiology
St. Xavier's College (Autonomous)
Kolkata

Sanjana Ghosh
Department of Microbiology
St. Xavier's College (Autonomous)
Kolkata
West Bengal, India

Bidisha Chatterjee
Department of Microbiology
St. Xavier's College (Autonomous)
Kolkata
West Bengal, India

Debapriya Maitra
Department of Microbiology
St. Xavier's College (Autonomous)
Kolkata
West Bengal, India

Ayan Chandra
Department of Statistics
St.Xavier's College(Autonomous)
Kolkata
West Bengal, India

Rajeshwari Podder
Department of Biotechnology
St. Xavier's College (Autonomous)
Kolkata
West Bengal, India

Tannishtha Das
Department of Biotechnology
St. Xavier's College (Autonomous)
Kolkata
West Bengal, India

Snata Pandey
Department of Microbiology
St. Xavier's College (Autonomous)
Kolkata
West Bengal, India

Jaydip Ghosh
Department of Microbiology
St. Xavier's College (Autonomous)
Kolkata
West Bengal, India

Sophiya Alexander
Amity Institute of Biotechnology
Amity University Maharashtra
Mumbai-Pune Expressway
Bhatan, Panvel
Mumbai 410206
Maharashtra, India

Siddant Ratanpal
Amity Institute of Biotechnology
Amity University Maharashtra
Mumbai-Pune Expressway
Bhatan, Panvel
Mumbai 410206
Maharashtra, India

Nilesh S. Wagh
Amity Institute of Biotechnology
Amity University Maharashtra
Mumbai-Pune Expressway
Bhatan, Panvel
Mumbai 410206
Maharashtra, India

Jaya Lakkakula
Amity Institute of Biotechnology
Amity University Maharashtra
Mumbai-Pune Expressway
Bhatan, Panvel
Mumbai 410206
Maharashtra, India
And
Centre for Computational Biology and
Translational Research
Amity Institute of Biotechnology
Amity University Maharashtra
Mumbai-Pune Expressway
Bhatan, Panvel
Mumbai 410206
Maharashtra, India

Nairouz H. Mahmoud, Nourhan A. Abozahra and Bahaa Abdella*

Chapter 1
Response of microbial communities to environmental changes

Abstract: Microbes are the principal role players in nutrient and carbon cycles in food web interactions. Furthermore, it is important for animal, plant, and human health. Their physiological activities are being employed in wastewater treatment via bioremediation technologies. Different environments harbor different compositions and structures of microbial communities that are extraordinarily complex with thousands of "species," including many with unknown functions. Environmental factor changes, such as temperature fluctuation, nutrient enrichment or depletion, discharge of synthetic chemicals, variation in pH value, and drought, influence the physiological activities of these microbial groups. These changes occur either naturally or through anthropogenic interventions. Copious factors govern the resistance and resilience of the microbial communities to any environmental change, such as prior conditions and whether the environment is stable or fluctuating. The more fluctuating the environment, the more resilient the microbial communities. The microbial communities might respond to the environmental factors through a different mechanism, such as a variation in the physiological response, community composition shift, or evolution. A closer look at how microbial communities' structure and function adapt to the changes in environmental conditions is essential to advance our understanding of their environmental impact.

1.1 Introduction

Microbes are the backbone of all ecosystems on the Earth that drive biochemical cycles. They are taxonomically rich, with the intricate interplay of environmental conditions and ecological interactions determining the composition of their communities (Mahmoud et al., 2022). The makeup of microbial community changes across time and geography (Abdella et al., 2017; Martiny et al., 2006). The mechanisms of adaptation and mitigation of the severe impact of climate change are a necessity to overcome its

***Corresponding author: Bahaa Abdella**, Faculty of Aquatic and Fisheries Sciences, Kafrelsheikh University, Kafrelsheikh, Egypt, e-mail: ba.abdella@gmail.com
Nairouz H. Mahmoud, Botany and Microbiology Department, Faculty of Science, Alexandria University, Alexandria, Egypt
Nourhan A. Abozahra, Faculty of Aquatic and Fisheries Sciences, Kafrelsheikh University, Kafrelsheikh, Egypt

https://doi.org/10.1515/9783110793635-001

effects (Sjøgaard et al., 2018). As a result of increasing the seawater level, many of the coastal lines are flooded with seawater either permanently or on intermittent bases. The introduction of seawater to agricultural soil dramatically affects the microbial communities present. It makes the soil more anaerobic, increases the salt content, and supplies the sulfur compounds. This alteration of the soil conditions will result in a shift in the microbial community structure and function in the response to the contact with seawater (Sjøgaard et al., 2018). The flooded agricultural soils with seawater have shown an increase in the abundance of Proteobacteria and detection of sulfur-oxidizing bacteria (Asano et al., 2013). Moreover, seawater-exposed soil tends to develop more salt-tolerant microbes (Zhang et al., 2014).

Additionally, the demand for increased productivity in the industrial sectors will rise as the world's population grows. This, however, will be accompanied by an increase in industrial waste and by-products in the manufacturing process (Y et al., 1998). As mentioned earlier, the behavior of complex microbial communities is at the heart of modern wastewater treatment technology (Jatowiecki et al., 2016). The improvement of the existing technologies and the innovation of new ones are vital for determining the performance of the entire system as well as the flow and composition of the wastewater, especially in rural areas and developing countries (Massoud et al., 2009). Practically, dynamic microbial populations and diversity are at the heart of successful wastewater treatment (Kapley et al., 2007). The amount and structure of organic compounds in the effluent, the capacity of microbial communities to digest these organic compounds, and the microbial density all influence the efficacy of microbial treatment systems (Jatowiecki et al., 2016). The micropollutant content of wastewater has a significant influence not just inside the wastewater treatment plant but also in the streams and the surrounding environment where the treated water is discharged (Tamminen et al., 2022).

In another explanation, these micropollutants affect the microbial community structure and ecological function. The treatment plants are built to preserve and maintain the maximum possible density of microbial cells capable of meeting the cleanness and purity requirements of treated water (Y et al., 1998). However, this is difficult to be maintained as the microbial populations are extremely sensitive to changes in their environment (Ruprecht et al., 2021). Global changes in environmental stressors such as climate change, including rising seawater temperatures and ocean acidification besides eutrophication, have had serious environmental implications on natural microbial communities, which are connected to a significant reduction in ecosystem productivity (Ruprecht et al., 2021). The composition of these microbial communities is complex and not easy to be examined and most of the participating species are with unknown functions. These microbial communities are governed by natural selection. Thus, understanding how environmental variables control the community structure is critical because the richness and diversity of the microbial community in the treatment system determine the overall performance and efficiency of the system. In other words, more research on the responses of microbial communities to environmental stresses might reveal turning points that could lead to ecological changes. Moreover, understanding the microbial ecology of the different populations in these systems and creating relevant environmental biotechnology methods for forecasting their functional roles and reactions to environmental stress are critical for improving these biological processes. To gain a better understanding of the dynamic response to different environmental stressors, this chapter discusses the structural diversity of the microbial communities responsible for wastewater treatment, aquaculture recirculatory systems and how these microbial communities are formed, and several factors that affect their richness and abundance, in addition to the effect of environmental change on the microbial communities beyond wastewater treatment.

1.2 Impact on environmental factors

Microbial populations control environmental factors such as pH value, temperature, and nutrient availability of the environment in every habitat, while also these communities themselves are being influenced by such parameters. It is a dynamic process that maintains a harmonious balance between resources and living creatures. For example, the presence of sulfur compounds encourages microorganisms that can use sulfur compounds to generate sulfuric acid by oxidation; but, once this acid is formed, the pH decreases, making the environment only suited for acid-tolerant microbes. It is a series of extremely complex interactions that occur regularly. The nature and complexity of the microbial communities that live locally, as well as other environmental factors, have influenced these interactions. These factors can positively or negatively affect the natural geochemical cycles of the major elements such as phosphorus, nitrogen, and carbon cycles (Arayes et al., 2021; Bi et al., 2022; Dong et al., 2021). In a study of the effect of the global warming on the microbial community structure, Dong et al. (2021) have studied the effect of artificial warming on the soil microbial communities in forest-wetland samples from China, and the results showed that the warming was favored by Actinobacteria and Proteobacteria. However, Chloroflexi and Acidobacteria were negatively affected. The warming effect extends to the marine environment and bacteria affiliated with marine life. Cold-water corals are a good example: by raising the temperature to only 2 °C, the coral *Lophelia pertusa* has shown changes in microbial halobiont before changing its behavior, and by raising the temperature to 4 °C above its normal temperature, *L. pertusa* and *Madrepora oculata* have lost their associated microbes before their death (Chapron et al., 2021). Any minor change in the environmental parameters reflects a severe impact on the system and the wastewater treatment plants are no exception; they are severely affected by the change in the operating conditions and environmental parameters.

1.3 Building up microbial community in wastewater treatment

During the start-up operation of a wastewater treatment plant, there is a period in any microbial-based treatment plant when microbial communities capable of degrading and eliminating contaminants from wastewater are being formed. In a study by Yu et al. (2021), the selection and development of microbial communities in the treatment plants were investigated. The results of the specimens taken from 10 activated-sludge and five anaerobic digestion systems demonstrated that the microbial community's richness and complexity are highly dependent on the effluent composition (Yu et al., 2021). Most of the bacteria in the activated-sludge system belonged to Proteobacteria, Planctomycetes, Bacteroidetes, and Chloroflexi. Proteobacteria, Bacteroidetes, and Euryarchaeota are the dominant groups in the anaerobic digestion system. However, the overall dominance for both systems was from Proteobacteria and Bacteroidetes (Yu et al., 2021). Members of Proteobacteria and Bacteroidetes are primarily responsible for carbon and nutrient removal, which is dictated by the wastewater composition to be treated as established earlier in similar studies (Weissbrodt et al., 2014; Wu et al., 2019; Yu et al., 2021). However, the dominance of Planctomycetes in the activated sludge might be attributed to the presence of ammonia and aromatic compounds in the effluent (Yu et al., 2021).

1.3.1 Impact on wastewater composition

The microbial community structure of a biofilm in contact with wastewater treatment plant effluents is affected by the constituent of the effluent. Moreover, the nutrients in treated effluent contribute to eutrophication in streams and water bodies generating a numerous ecological problems (Ruprecht et al., 2021). Tamminen et al. (2022) study discovered a noticeable change in the microbial community structure in the water stream in Switzerland where the treated water from a wastewater treatment facility is released. This treated water contains microorganisms, which escaped from the treatment unit plus micropollutants/nutrients resulting from the treatment process. The results demonstrated that the presence of herbicides and insecticides particularly affects the presence of species belonging to the genus *Rhodobacter*. These herbicides found in the wastewater include chlortoluron, atrazine, and terbuthylazine, while insecticides were represented by pirimicarb and fungicide tebuconazole (Tamminen et al., 2022). The shift in microbial communities was attributed to the effluent concentration in research conducted by Ruprecht et al. (2021). The less diluted the effluent, the greater the change in the microbial communities' structure and function. The high concentration of the effluent increased the abundance of the nitrate reduction genes (*narG*/*napA*), indicating the shift in the microbial communities' function to nitrate reduction (Ruprecht et al., 2021).

1.3.2 Impact of seasonal change

The microbial flora present in the wastewater influent, the kind of contaminant present, and the method used all influence the microbial community structure and diversity in the activated-sludge treatments. Nonetheless, in the most stable treatment systems, seasonal fluctuations in environmental conditions may make microbial populations vulnerable to that change and might affect the efficiency of the entire treatment process. To understand the dynamics of microbial communities' responses and functional revamps of the communities toward the change in the environment, a study was conducted in Shenyang, Liaoning province of China, for monitoring the restructure of microbial communities using 16s rRNA gene sequencing and functional gene prediction using Phylogenetic Investigation of Communities by Reconstruction of Unobserved States (PICRUSt) (F. Liu et al., 2019). The results manifested that the efficiency of NH^{4+}–N removal was less in winter than in summer, which was attributed to the average lower temperature in the winter compared to summer. The average temperature was 21 °C and 11 °C in the summer and winter, respectively. The genera responsible for ammonia removal were dominated by *Nitrosomonas* and *Nitrospira* (F. Liu et al., 2019). This observation was also supported by Downing and Hopwood (1964), who recognized that the lower temperature has a negative effect on the ammonia removal and nitrification process. The comparison of microbial richness in the summer and winter shows that the microbial communities in the summer are richer than in the winter (F. Liu et al., 2019). Moreover, 29.5% and 22.7% of the detected microbes were not classified in the summer and winter, respectively, as reported by Liu et al. (2019). This adds to the evidence that the seasonal transition of higher average summer temperatures plays a role in increasing species richness and abundance in summer. The most abundant groups that were identified in the summer were *Geothrix*, *Intrasporangium*, *Phaeodactylibacter*, Parcubacteria, *Aridibacter*, *Terrimonas*, *Jiella*, Gp7, and *Tepidisphaera*. The most abundant groups that were identified in winter were *Dechloromonas*, Parcubacteria, *Haliscomenobacter*, *Phaeodactylibacter*, *Rhodoferax*, *Gemmobacter*, *Dokdonella*, Saccharibacteria, and *Sideroxydans* (F. Liu et al., 2019). The variation in abundance and order was ascribed to changes in temperature and the quantities of influent contaminants. The location is also very crucial in determining the microbial community structures. According to Tamminen et al. (2022), these community organizations may vary depending on the geographical location. Certain genera will completely disappear and be undetectable in a particular season, for example, *Exiguobacterium* that was detectable in summer and completely disappeared in winter (F. Liu et al., 2019). Furthermore, the significant prevalence of *Dechloromonas* during the winter months might be ascribed to the high level of aromatic chemicals in the wastewater during this time (F. Liu et al.,

2019). Based on the interpretation above, we may deduce that changes in influent composition and environmental parameter states influence the richness and abundance of microbial species in wastewater treatment plants and that the richness and abundance of microbial species will vary by geographical location. The changes in the environment affected the whole treatment process, for example, the observed reduction in ammonia removal at lower temperatures.

1.4 Effects beyond wastewater treatment

1.4.1 Beyond the conventional life cycle inventory in wastewater treatment plants

The biosolid generated by the wastewater treatment facility is high in minerals such as phosphorus, nitrogen, and potassium, as well as other organic compounds, making it an excellent alternative to synthetic fertilizers (Roig et al., 2012). Heavy metals, microorganisms, and potentially harmful chemical components are still present in this biosolid. Thus, the use of these biosolids may disturb and alter the microbial diversity in agricultural soils. In a research conducted in Tangshan, China, the effect of applying biosolid created from a municipal wastewater treatment facility was investigated. The results revealed that biosolid application in agricultural soil altered the structure and population abundance of microbial communities in the soil. The results showed a decrease in the abundance of total bacteria (gram-positive and gram-negative), soil fungi, arbuscular mycorrhiza, and saprophytic fungal communities (Hu et al., 2019). Even though the bacteria-fungi ratio remained unchanged following the usage of the biosolid, the gram-positive-to-gram-negative bacteria ratio grew dramatically (Hu et al., 2019). In a previous study, the microorganisms and micropollutants present in wastewater effluent modified the organization of microbial communities in rivers/streams (Drury et al., 2013). The wetting and drying phases influence the microbial biofilm sorption and desorption properties of metal ions from water streams. During the wetting stage after the drying period, the biofilm tends to release metals such as Zn, Cd, and As (Luo et al., 2022). However, the relative richness of phosphate-solubilizing bacteria and fungi was increased after the wetting and drying in the corn field with a notable change in the genera *Pseudomonas*, *Arthrobacter*, and *Penicillium* (Bi et al., 2022).

1.4.2 Microbial anticipation of aquatic environment fluctuations

One of the main challenges in microbial ecology is understanding the relationship and function between microbial communities and the surrounding environment

(Widder et al., 2016). and to what extent do ecosystems rely on a group of organisms or particular organisms for any specific function and process (Antwis et al., 2017). The environment has a powerful influence on microbial communities' structure, where any fluctuation in environmental factors will correspondingly affect the microbial communities. Those microorganisms are considered natural indicators of biogeochemical processes that link the environment and ecosystem functioning and indicate that there is a reciprocal relationship between microbes and the environment (Gutknecht et al., 2012).

The environment of aquatic animals consists of water, sediment, feed, and microorganisms present in each part of this environment and in the body of the aquatic animal itself. Microbial communities play a key role in water quality and aquatic animals' health. Microbial communities' structure and their variations (expressed by operational taxonomic units (OTUs)) are affected by several factors in the environment such as water temperature, climate, regional distribution, type of aquaculture production, pH, the salinity of water, stress factors, and the degree of human intervention (Yannarell et al., 2005; Liu et al., 2019; Zhang et al., 2020). Studies focused on fish species as they represent the highest presence of all vertebrates. In the aquaculture environment, there are various microbial communities' compositions and species that differ from those in the natural environment. Fish eggs in the natural environment are free from any microbes, and after hatching, larvae are exposed to microbes in the environment, as the surfaces of the external mucosa are colonized immediately, including the gastrointestinal (GI) tract that forms gut microbiota responsible for GI tract development and this colonization is also important for developing the immune system.

Gut microbiota play a role in fish health as follows: synthesis of hormones and vitamins, which are not formed by the host itself, producing energy from food, production of short-chain fatty acids, and increased resistance to diseases, and these microbes compete with opportunistic microbes in nutrient and other resources. In aquaculture rearing environments, microbes are introduced to fish through feed, water, and fish excrement. The microbial load depends on feed, organic matter production, and the design of the aquaculture system. The microbial community structure varies with the bacterial loads. For example, using a biofilter in a recirculating aquaculture system (RAS) affects the microbial community composition, load, and general fish health (Derome, 2019).

1.4.3 Impact of salinity change on aquatic microbial communities

Fish have an immune system that is divided into innate and adaptive defense mechanisms, which are further subdivided into cell-mediated and humoral defenses. Gut microbiota contribute to this immune system, protecting fish against sickness and keeping them healthy (Smith et al., 2019). However, any disturbance in the surrounding environment stresses the fish. These stresses expose the fish to pathogenic or opportunistic bacteria-caused diseases, necessitating the use of external factors to increase immunity and cure the fish (El-Wazzan et al., 2020; Magnadottir, 2010). Furthermore, RAS has been created, which is based on the use of biofilters (bacteria) that improve the water quality and fish health by converting ammonia (toxic compounds) into nitrate and nitrite (nontoxic products). Similarly, instead of using antibiotics, which lead pathogenic bacteria to develop resistance to them while also killing the beneficial bacteria, biofloc and probiotics are used in three forms such as soil probiotics, feed probiotics, and water probiotics to boost the immune system, improve water quality, and serve as a source of feed for fish (Shitu et al., 2022).

Bacteroidetes and Firmicutes phyla make up between 80% and 90% of all gut microbiota, and in a normal healthy state, all gut microbiota and other microorganisms work in conjunction to place the fish in hemostasis. Any stress or disturbance leads to an imbalance between beneficial and pathogenic bacteria, resulting in dysbiosis (hemostasis disturbance). Salinity is the most powerful environmental element influencing the microbial community in aquaculture systems and water treatment. High salt concentrations inactivate the microbial metabolic processes as well as extracellular enzymes essential for food and carbon metabolism.

According to Zhang et al. (2020), Firmicutes superseded Bacteroidetes as the dominant microbes when salinity increased gradually. *Bacteroidetes* had a weaker taxonomic link with the bacterial population in the saline environment, and the presence of the freshwater bacterial group GKS98 in higher salinity with a change in its symbiotic relationship with other strains exhibits that salt played an important role in bacterial community selection.

A decline in some microbial community species when salinity increases due to the high osmotic pressure leads the cells to shrivel (Newton et al., 2011). In contrast, Zhang et al. (2013) stated that bacterial diversity was in a dwindling number at an intermediate level of salinity (6.86 g/L). Moreover, Zhang et al. (2020) demonstrated that bacterial community diversity, OTUs, and abundance are higher in fresh and brackish water groups than in oligohaline and saline water groups, confirming that bacterial diversity decreases with high salinity. Firmicutes, Actinobacteria, and Proteobacteria phyla were increased in abundance in high saline water. Alternatively, Firmicutes were low in abundance in brackish water, whereas Actinobacteria and Proteobacteria were plentiful. In oligohaline water, Bacteroidetes were dominant. Generally speaking, microbial diversity was reduced in the high salinity group compared with the low salinity group, owing to suboptimal salinity levels for fresh and marine water bacteria.

To conclude, the bacterial community structure and diversity are adapted in response to variation in the salinity levels, indicating that they can adjust their composition gradually in response to environmental changes, and some bacteria exhibited a high degree of adaptability and plasticity. For example, *Pseudomonas* genus and GKS98 are known as "typical freshwater genus" (Evans, 2012).

1.5 Approaches for studying microbial diversity

Usually, culture-dependent and -independent techniques are utilized to investigate the microbial diversity in wastewater treatment facilities (Cydzik-Kwiatkowska & Zielińska, 2016). Enumeration of culturable aerobic bacteria employing selective/nutrient medium for the target/total microbial group or community-level substrate consumption using an ecologically appropriate substrate are examples of culture-dependent approaches (Jatowiecki et al., 2016; Preston-Mafham et al., 2002). According to the research done by Jatowiecki et al. (2016), Biolog EcoPlates is a valuable instrument for determining the factors impacting the microbial community structure in wastewater treatment facilities based on changes in the specified substrate's usage behavior. The use of next-generation DNA sequencing, on the other hand, accurately represents the culture-independent techniques through the sequencing of 16s rRNA and 18s rRNA genes from prokaryotic and eukaryotic microorganisms, respectively, from DNA isolated from environmental samples (Tamminen et al., 2022). Other attempts to identify the microbial communities or species in wastewater treatment systems using chemotaxonomy have also been made; however, owing to the tremendous variety and complexity of the bacteria in the treatment plant, this technique has several flaws (Y et al., 1998). This approach is based on identifying a specific substance generated by a specific group of bacteria and using that component as an indicator/marker for that group. Although the physiological condition of the cell affects the concentration of these substances, it is difficult to quantify the absolute number of cells using chemical concentrations alone. Immunofluorescence is another approach that has proven successful in detecting the existence, quantity, and the function of slow-growing bacteria in wastewater treatment, such as ammonia-oxidizing bacteria like *Nitrosomonas* spp. (Fiencke & Bock, 2004). It is worth noting that phospholipid fatty acid (PLFA) is also used to analyze microbial communities in environmental samples using an automated identification approach based on PLFA's unique structure, which varies between microbial groups (Hu et al., 2019).

1.6 Conclusion

Microbes are the keystones of all ecosystems and the engines of biochemical cycles on the Earth. The composition of microbial communities varies across different seasons and locations. Adaptation and mitigation methods are essential for mitigating the severity of the consequences of environmental parameter change. Microbial populations have an impact on factors such as pH, temperature, and nutrient availability in every habitat; yet these communities are also influenced by such parameters. This applies to the first stages of establishing microbial communities in a wastewater treatment facility, aquaculture bonds and cages, agricultural soil, and adjacent habitats. The biological and chemical composition of wastewater dominates not only the microbial populations in the treatment plant, in spite of the fact that the streams and the surrounding habitats where the treated water is discharged. Furthermore, seasonal and temperature fluctuations alter the functional and structural features of the active microbes in industrial processes and nature. Restructures in the microbial populations caused by fluctuation of the environmental factors are unpredictable and impact the whole process. More research is required for a better understanding of microbial communities' evolved process in response to various environmental conditions, and it is censorious to turn down their impacts.

References

Abdella, B., El-Wazzan, E., El-Sersy, N. A., Sabry, S. A., & El-Helow, E. R. (2017). Pathogenicity and antibiotic susceptibility of two bacterial pathogens associated with the clam Tapes decussatus in some Egyptian fisheries. *Ege Journal of Fisheries and Aquatic Sciences*, *34*(4), 383–389. https://doi.org/10.12714/egejfas.2017.34.4.04

Antwis, R. E., Griffiths, S. M., Harrison, X. A., Aranega-Bou, P., Arce, A., Bettridge, A. S., Brailsford, F. L., de Menezes, A., Devaynes, A., Forbes, K. M., Fry, E. L., Goodhead, I., Haskell, E., Heys, C., James, C., Johnston, S. R., Lewis, G. R., Lewis, Z., Macey, M. C., & Sutherland, W. J. (2017). Fifty important research questions in microbial ecology. *FEMS Microbiology Ecology*, *93*(5), 1–10. https://doi.org/10.1093/femsec/fix044

Arayes, M. A., Mabrouk, M. E. M., Sabry, S. A., & Abdella, B. (2021). Diversity and characterization of culturable haloalkaliphilic bacteria from two distinct hypersaline lakes in northern Egypt. *Biologia*, *76*(2), 751–761. https://doi.org/10.2478/s11756-020-00609-5

Asano, R., Nakai, Y., Kawada, W., Shimura, Y., Inamoto, T., & Fukushima, J. (2013). Seawater inundation from the 2011 Tohoku Tsunami continues to strongly affect soil bacterial communities 1 year later. *Microbial Ecology*, *66*(3), 639–646. https://doi.org/10.1007/s00248-013-0261-9

Bi, W., Weng, B., Yan, D., Wang, H., Wang, M., Yan, S., Jing, L., Liu, T., & Chang, W. (2022). Responses of phosphate-solubilizing microorganisms mediated phosphorus cycling to drought-flood abrupt alternation in summer maize field soil. *Frontiers in Microbiology*, *12*. https://doi.org/10.3389/fmicb.2021.768921

Chapron, L., Galand, P. E., Pruski, A. M., Peru, E., Vétion, G., Robin, S., & Lartaud, F. (2021). Resilience of cold-water coral holobionts to thermal stress. *Proceedings of the Royal Society B: Biological Sciences*, *288*(1965). https://doi.org/10.1098/rspb.2021.2117

Cydzik-Kwiatkowska, A., & Zielińska, M. (2016). Bacterial communities in full-scale wastewater treatment systems. *World Journal of Microbiology and Biotechnology*, *32*(4), 1–8. https://doi.org/10.1007/s11274-016-2012-9

Derome, N. (2019). Microbial Communities in Aquaculture Ecosystems. In N. Derome (Ed.), *Microbial Communities in Aquaculture Ecosystems*. Springer International Publishing. https://doi.org/10.1007/978-3-030-16190-3

Dong, X., Liu, C., Ma, D., Wu, Y., Man, H., Wu, X., Li, M., & Zang, S. (2021). Organic carbon mineralization and bacterial community of active layer soils response to short-term warming in the Great Hing'an Mountains of Northeast China. *Frontiers in Microbiology*, *12*. https://doi.org/10.3389/fmicb.2021.802213

Downing, A. L., & Hopwood, A. P. (1964). Some observations on the kinetics of nitrifying activated-sludge plants. *Schweizerische Zeitschrift Für Hydrologie*, *26*(2), 271–288. https://doi.org/10.1007/BF02504050

Drury, B., Rosi-Marshall, E., & Kelly, J. J. (2013). Wastewater treatment effluent reduces the abundance and diversity of benthic bacterial communities in urban and suburban rivers. *Applied and Environmental Microbiology*, *79*(6), 1897–1905. https://doi.org/10.1128/AEM.03527-12

El-Wazzan, E., Ghareeb, D. A., & Abdella, B. (2020). Pre-induction of Hsp70 expression to protect the grooved carpet shell clam, Ruditapes decussatus, against Micrococcus luteus: A trained immunity strategy. *Egyptian Journal of Aquatic Research*, *46*(1), 79–84. https://doi.org/10.1016/j.ejar.2019.10.004

Evans, T. G., & Hofmann, G. E. (2012). Defining the limits of physiological plasticity: How gene expression can assess and predict the consequences of ocean change. *Philosophical Transactions of the Royal Society B: Biological Sciences*, *367*(1596), 1733–1745. https://doi.org/10.1098/rstb.2012.0019

Fiencke, C., & Bock, E. (2004). Genera-specific immunofluorescence labeling of ammonia oxidizers with polyclonal antibodies recognizing both subunits of the ammonia monooxygenase. *Microbial Ecology*, *47*(4), 374–384. http://www.jstor.org/stable/25153069

Gutknecht, J. L. M., Field, C. B., & Balser, T. C. (2012). Microbial communities and their responses to simulated global change fluctuate greatly over multiple years. *Global Change Biology*, *18*(7), 2256–2269. https://doi.org/10.1111/j.1365-2486.2012.02686.x

Hu, Y., Pang, S., Yang, J., Zhao, X., & Cao, J. (2019). Changes in soil microbial community structure following amendment of biosolids for seven years. *Environmental Pollutants and Bioavailability*, *31*(1), 24–31. https://doi.org/10.1080/26395940.2019.1569478

Jatowiecki, Ł., Chojniak, J. M., Dorgeloh, E., Hegedusova, B., Ejhed, H., Magnér, J., & Płaza, G. A. (2016). Microbial community profiles in wastewaters from onsite wastewater treatment systems technology. *PLoS ONE*, *11*(1), 1–15. https://doi.org/10.1371/journal.pone.0147725

Kapley, A., De Baere, T., & Purohit, H. J. (2007). Eubacterial diversity of activated biomass from a common effluent treatment plant. *Research in Microbiology*, *158*(6), 494–500. https://doi.org/10.1016/j.resmic.2007.04.004

Liu, F., Hu, X., Zhao, X., Guo, H., & Zhao, Y. (2019). Microbial community structures' response to seasonal variation in a full-scale municipal wastewater treatment plant. *Environmental Engineering Science*, *36*(2), 172–179. https://doi.org/10.1089/ees.2018.0280

Liu, J., Meng, Z., Liu, X., & Hua, X. (2019). Microbial assembly, interaction, functioning, activity and diversification : a review derived from community compositional data. *Marine Life Science & Technology*, *1*(1), 112–128. https://doi.org/10.1007/s42995-019-00004-3

Luo, X., Yang, Y., Xie, S., Wang, W., Li, N., Wen, C., Zhu, S., & Chen, L. (2022). Drying and rewetting induce changes in biofilm characteristics and the subsequent release of metal ions. *Journal of Hazardous Materials*, *433*, 128832. https://doi.org/10.1016/j.jhazmat.2022.128832

Magnadottir, B. (2010). Immunological control of fish diseases. *Marine Biotechnology*, *12*(4), 361–379. https://doi.org/10.1007/s10126-010-9279-x

Mahmoud, N. H., Elsherbiny, B. A., Moffit, S. M., Mohamed, J. H., Abouelkheir, S. S., & Abdella, B. (2022). Cell Biology and Microbial Interactions in Algal Cells. In M. M. El-Sheekh, N. Abdullah, & I. Ahmad (Eds.), *Handbook of Research on Algae as a Sustainable Solution for Food, Energy, and the Environment* (pp. 84–108). IGI. https://doi.org/10.4018/978-1-6684-2438-4.ch004

Martiny, J. B. H., Bohannan, B. J. M., Brown, J. H., Colwell, R. K., Fuhrman, J. A., Green, J. L., Horner-Devine, M. C., Kane, M., Krumins, J. A., Kuske, C. R., Morin, P. J., Naeem, S., Øvreås, L., Reysenbach, A.-L., Smith, V. H., & Staley, J. T. (2006). Microbial biogeography: putting microorganisms on the map. *Nature Reviews Microbiology*, *4*(2), 102–112. https://doi.org/10.1038/nrmicro1341

Massoud, M. A., Tarhini, A., & Nasr, J. A. (2009). Decentralized approaches to wastewater treatment and management: Applicability in developing countries. *Journal of Environmental Management*, *90*(1), 652–659. https://doi.org/10.1016/j.jenvman.2008.07.001

Newton, R. J., Jones, S. E., Eiler, A., McMahon, K. D., & Bertilsson, S. (2011). A Guide to the natural history of freshwater lake bacteria. *Microbiology and Molecular Biology Reviews*, *75*(1), 14–49. https://doi.org/10.1128/MMBR.00028-10

Preston-Mafham, J., Boddy, L., & Randerson, P. F. (2002). Analysis of microbial community functional diversity using sole-carbon-source utilisation profiles – a critique. *FEMS Microbiology Ecology*, *42*(1), 1–14. https://doi.org/10.1111/j.1574-6941.2002.tb00990.x

Roig, N., Sierra, J., Martí, E., Nadal, M., Schuhmacher, M., & Domingo, J. L. (2012). Long-term amendment of Spanish soils with sewage sludge: Effects on soil functioning. *Agriculture, Ecosystems & Environment, 158*, 41–48. https://doi.org/10.1016/j.agee.2012.05.016

Ruprecht, J. E., Birrer, S. C., Dafforn, K. A., Mitrovic, S. M., Crane, S. L., Johnston, E. L., Wemheuer, F., Navarro, A., Harrison, A. J., Turner, I. L., & Glamore, W. C. (2021). Wastewater effluents cause microbial community shifts and change trophic status. *Water Research, 200*, 117206. https://doi.org/10.1016/j.watres.2021.117206

Shitu, A., Liu, G., Muhammad, A. I., Zhang, Y., Tadda, M. A., Qi, W., Liu, D., Ye, Z., & Zhu, S. (2022). Recent advances in application of moving bed bioreactors for wastewater treatment from recirculating aquaculture systems: A review. *Aquaculture and Fisheries, 7*(3), 244–258. https://doi.org/10.1016/j.aaf.2021.04.006

Sjøgaard, K. S., Valdemarsen, T. B., & Treusch, A. H. (2018). Responses of an agricultural soil microbiome to flooding with seawater after managed coastal realignment. *Microorganisms, 6*(1). https://doi.org/10.3390/microorganisms6010012

Smith, N. C., Rise, M. L., & Christian, S. L. (2019). A comparison of the innate and adaptive immune systems in cartilaginous fish, ray-finned fish, and lobe-finned fish. *Frontiers in Immunology, 10*. https://doi.org/10.3389/fimmu.2019.02292

Tamminen, M., Spaak, J., Tlili, A., Eggen, R., Stamm, C., & Räsänen, K. (2022). Wastewater constituents impact biofilm microbial community in receiving streams. *Science of the Total Environment, 807*. https://doi.org/10.1016/j.scitotenv.2021.151080

Weissbrodt, D. G., Shani, N., & Holliger, C. (2014). Linking bacterial population dynamics and nutrient removal in the granular sludge biofilm ecosystem engineered for wastewater treatment. *FEMS Microbiology Ecology, 88*(3), 579–595. https://doi.org/10.1111/1574-6941.12326

Widder, S., Allen, R. J., Pfeiffer, T., Curtis, T. P., Wiuf, C., Sloan, W. T., Cordero, O. X., Brown, S. P., Momeni, B., Shou, W., Kettle, H., Flint, H. J., Haas, A. F., Laroche, B., Kreft, J. U., Rainey, P. B., Freilich, S., Schuster, S., Milferstedt, K., & Wilmes, P. (2016). Challenges in microbial ecology: Building predictive understanding of community function and dynamics. *ISME Journal, 10*(11), 2557–2568. https://doi.org/10.1038/ismej.2016.45

Wu, L., Ning, D., Zhang, B., Li, Y., Zhang, P., Shan, X., Zhang, Q., Brown, M. R., Li, Z., Van Nostrand, J. D., Ling, F., Xiao, N., Zhang, Y., Vierheilig, J., Wells, G. F., Yang, Y., Deng, Y., Tu, Q., Wang, A., & Zhou, J. (2019). Global diversity and biogeography of bacterial communities in wastewater treatment plants. *Nature Microbiology, 4*(7), 1183–1195. https://doi.org/10.1038/s41564-019-0426-5

Y, R. A., Lemmer, H., & Wagner, M. (1998). Monitoring the community structure of wastewater treatment plants : a comparison of old and new techniques. *FEMS Microbiology Ecology, 25*, 205–215.

Yannarell, A. C., & Triplett, E. W. (2005). Geographic and environmental sources of variation in lake bacterial community composition. *Applied and Environmental Microbiology, 71*(1). https://doi.org/10.1128/AEM.71.1.227

Yu, J., Tang, S. N., & Lee, P. K. H. (2021). Microbial communities in full-scale wastewater treatment systems exhibit deterministic assembly processes and functional dependency over time. *Environmental Science and Technology, 55*(8), 5312–5323. https://doi.org/10.1021/acs.est.0c06732

Zhang, L., Gao, G., Tang, X., & Shao, K. (2014). Can the freshwater bacterial communities shift to the “marine-like” taxa? *Journal of Basic Microbiology, 54*(11), 1264–1272. https://doi.org/10.1002/jobm.201300818

Zhang, L., Gao, G., Tang, X., Shao, K., Bayartu, S., & Dai, J. (2013). Bacterial community changes along a salinity gradient in a Chinese wetland. *Canadian Journal of Microbiology, 59*(9), 611–619. https://doi.org/10.1139/cjm-2013-0212

Zhang, L., Hu, Y., Li, X., Lu, W., & Li, J. (2020). Function prediction and network analysis to investigate the response of microbial communities to a single environmental factor. *Journal of Freshwater Ecology, 35*(1), 271–289. https://doi.org/10.1080/02705060.2020.1791269

Susmita Paul, Trishnamoni Gautom and Santa Ram Joshi*

Chapter 2
A perspective on bacterial metagenomics in wastewater research

Abstract: The evolving field of research and development exploring novel areas in environmental genomics is growing at a fast pace. Metagenomics, also referred to as eco-genomics, has provided an opportunity for unprejudiced exploration and discovery of microorganisms within their ecological niches. The metagenomics approach is used to study microbial communities, especially to decipher distinction between microbes and groups and their functional genomics. This emerging area has offered vistas for the discovery of unknown groups and genes that can be explored and utilized in remediation of environmental problems. One such area is wastewater and its treatment for restoration and rehabilitation of water bodies. In this chapter, an overview of emerging potential applications of metagenomics used in wastewater research is presented with the focus of novel genome(s) and gene(s) available through eco-genomics profiling of environmental DNA.

2.1 Introduction

Wastewater treatment through biological means utilizes the metabolism process of microbes belonging to the groups that convert contaminants into products that are not lethal. Since there is little or no comprehensive understanding regarding the multiplicity and density of groups of microbes, it is expected that the potential processes involving groups of microbes in the process of wastewater treatment would be ignored. The technology of metagenomics is developed in the field of molecular biology, a field with huge sequences of genes attained from the environmental specimens, and these are studied using bioinformatic tools to govern the alignment and function of the community of microbes. Metagenomics, when compared with traditional molecu-

*Corresponding author: Santa Ram Joshi, Microbiology Laboratory, Department of Biotechnology and Bioinformatics, North-Eastern Hill University, Shillong 793022, Meghalaya, India, e-mail: srjoshi2006@yahoo.co.in
Susmita Paul, Royal School of Biosciences, The Assam Royal Global University, Guwahati 781035, Assam, India; Microbiology Laboratory, Department of Biotechnology and Bioinformatics, North-Eastern Hill University, Shillong 793022, India
Trishnamoni Gautom, Department of Industrial Biotechnology, KTH Royal Institute of Technology, Stockholm, Sweden

https://doi.org/10.1515/9783110793635-002

lar methods, were found to be more efficient in identifying potential microorganisms in their inherent surroundings (Singh et al., 2020).

To discover the bionetwork of microbes, in terms of genetics, metabolics, diversity developments, and their relationships in the treatment of wastewater systems through biological means, several methods that have been observed were applied. It has been observed that the traditionally used techniques that are cultivation-dependent, utilized for the investigation of ecology of microbes, are unable to expansively categorize the natural world of intricate microbiological communities (Wommack & Ravel, 2013).

Handelsman et al. (1998) first gave the definition of the term "metagenomics," and it was explained that it is the fortitude of entire genetic makeup in the community of microbes for the analysis of multiplicity of such microbes as well as their genetics, and metabolic developments of microorganisms in a particular environment (Bengtsson-Palme, 2018; Boulund et al., 2018; Mardanov et al., 2018; Ramazzotti & Bacci, 2018). Metagenomics fetches multiple nucleotide fragments from environmental samples and utilizes the sequencing of DNA method to attain the targeted evidence straight up, which can more effectually classify the position of microbes in a particular surrounding, and this is not observed in traditional cultivation-dependent techniques utilized for researches in microbiology. Furthermore, metagenomics does not lead to any biasness in polymerase chain reaction, which is observed to be more beneficial compared to 16S amplicon sequencing (Mardanov et al., 2018).

Functional and sequence-based metagenomics are the two approaches that are presently used to attain metagenomic data from environmental samples (Mardanov et al., 2018) (Figure 2.1).

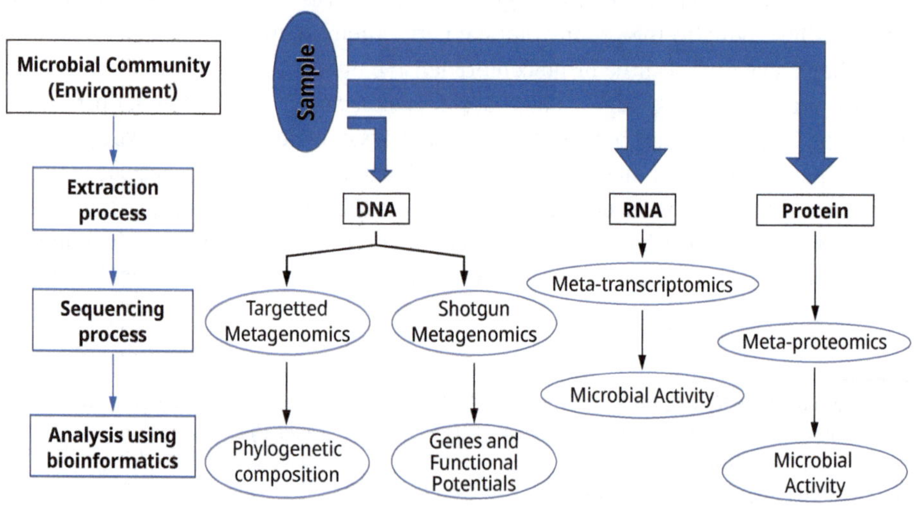

Figure 2.1: Metagenomic strategies to understand the communities of microbes.

Function-based metagenomic data primarily utilizes an expression library designed by heterologous expression of numerous replicated strands of DNA in vectors to get hold of enzymatic functions and actions interrelated to microbes as well as their route of metabolism (Mardanov et al., 2018; Allen et al., 2009; Felczykowska et al., 2015; Liebl et al., 2014). Consequently, for the identification of function of genes, the metagenomic approaches of function is commonly applied; for instance, antibiotic resistance genes (ARGs) are applicable for the resistivity of heavy metals, and genes associated with metabolic activity of phosphorus or nitrogen and utilisation of contaminants (Allen et al., 2009; Guo et al., 2017). Additionally, it has been observed that the function-based metagenomics can distinguish the genes that code for an enzyme with a specific function, allowing the detection of new enzymes whose roles might not be anticipated by utilizing the DNA sequence (Ferrer et al., 2007). Metagenomics utilizes next-generation sequencing approaches to study the sequences of large-sized DNA fragments, and also interprets genes by bioinformatic analysis (Bharagava et al., 2019). Metagenomics, which is sequence based, can offer genetic information on the entire community, and consequently, can be used in designing metabolic procedures as well as predicting the function of possible genes (Chong et al., 2020).

This chapter outlines the application of metagenomics to evaluate the groups of microbes in the biological treatment of wastewater, for example, the genetic amputation of nitrogen and phosphorus by means of bacteria, the exploration of ARGs, and the decrease

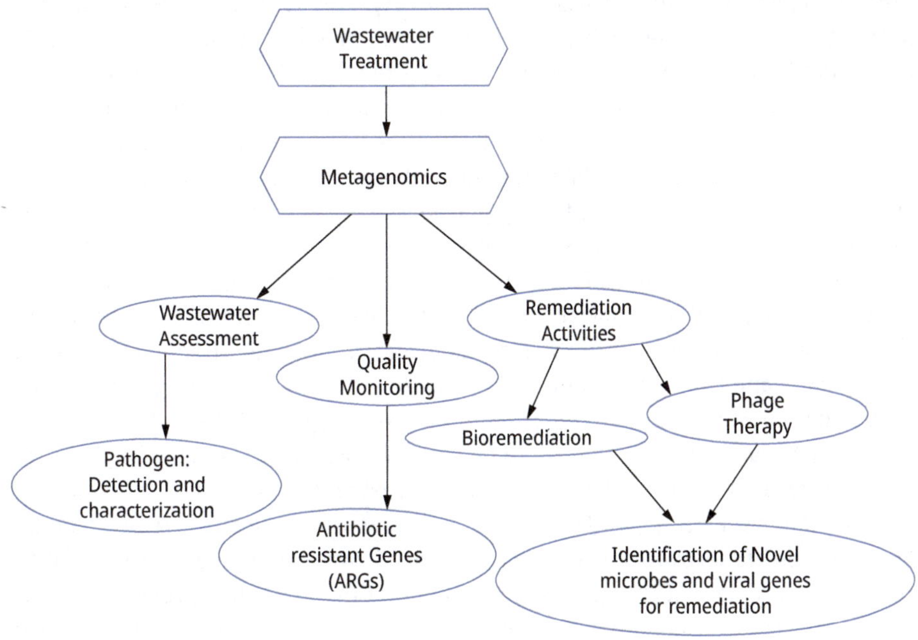

Figure 2.2: Metagenomics in wastewater treatment and its applications.

of heavy metals through involvement of the groups of microbes, with a prominence on the involvement of microbial multiplicity as well as variety in metabolism, and in detecting and analyzing viruses present in wastewater (Figure 2.2).

2.2 Metagenomics in biological phosphorus removal

Eutrophication, which is majorly caused by dispensing high-concentrate phosphorus wastewater into natural waters, results in destruction of ecological structures and functions (Conley et al., 2009). Biological means of anthropogenic phosphorus removal are extensively utilized when treating wastewater to tackle eutrophication, owing to its environmental friendliness and low cost. Phosphate-accumulating organisms (PAOs) are vital bacteria employed in the elimination of phosphates biologically, and therefore studies on PAOs can aid in optimizing this procedure. Removal of phosphorus through biological means processsing the PAO-stored phosphate in sewage via metabolism of bacteria by converting it into polyphosphate, a substance that is ultimately released as elevated-concentrate phosphorus sludge (Mino et al., 1998). In spite of decades of progress, certain constraints on the biological elimination of phosphorus, including the association between PAOs and glycogen-accumulating organisms (GAOs), that is a critical aspect monitoring the reactor function, still remain unclear. Also, an inadequate comprehension of biodiversity, interactions between functioning microorganisms, and the distinction of niches and metabolic function influence the total phosphorus removal competency and constancy within the system. Thus, PAO biodiversity ought to be taken into account during wastewater treatment for elimination of phosphorus (Mino et al., 1998). Although the PAOs are fewer significantly more varied in lab cultures than in the nature but only the genera *Thioploca, Tetrasphaera*, and *Accumulibacter* are reported to be phosphorus stores (Holmkvist et al., 2010; Fernando et al., 2018). This limited knowledge considerably restricts our comprehension of the phylogeny and variety of PAOs. However, the use of metagenomic binning techniques has improved our comprehension on the main clade of PAOs, which was revealed to be *Candidatus Accumulibacter* (Zhang et al., 2011; Albertsen et al., 2016). Although PAOs and GAOs were known not to share any homology phylogenetically, *Ca. Propionivibrio* was eventually reported to be a new type of GAOs (Albertsen et al., 2016), which was previously thought to be a PAO favorably coexisting through *Ca. Accumulibacter* in sequencing batch reactors on a laboratory scale. *Ca. Propionivibrio* was thereafter utilized as a PAOmix FISH probe to assess the richness of *Accumulibacter*, leading to its overestimation in situ. This discovery presented a better and accurate measurement of the species-richness and at the same time provided the estimation of reactor's potential to remove phosphorus. Like PAOs, GAOs are also considered a significant microbiological member species in enhancing the performance of method for biological elimination of phosphorus denoting the significance of also exploring GAO diversity. The applications of metagenomics will eventually simplify the underlying challenge of elaborating and strengthening the mechanism of competition in biological phosphorus removal systems between PAOs and GAOs (Albertsen et al., 2016).

2.3 Metagenomics in nitrogen removal

The advancement of modern agriculture and industries has contributed to the release of substances containing nitrogen, which far exceeds the environmental capacity, subsequently leading to numbers of environmental concerns such as acidic rain, acidifying the soil and eutrophication. Thus, the competent elimination of nitrogen extracted from sewage grave matter aimed at recent handling of structures. Elimination of nitrogen through biological means is inexpensive and hence extensively used to condense nitrogen discharge from wastewater treatment structures. Since nitrogen cycling microbes is an important element in elimination of nitrogen biologically, information on microbial nitrogen cycling systems will be helpful in accurately controlling these microorganisms and thereby increasing their operational efficiency. Knowledge on denitrification of microbial diversity was expanded by metagenomics by analyzing the process of partial nitrification-anammox (PNA) that yielded 23 metagenomes – a mass of genomes of denitrification microbes that were then categorized on the basis of their functional genes within genomes (Speth et al., 2016). Later, in 2020, metagenomics was used to govern the consequence of particle size on granule-based PNA presentation, and it was observed that larger molecules shaped high biodiversity based on the reactor's function. This could be the reason behind reasonable denitrification presentation of big granule-based PNA. Metagenomic binning technology was used to gather the genetic data of comammox *Nitrospira*, primarily found in structures where wastewater handling is performed. *Nitrospira* exhibited the capability of complete nitrification by expressing the oxidation of ammonia as well as nitrite genes concurrently throughout the process of nitrification (Daims et al., 2015). Such exploration directed to the finding of an entirely innovative procedure of metabolism of comammox that expanded our knowledge of the outmoded nitrification-denitrification development. Metagenomics has the potential not only to enrich our knowledge on the diversity of nitrogen-cycling microorganism but also to pave a fresh system of organizing the ecology of microbes.

2.4 Metagenomics including genes linked to antibiotic resistance in wastewater treatment plants

It has been observed that occurrence of medication resistance and ARGs of bacteria is very common in the biological treatment of wastewater, which is mostly produced by stock farming, medications, and other endeavors (Kim & Carlson, 2007). Wastewater treatment plants (WWTPs) gives a platform for resistance genes to mix in the nature (Munir et al., 2011; Berendonk et al., 2015).

The advancement in gene resistance study is a dynamic research area and metagenomics play a vital role to explicate the horizontal transfer, circulation, and ARG deterioration in WWTPs.

One of the significant means of microbial evolution is the horizontal gene transfer (HGT) fragments and is also a key approach for distribution of ARGs. In the meantime, HGT of ARGs aids microbes to harvest multiple drug resistance genes, which have the potential to prime various environmental health complications, and has gained a progressively prevalent pedestal in research (Bellanger et al., 2014; Jin et al., 2020). In order to assemble the genetic components that are mobile and permit investigators to discover the variability in such components across and within genomes, metagenomics provides the platform (New & Brito, 2020). Thus, for expounding ARGs' horizontal movement in wastewater, metagenomics is utilized. It has been observed that numerous ARGs in WWTPs are transported by plasmids and thus the maximum ARGs' relative abundance was observed in plasmids followed by conjugate transposons, which could be the key areas of endorsing HGT of antibiotic resistance in wastewater (Che et al., 2019).

Analogous phenomenon was observed in medical wastewaters too, where multiple mobile ARGs have been identified and presented, indicating that an intricate wastewater environment might persuade numerous diverse resistance genes that are supported by microbes (Marathe et al., 2019) that could speed up the spread of super-resistant microorganisms.

The research areas that are at present very much active include elimination of ARGs as well as their spread in wastewater and thus to the environment (Krzeminski et al., 2019; Zhu et al., 2021). It is recorded that ARGs present in wastewater can be partly removed by means of ultraviolet disinfection (Munir et al., 2011; Aminov, 2011; McKinney & Pruden, 2012). Several other assumptions that are analogous were also attained. In such similar works which were studied, combination of metagenomics along with technology of molecular biology were carried out for examining the multiplicity as well as copiousness of ARGs in wastewater. It was observed that although exposure to UV light significantly condensed the microbial count along with ARGs, at the same time it showed the high number of multiple resistance genes carrying microbes (Hu et al., 2016).

One of the operative approaches to eliminate ARGs is the use of anaerobic digestion, that is, the method of waste activated sludge. In a study, anaerobic digestion reactor was used for metagenomics, which was then utilized to discover resistance gene degradation, and it was carried out under different temperature systems. From the study, thermophilic-thermophilic systems reasonably eliminated heavy metal resistance genes and ARGs. On the other hand, a thermophilic-mesophilic system was observed to be further most appropriate for eliminating resistance genes carried by class 1 integrons as well as pathogenic bacteria (Jang et al., 2018).

Metagenomics has significantly improved our understanding on ARGs, empowering a detailed learning on the influence of HGT as well as the transmission pathway of antibiotic resistance, influence of antibiotics on different communities of microorganisms, and development of multidrug-resistant bacteria.

2.5 Metagenomics and heavy metals

With the hasty growth of current industry, heavy metal effluence has become one of the most acute problems in the environment. Wastewater from industries, including metals, turns out to be highly biotoxic and bioaccumulative and can lead to drastic destruction of the ecology upon the access into the atmosphere as well as the food chain. Hence, a sequence of chemical and physical procedures has been established to eliminate heavy metals that are toxic (Barakat, 2011; Fu & Wang, 2011; Kurniawan et al., 2006), and a few of them are membrane filtration, adsorption, ion exchange, and chemical precipitation. While such kind of methods attain reasonable elimination in definite pollution conditions, high maneuver and preservation charges prevent large-scale applications. Heavy metal remediation by utilizing microbes has been broadly premeditated for the reason that it is cost-effective and at the same time provides satisfactory results. Remediation utilizing microbes runs on a significant principle and that is the decreased level of lethal heavy metal valences to atoxic or less toxic levels through metabolism of microbes, which is for resource recycling. Over the years, numerous studies have focused on the reduction of heavy metals; however, comprehensive understanding of microbial remediation techniques, as well as the physiological and biochemical responses of microbial communities to heavy metal-induced stress, remains limited. A study of metagenome could be utilized to suggestively promote the progress of heavy metal redress using microbes.

For explicating the retort methods of groups of microbes under heavy metal stress, metagenomics turned out to be acting as the most important part. It has been observed that a few outcomes discovered that the community of microbes, together with the structure, function, and processes involved in metabolism, varies to a certain extent under stress from heavy metals (Sharma et al., 2021). In a study, application of metagenomics to evaluate the anammox's level of toxicity (anaerobic ammonium oxidation) procedure utilized for the treatment of wastewater by means of elevated levels of nitrogen comprising Ni and Zn presented that the microorganism's group constraint is associated with valency states of heavy metals and bacterial organisms that are dominant (Aktan et al., 2018). It is also studied that during Cr(VI) stress condition, the community of microbes may show dissimilar response mechanisms during distinct development procedures. In another study, metagenomics was utilized along with grids to examine diverse amounts of Cr(VI) on the execution of the A2O (anaerobic-anoxic-aerobic) method and presented that a great amount of Cr(VI) reserved the expressing capabilities of the genes that are associated with nitrogen metabolism, and at the same time lowered the diversity of microbes associated with nitrogen metabolism, ultimately dropping the elimination presentation of NH3–N as well as total nitrogen (TN) (Sun et al., 2019). Nevertheless, in an observation, researchers attained granulated sludge proficient of denitrification and decrease of nitrate as well as chromate during a condition of extended exposure to hexavalent chromium. Also, the metagenomic investigation presented that *Halomonas* sp., which required a good skill to eliminate nitrate and chromate, was considerably augmented across activated sludge (Kiran Kumar Reddy & Nancharaiah, 2018). The preparatory assessment of the communities' reaction system of microbes under stress from heavy metals is being carried out and understood by the researchers; nevertheless, wastewater from industries was frequently found to hold numerous or even masses of heavy metals. Thus, further analysis on the response strategies of groups of microorganisms at such intricate environments is looked for.

2.6 Viral metagenomics and bioremediation

Sewage contaminated water harbors several pathogenic bacteria such as *E. coli, Salmonella, Vibrio cholerae,* and viruses, which may result in severe gastrointestinal illnesses caused by water and hepatitis (Jassim et al., 2016; Bofill-Mas & Rusiñol, 2020). Bacterial indicators like coliforms that are traditionally used to evaluate water's microbiological safety frequently do not succeed in detecting infective viruses that endure longer in the water (Aw et al., 2014). Several enteric viruses in humans have been found in samples of ambient water indicating their presence and potential transmission by water sources. Therefore, efficient pathogen monitoring and management levels are vital steps in wastewater treatment procedure as well as public safety, where an account of pathogens in water sources plays a crucial role. Metagenomics is one such modern strategy used to recognize and describe viruses and bacteria by sequencing and analyzing the genome, eliminating the need for cultivating them. The field of metagenomics has produced large data from the viral genome and related sequences of which stand unavailable inside databases, indicating that viral communities in the environment are considerably dissimilar from the well-studied viruses

(Haramoto et al., 2018). This novel genetic data has fetched noteworthy phages' significance as a source of untapped genomic data (Rosario & Breitbart, 2011). Phages are the most prevalent living organisms in freshwaters as well as oceans. Earlier studies have investigated phages' possibility to be used as potential biological tools in bacterial control and genome manipulations (Hayes et al., 2017). An efficient and successful application of phages in wastewater treatment requires a thorough comprehension of the interactions and dynamics of the microbial community in wastewater (Jassim et al., 2016). Metagenome study using marker gene amplification and shotgun metagenomics are prime lines that permit the examination of uncultured viral and microbial inhabitants using both qualitative and quantitative methods. Viruses are generally contaminated using DNA from bacteria that must be detached before sequence analysis. Although the small size of viruses favors the sequencing-based methods, still the use of metagenome study in virology becomes a bit complicated due to the huge multiplicity in viruses and the absence of common universal marker genes. Till date, from various wastewaters, viruses with DNA and RNA have been found by means of viral metagenome studies (Figure 2.3).

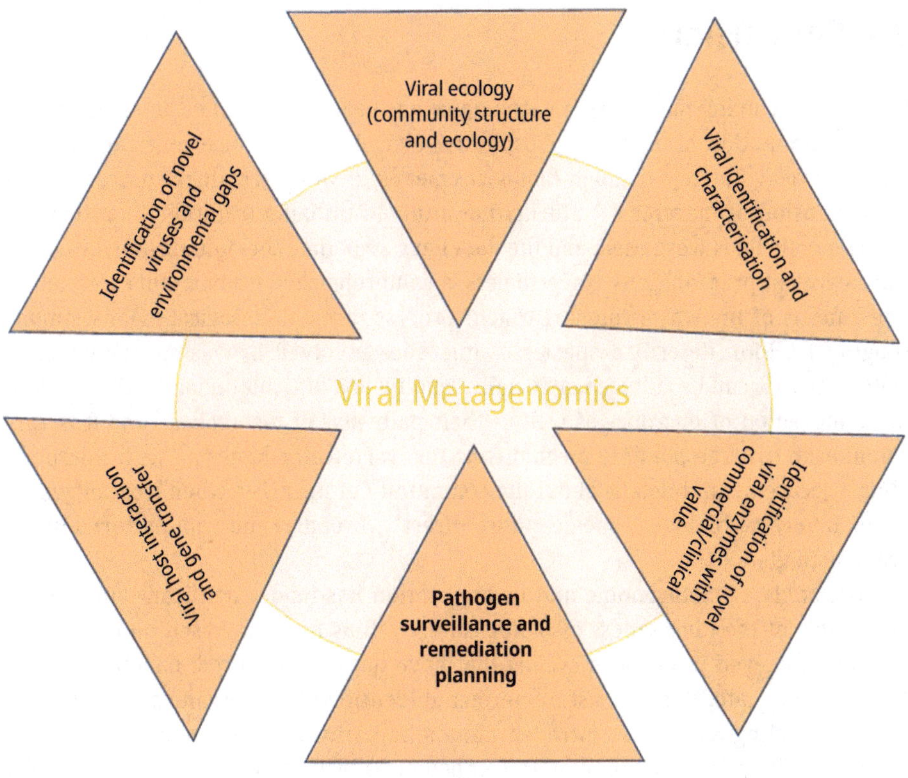

Figure 2.3: Applications of viral metagenomics.

In addition to detecting emerging viruses/novel strains in wastewater, next-generation sequencing also aids in finding innovative information on emerging viral strains that were previously identified (Bofill-Mas & Rusiñol, 2020; Höper et al., 2017).

Another exciting environmentally friendly technology is bioremediation, which explores the capacity of microorganisms to restore the polluted water. It utilizes the capability of microorganisms to degrade or transform the pollutants in wastewater through bioaugmentation, bioattenuation, or biostimulation. Metagenomics, therefore, can be applied to explore the genes expressed in such degradation processes of contaminants (Bharagava et al., 2019). This knowledge has to be exploited to examine microorganisms having strong ability to degrade a particular contaminant of interest. These genes could be recognized and used to their advantage for bioremediation.

Metagenomic libraries are able to be filtered to explore the current collection of novel gene families as well as enzymes entangled in the process of pollutant degradation. Superior viral genomes will help clarify the evolution of viruses and make the foundation of numerous forthcoming pathogen surveillance studies.

2.7 Conclusion

In studies involving microbes, metagenomics has been developed as an essential tool for the elimination of nitrogen, phosphorus, as well as resistance genes and the decrease in heavy metals through biological means in water treatment processes. For the expansion of margins of information attained utilizing traditional approaches, metagenomic tools were used, and the data have smoothed the optimization of biological treatment presentations. Nevertheless, a comprehensive study is still essential for a few facets of the wastewater treatment process through biological means. Among roughly 1 billion described species of microbes involved in wastewater which is global, only 1% can be cultured artificially, leading to a difficult understanding toward the composition of microbes as well as their pathways of metabolism, and it is very common to oversee possible mechanisms for wastewater handling with microbes. Thus, procurement of details about the community of microbes when treating wastewater involving biological means, more directly, broadly, and rapidly turns into a pressing matter.

The study of metagenome and its application has made significant alms to our knowledge of the diverseness of biological as well as function-based multiplicity of microbes involved in nitrogen cycling that have not been cultured; nonetheless, it is far from adequate. Here persist numerous unidentified microorganisms involved in nitrogen cycling as well as nitrogen metabolic pathways, and their investigation might benefit us in developing more competent as well as environmentally friendly nitrogen elimination systems by biological means.

Not only does wastewater contaminate receiving bodies of water but also it contaminates the soil. Microbes has reservoirs as water, also soil. Metagenomics can be utilized in a complete valuation of the whole community of microbes, in comprehending the natural association and the interchange of biodiversity among soil and water. Work in the future can be attempted to classify new viruses utilizing third-generation technologies of sequence analysis that can examine the number of nucleotides.

References

Aktan, C., Uzunhasanoglu, A. E., & Yapsakli, K. (2018). Speciation of nickel and zinc, its short-term inhibitory effect on anammox, and the associated microbial community composition. *Bioresource Technology, 268*, 558–567. https://doi.org/10.1016/j.biortech.2018.08.011

Albertsen, M., McIlroy, S. J., Stokholm-Bjerregaard, M., Karst, S. M., & Nielsen, P. H. (2016). "Candidatus Propionivibrio aalborgensis": A novel glycogen accumulating organism abundant in full-scale enhanced biological phosphorus removal plants. *Frontiers in Microbiology, 7*, 1033. https://doi.org/10.3389/fmicb.2016.01033

Allen, H. K., Moe, L. A., Rodbumrer, J., Gaarder, A., & Handelsman, J. (2009). Functional metagenomics reveals diverse beta-lactamases in a remote Alaskan soil. *ISME Journal, 3*, 243–251. https://doi.org/10.1038/ismej.2008.86

Aminov, R. (2011). Horizontal gene exchange in environmental microbiota. *Frontiers in Microbiology, 2*(158). https://doi.org/10.3389/fmicb.2011.00158

Aw, T. G., Howe, A., & Rose, J. B. (2014). Metagenomic approaches for direct and cell culture evaluation of the virological quality of wastewater. *Journal of Virological Methods, 210*, 15–21. https://doi.org/10.1016/j.jviromet.2014.09.017

Barakat, M. A. (2011). New trends in removing heavy metals from industrial wastewater. *Arabian Journal of Chemistry, 4*, 361–377. https://doi.org/10.1016/j.arabjc.2010.07.019

Bellanger, X., Guilloteau, H., Bonot, S., & Merlin, C. (2014). Demonstrating plasmid-based horizontal gene transfer in complex environmental matrices: A practical approach for a critical review. *Science of the Total Environment, 493*, 872–882. https://doi.org/10.1016/j.scitotenv.2014.06.070

Bengtsson-Palme, J. (2018). Strategies for Taxonomic and Functional Annotation of Metagenomes. In *Metagenomics* (pp. 55–79). Academic press.

Berendonk, T. U., Manaia, C. M., Merlin, C., Fatta-Kassinos, D., Cytryn, E., Walsh, F., Bürgmann, H., Sørum, H., Norström, M., Pons, M.-N., Kreuzinger, N., Huovinen, P., Stefani, S., Schwartz, T., Kisand, V., Baquero, F., & Martinez, J. L. (2015). Tackling antibiotic resistance: The environmental framework. *Nature Reviews Microbiology, 13*, 310–317. https://doi.org/10.1038/nrmicro3439

Bharagava, R. N., Purchase, D., Saxena, G., & Mulla, S. I. (2019). Chapter 26 – Applications of Metagenomics in Microbial Bioremediation of Pollutants: From Genomics to Environmental Cleanup. In S. Das, & H. R. Dash (Eds.), *Microbial Diversity in the Genomic Era* (pp. 459–477). Academic Press. https://doi.org/10.1016/B978-0-12-814849-5.00026-5

Bofill-Mas, S., & Rusiñol, M. (2020). Recent trends on methods for the concentration of viruses from water samples. *Current Opinion in Environmental Science and Health, 16*, 7–13. https://doi.org/10.1016/j.coesh.2020.01.006

Boulund, F., Pereira, M. B., Jonsson, V., & Kristiansson, E. (2018). Chapter 4 – Computational and Statistical Considerations in the Analysis of Metagenomic Data. In *Metagenomics* (pp. 81–102). https://doi.org/10.1016/b978-0-08-102268-9.00004-5

Che, Y., Xia, Y., Liu, L., Li, A.-D., Yang, Y., & Zhang, T. (2019). Mobile antibiotic resistome in wastewater treatment plants revealed by Nanopore metagenomic sequencing. *Microbiome*, *7*, 44. https://doi.org/10.1186/s40168-019-0663-0

Chong, J., Liu, P., Zhou, G., & Xia, J. (2020). Using MicrobiomeAnalyst for comprehensive statistical, functional, and meta-analysis of microbiome data. *Nature Protocols*, *15*, 799–821. https://doi.org/10.1038/s41596-019-0264-1

Conley, D. J., Paerl, H. W., Howarth, R. W., Boesch, D. F., Seitzinger, S. P., Havens, K. E., Lancelot, C., & Likens, G. E. (2009). Ecology. Controlling eutrophication: Nitrogen and phosphorus. *Science*, *323*, 1014–1015. https://doi.org/10.1126/science.1167755

Daims, H., Lebedeva, E. V., Pjevac, P., Han, P., Herbold, C., Albertsen, M., Jehmlich, N., Palatinszky, M., Vierheilig, J., Bulaev, A., Kirkegaard, R. H., Von Bergen, M., Rattei, T., Bendinger, B., Nielsen, P. H., & Wagner, M. (2015). Complete nitrification by Nitrospira bacteria. *Nature*, *528*, 504–509. https://doi.org/10.1038/nature16461

Felczykowska, A., Krajewska, A., Zielińska, S., Łoś, J. M., Bloch, S. K., & Nejman-Faleńczyk, B. (2015). The most widespread problems in the function-based microbial metagenomics. *Acta Biochimica Polonica*, *62*, 161–166. https://doi.org/10.18388/abp.2014_917

Fernando, E. Y., McIlroy, S. J., Nierychlo, M., Herbst, F.-A., Schmid, M. C., Wagner, M., Nielsen, J. L., & Nielsen, P. H. (2018). Resolving the individual contribution of key microbial populations to enhanced biological phosphorus removal with Raman-FISH. *ISME Journal*, *13*(8), 1933–1946. https://doi.org/10.1101/387795

Ferrer, M., Golyshina, O., Beloqui, A., & Golyshin, P. N. (2007). Mining enzymes from extreme environments. *Current Opinion in Microbiology*, *10*, 207–214. https://doi.org/10.1016/j.mib.2007.05.004

Fu, F., & Wang, Q. (2011). Removal of heavy metal ions from wastewaters: A review. *Journal of Environmental Management*, *92*, 407–418. https://doi.org/10.1016/j.jenvman.2010.11.011

Guo, J., Li, J., Chen, H., Bond, P. L., & Yuan, Z. (2017). Metagenomic analysis reveals wastewater treatment plants as hotspots of antibiotic resistance genes and mobile genetic elements. *Water Research*, *123*, 468–478. https://doi.org/10.1016/j.watres.2017.07.002

Handelsman, J., Rondon, M. R., Brady, S. F., Clardy, J., & Goodman, R. M. (1998). Molecular biological access to the chemistry of unknown soil microbes: A new frontier for natural products. *Chemical Biology*, *5*(10), R245–R249. https://doi.org/10.1016/S1074-5521(98)90108-9

Haramoto, E., Kitajima, M., Hata, A., Torrey, J. R., Masago, Y., Sano, D., & Katayama, H. (2018). A review on recent progress in the detection methods and prevalence of human enteric viruses in water. *Water Research*, *135*, 168–186. https://doi.org/10.1016/j.watres.2018.02.004

Hayes, S., Mahony, J., Nauta, A., & Van Sinderen, D. (2017). Metagenomic approaches to assess bacteriophages in various environmental niches. *Viruses*, *9*(6), 127. https://doi.org/10.3390/v9060127

Holmkvist, L., Arning, E. T., Küster-Heins, K., Vandieken, V., Peckmann, J., Zabel, M., & Jørgensen, B. B. (2010). Phosphate geochemistry, mineralization processes, and Thioploca distribution in shelf sediments off central Chile. *Marine Geology*, *277*, 61–72. https://doi.org/10.1016/j.margeo.2010.08.011

Höper, D., Wylezich, C., & Beer, M. (2017). Chapter Two – Loeffler 4.0: Diagnostic Metagenomics. In M. Beer, & D. Höper (Eds.), *Loeffler's Footsteps – Viral Genomics Era High-Throughput Sequencing* (pp. 17–37). Academic Press. https://doi.org/10.1016/bs.aivir.2017.08.001

Hu, Q., Zhang, -X.-X., Jia, S., Huang, K., Tang, J., Shi, P., Ye, L., & Ren, H. (2016). Metagenomic insights into ultraviolet disinfection effects on antibiotic resistome in biologically treated wastewater. *Water Research*, *101*, 309–317. https://doi.org/10.1016/j.watres.2016.05.092

Jang, H. M., Lee, J., Choi, S., Shin, J., Kan, E., & Kim, Y. M. (2018). Response of antibiotic and heavy metal resistance genes to two different temperature sequences in anaerobic digestion of waste activated sludge. *Bioresource Technology*, *267*, 303–310. https://doi.org/10.1016/j.biortech.2018.07.051

Jassim, S. A. A., Limoges, R. G., & El-Cheikh, H. (2016). Bacteriophage biocontrol in wastewater treatment. *World Journal of Microbiology and Biotechnology*, *32*, 70. https://doi.org/10.1007/s11274-016-2028-1

Jin, M., Liu, L., Wang, D., Yang, D., Liu, W., Yin, J., Yang, Z., Wang, H., Qiu, Z., Shen, Z., Shi, D., Li, H., Guo, J., & Li, J. (2020). Chlorine disinfection promotes the exchange of antibiotic resistance genes across bacterial genera by natural transformation. *ISME Journal, 14*, 1847–1856. https://doi.org/10.1038/s41396-020-0656-9

Kim, S.-C., & Carlson, K. (2007). Temporal and spatial trends in the occurrence of human and veterinary antibiotics in aqueous and river sediment matrices. *Environmental Science and Technology, 41*, 50–57. https://doi.org/10.1021/es060737+

Kiran Kumar Reddy, G., & Nancharaiah, Y. V. (2018). Sustainable bioreduction of toxic levels of chromate in a denitrifying granular sludge reactor. *Environmental Science and Pollution Research, 25*, 1969–1979. https://doi.org/10.1007/s11356-017-0600-3

Krzeminski, P., Tomei, M. C., Karaolia, P., Langenhoff, A., Almeida, C. M. R., Felis, E., Gritten, F., Andersen, H. R., Fernandes, T., Manaia, C. M., Rizzo, L., & Fatta-Kassinos, D. (2019). Performance of secondary wastewater treatment methods for the removal of contaminants of emerging concern implicated in crop uptake and antibiotic resistance spread: A review. *Science of the Total Environment, 648*, 1052–1081. https://doi.org/10.1016/j.scitotenv.2018.08.130

Kurniawan, T. A., Chan, G. Y. S., Lo, W.-H., & Babel, S. (2006). Physico–chemical treatment techniques for wastewater laden with heavy metals. *Chemical Engineering Journal, 118*, 83–98. https://doi.org/10.1016/j.cej.2006.01.015

Liebl, W., Angelov, A., Juergensen, J., Chow, J., Loeschcke, A., Drepper, T., Classen, T., Pietruszka, J., Ehrenreich, A., Streit, W. R., & Jaeger, K.-E. (2014). Alternative hosts for functional (meta)genome analysis. *Applied Microbiology and Biotechnology, 98*, 8099–8109. https://doi.org/10.1007/s00253-014-5961-7

Marathe, N. P., Berglund, F., Razavi, M., Pal, C., Dröge, J., Samant, S., Kristiansson, E., & Larsson, D. G. J. (2019). Sewage effluent from an Indian hospital harbors novel carbapenemases and integron-borne antibiotic resistance genes. *Microbiome, 7*(1), 1–11. https://doi.org/10.1186/s40168-019-0710-x

Mardanov, A. V., Kadnikov, V. V., & Ravin, N. V. (2018). Metagenomics: A Paradigm Shift in Microbiology. In *Metagenomics* (pp. 1–13). Academic press.

McKinney, C. W., & Pruden, A. (2012). Ultraviolet disinfection of antibiotic resistant bacteria and their antibiotic resistance genes in water and wastewater. *Environmental Science and Technology, 46*, 13393–13400. https://doi.org/10.1021/es303652q

Mino, T., van Loosdrecht, M. C. M., & Heijnen, J. J. (1998). Microbiology and biochemistry of the enhanced biological phosphate removal process. *Water Research, 32*, 3193–3207. https://doi.org/10.1016/S0043-1354(98)00129-8

Munir, M., Wong, K., & Xagoraraki, I. (2011). Release of antibiotic resistant bacteria and genes in the effluent and biosolids of five wastewater utilities in Michigan. *Water Research, 45*, 681–693. https://doi.org/10.1016/j.watres.2010.08.033

New, F. N., & Brito, I. L. (2020). What is metagenomics teaching us, and what is missed? *Annual Review of Microbiology, 74*, 117–135. https://doi.org/10.1146/annurev-micro-012520-072314

Ramazzotti, M., & Bacci, G. (2018). Chapter 5 16S rRNA-Based Taxonomy Profiling in the Metagenomics Era. In *Metagenomics* (pp. 103–119). https://doi.org/10.1016/b978-0-08-102268-9.00005-7

Rosario, K., & Breitbart, M. (2011). Exploring the viral world through metagenomics. *Current Opinion in Virology, 1*, 289–297. https://doi.org/10.1016/j.coviro.2011.06.004

Sharma, P., Tripathi, S., & Chandra, R. (2021). Metagenomic analysis for profiling of microbial communities and tolerance in metal-polluted pulp and paper industry wastewater. *Bioresource Technology, 324*, 124681. https://doi.org/10.1016/j.biortech.2021.124681

Singh, R., Chirag, C., Neeta, C., & Sharma, R. (Eds.) (2020). *Metagenomics: Techniques, Applications, Challenges and Opportunities* (pp. 3–227). https://doi.org/10.1007/978-981-15-6529-8

Speth, D. R., In'T Zandt, M. H., Guerrero-Cruz, S., Dutilh, B. E., & Jetten, M. S. M. (2016). Genome-based microbial ecology of anammox granules in a full-scale wastewater treatment system. *Nature Communications*, *7*(1), 1–10. https://doi.org/10.1038/ncomms11172

Sun, F.-L., Fan, -L.-L., Wang, Y.-S., & Huang, L.-Y. (2019). Metagenomic analysis of the inhibitory effect of chromium on microbial communities and removal efficiency in A2O sludge. *Journal of Hazardous Materials*, *368*, 523–529. https://doi.org/10.1016/j.jhazmat.2019.01.076

Wommack, K. E., & Ravel, J. (2013). Microbiome, demystifying the role of microbial communities in the biosphere. *Microbiome*, *1*, 1–2. https://doi.org/10.1186/2049-2618-1-1

Zhang, T., Zhang, -X.-X., & Ye, L. (2011). Plasmid metagenome reveals high levels of antibiotic resistance genes and mobile genetic elements in activated sludge. *PLoS One*, *6*, e26041–e26041. https://doi.org/10.1371/journal.pone.0026041

Zhu, T., Su, Z., Lai, W., Zhang, Y., & Liu, Y. (2021). Insights into the fate and removal of antibiotics and antibiotic resistance genes using biological wastewater treatment technology. *Science of the Total Environment*, *776*, 145906. https://doi.org/10.1016/j.scitotenv.2021.145906

Hiren K. Patel* and Divyesh K. Vasava

Chapter 3 Microbial metagenomics: a paradigm shift in the new era of environmental biotechnology

Abstract: Nature has its own ways of resolving unusual traits in its current situation, and microbes are probably the finest mechanism for removing dangerous chemicals. “Bioremediation” refers to the process of removing contaminants using microorganisms. Rapid improvements in sequencing innovations have significantly reduced the sequencing costs and correspondingly increased the amount of sequencing programs. This has allowed the routine use of sequencing tools in wildlife studies. Although metagenomic applications have long been used to better understand the properties and diversity of organisms, their use in environmental monitoring and bioremediation is becoming increasingly common. Metagenomics is a critical technique for dissecting microbial networks at the genomic level. It stands out among previous mechanical bioremediation upgrades. In a metagenomic investigation, it is essential to identify and filter out metagenomes from unclean media. Metagenomics, in this case, uses sorting-based screening techniques and the ability to reveal information about the microbial networks of organisms that cannot be cultured in natural niches. The discoveries of new microorganisms or quality bundles encoding compounds that debase contaminations are examples of applications. Metagenomics also aids in the analysis and screening of microbes.

3.1 Introduction

In the field of metagenomics, DNA is extracted from ambient samples and studied without the microbial populations being isolated and cultured. This method was primarily used to sample the diversity of microorganisms from different environmental niches and to uncover novel microbial products. The metagenome is the collective reservoir of DNA isolated from each environmental sample and acts as a source of genetic data that may be examined for the goal of identifying novel species and previously unknown biochemical pathways and activities (Alves et al., 2018). There were

***Corresponding author: Hiren K. Patel,** School of Science, P. P. Savani University, Surat 394125, Gujarat, India, e-mail: drhkpatel1@gmail.com
Divyesh K. Vasava, R&D Department, Organic Pesticides Industries, Ankleshwar 393002, Gujarat, India

https://doi.org/10.1515/9783110793635-003

numerous environmental microorganisms that made it difficult to culture them in a laboratory. Additionally, it has been claimed that the bacteria formed more than 109 new small compounds that had practical applications (Stewart, 2012). Therefore, one of the biggest advantages of metagenomics was that it assisted in analyzing the genetic data of organisms that had never been formed before or were challenging to extract into pure cultures (Simon & Daniel, 2011). Microbial metagenomes have contributed significantly to the discovery of numerous novel enzymes, antimicrobials, medicinal chemicals, and various biochemically active substances over the past 20 years.

3.2 The focus of metagenomics research

A better knowledge of the microbiomes in various environmental settings can also be facilitated by the study of the microbial metagenome. In this chapter, we discuss the numerous current applications of metagenomics and how they have influenced the socioeconomic environment of the contemporary world. To provide light on the socioeconomic evolution, Figure 3.1 provides a graphical overview of many of the discoveries made by metagenomics over the past few years (Datta et al., 2020).

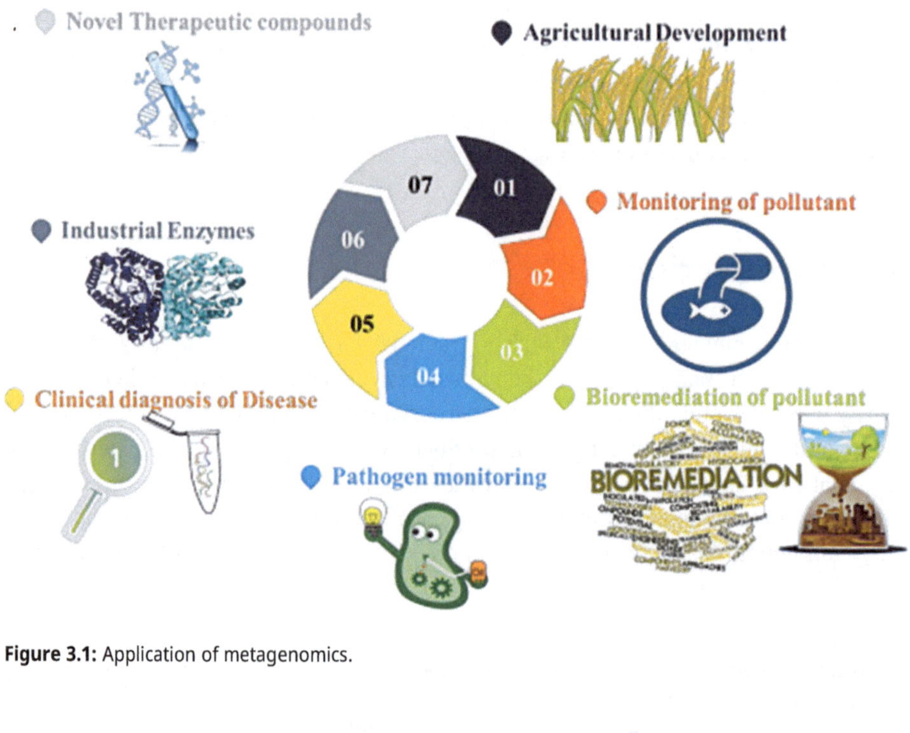

Figure 3.1: Application of metagenomics.

Metagenome-based studies often entail a workflow with several steps, including DNA extraction from the source, library generation, sequencing, post-metagenomic sequencing, data analysis, and experimentation. A general outline of the metagenomic study procedure is shown in Figure 3.2 (Datta et al., 2020). The most difficult and responsible step of metagenome screening was the isolation of DNA from environmental samples. Obtaining a pure, undamaged DNA sample containing DNA from every cell that predominated in the environmental sample proved to be quite difficult. For the extraction of the DNA material, a variety of physical, chemical, enzymatic, and mechanical techniques have been used, and they can change depending on the source.

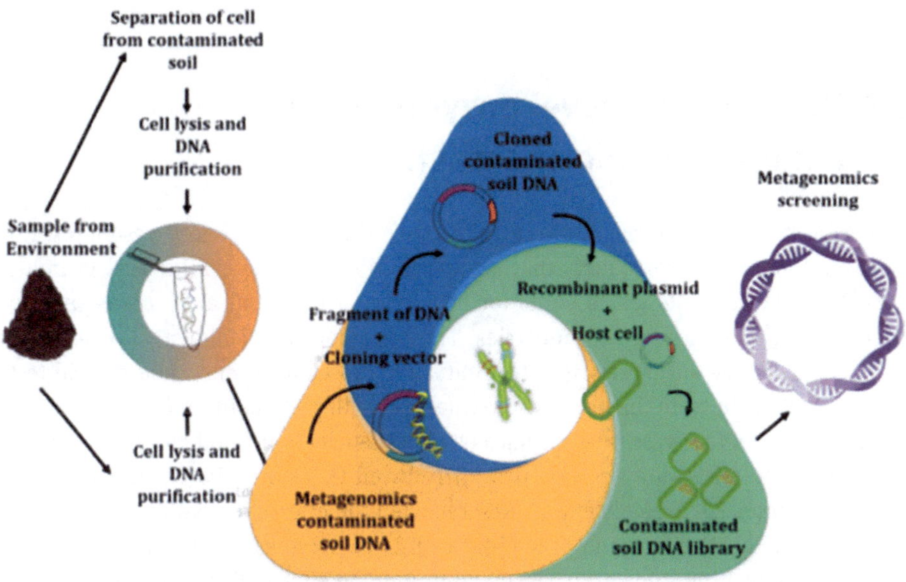

Figure 3.2: A general overview of the procedure for metagenomics of microorganism.

3.3 Metagenomic research on microbial communities

The microbial community is the world's largest one and plays a critical role in the biogeochemical cycle (Jansson & Hofmockel, 2018). Microbial communities have the most diverse species on the Earth, accounting for 60% of the Earth's biomass (UOE, 2017). The total number of microorganisms on the Earth is even predicted to reach 1,030 (Simon & Daniel, 2011). Given the importance of microbes to this terrestrial "balance of life," there is a need to better understand microbial communities to avoid the ecosystem damage (Prayogo et al., 2020). A fundamental understanding of the interactions between microbial communities enables better ecosystem management and advances in bioprospecting (Jansson & Hofmockel, 2018). Microbial communities with diverse biochemical reactions are a mystery that researchers are still trying to solve. The existence of metagenomics allows researchers to unravel its mysteries. It is fascinating to continue studying microbial communities. Furthermore, microbial communities have advantages in industrial processes. Cellulases (Cui et al., 2019), proteases (Biver et al., 2013), and esterases (Pereira et al., 2015) are examples of enzymes found in several publications and are derived from microbial communities. Enzymes from microbial communities are superior to those from animal and plant sources in terms of stability, yield, and ease of development (Liu & Kokare, 2017).

3.4 Discovery of novel enzymes using a metagenomic approach

Exploring natural resources is one way to make the most of genetic resources. Enzymes are one of the substances that are found in nature and can be used in industries. These biocatalysts are essential for both cellular biochemical processes and applications in today's modern industries. The pharmaceutical industry, which uses enzymes to change antibiotics, the soap industry, as well as forensic and clinical testing could all profit from the usage of enzymes, according to Robinson (2015).

It is not new to investigate enzymes using a metagenomic method. Direct cloning from environmental samples was first introduced in 1985 by Pace and colleagues (Simon & Daniel, 2011). Healy et al. (1995) reported the effective isolation from the environment of functional genes encoding cellulase enzymes in the first functional gene-based screening studies. Five years later, Rondon et al. (2000) used bacterial artificial chromosomes as vectors to generate metagenomic libraries from soil samples. Rondon et al. (2000) discovered specific enzymes, including lipases, amylases, and nucleases (Prayogo et al., 2020).

3.5 Recent studies using metagenomic approaches to explore new enzymes

In the past, enzymatic research still relied on traditional techniques: B. Cultivation of microorganisms on growth media (Prayogo et al., 2020). However, as technology has advanced, scientists have been able to study new enzymes without culturing them in growth media. Cellulase-, lipase-, and protease-type enzymes are essential for industrial operations (Bashir et al., 2014).

3.5.1 Cellulases

A set of enzymes known as cellulases catalyze the breakdown of cellulose polymers into simpler sugars (Escuder-Rodríguez et al., 2018). Detergents, the processing of cotton, and the paper sector can all benefit from this enzyme (Amroun et al., 2017). *Aspergillus* sp. has been identified as an entity with strong cellulase activity through standard exploration of cellulase enzymes (Paula et al., 2019). However, metagenomic techniques show that cellulase enzymes are broadly distributed in a variety of creature types. According to Cui et al. (2019), a number of species, including *Cloacibacterium, Paludibacter, Exiguobacterium, Acetivibrio, Tolumonas,* and *Clostridium,* have been identified as cellulolytic bacteria with the ability to manufacture cellulase enzymes. These six genera were discovered in bamboo paper manufacturing facilities with high cellulose conditions.

Previous studies have shown that genes producing cellulase enzymes are also present in biogas waste microbial communities (Kanokratana et al., 2013) and human gut microbial communities (Aggarwala et al., 2017). Research done on cellulase enzymes now leads to hot springs and other high-temperature situations (Schröder et al., 2014). Stringent conditions are chosen to obtain enzymes with high temperature stability (thermostability) (Prayogo et al., 2020).

3.5.2 Proteases

Enzymes called proteases cleave peptide bonds in amino acid chains. The enzyme Tavano is used in detergent, pharmaceutical, food, and beverage sectors. Numerous creatures, including plants, animals, and microbes, are sources of protease. *Bacillus* sp. is currently the most well-known protease producer in the sector (Razzaq et al., 2019). The growth of metagenomic technology makes it possible to look for additional organisms that may be more successful. According to Biver et al. (2013), a novel protease-coding gene was found in a bacterium with a similarity rate of 69% to *Desulfobacter postgatei 2ac9.* Devi et al. (2016) provide the results for the organic waste from protease enzyme encoded by the *Prt1A* gene. The *Prt1A* gene produces a protease enzyme that functions optimally at a temperature of 55 °C. In the following year, Pessoa et al. (2017) identified genes that produce the most active proteases at 60 °C.

3.5.3 Lipases

Hydrolytic cleavage of ester bonds between carboxylic acid and alcohol groups is catalyzed by enzymes known as lipases (López-López et al., 2014). Lipases are used in detergents, foods, and biodiesel, and in the bioremediation process. The best known bacterial lipase producers today are *Bacillus species. B. alcalophilus, B. licheniformis,*

B. pumilus, and *B. subtilis* (Hassan et al., 2018a; Jia et al., 2013; van Wageningen-Kessels et al., 2015). Currently, scientists are competing to investigate more organisms that may be able to produce stronger lipase. Hårdeman and Sjöling (2007) used a functional metagenomic approach to find an h1Lip1 gene that shares 54% similarity with the *Pseudomonas putida* lipase. The *h1Lip1* gene produces lipase enzyme at 35 °C (low temperature). According to López-López et al. (2014), this is often ideal for detergent cold wash processes. The metagenomic library also contains a large number of additional enzymes, and many of which have special biochemical characteristics that make them useful for industrial applications (Prayogo et al., 2020). A solvent-resistant enzyme is one illustration, and soil detergents are tainted with petroleum hydrocarbons (Pereira et al., 2015).

3.5.4 Other enzymes

Researchers have discovered several enzymes from metagenomic sources that might be made commercially available. A new bleomycin-resistant dioxygenase from contaminated agricultural soil was recently identified by Sharma et al. (2020). It serves its purpose in bioremediation processes by catalyzing the breakdown of hydrocarbon substrates such as pesticides. Furthermore, Berini et al. (2019) discovered the *53D1* gene, which encodes chitinases that may be exploited to manage plant pests. They looked at the chitinases that regulate the Lepidoptera *Bombyx mori*. The outcome demonstrated that the chitinase (*53D1* gene) enzyme was a promising pesticide. *AHL*-lactonase (Fan et al., 2017) and transaminases (Leipold et al., 2019) are employed in the pharmaceutical field, and oxoflavin-degrading enzyme (Choi et al., 2018) is a few other enzymes recently discovered to be produced from metagenomic sources.

3.6 Challenges in discovering new enzymes using the metagenomic approach

The choice of sampling sites is a hurdle for scientists looking into new enzymes. The function of enzyme characterization depends on the location. The investigation of novel enzymes can take place in each location's specific ecological niche. Functional interactions between the microbial population and its surroundings produce distinctive niches (Mhuantong et al., 2015; Staley & Sadowsky, 2018). A high concentration of cellulolytic bacteria was found in sugarcane bagasse samples, according to Mhuantong et al. (2015). When compared with noncellulolytic bacteria, the quantity of cellulolytic bacteria even appears to be substantial. Additionally, Nie et al. (2016) found that microbial populations in petroleum environments harbored genes encoding enzymes that degrade hydrocarbons. It is impossible to isolate the existence of hydrocarbon breakdown enzymes from the soil environment, which is abundant in hydrocarbon compounds. The findings support the idea that the environment affects how enzymes are characterized; hence, the location of the sampling needs to be chosen with respect to the type of enzyme being studied. Researchers have difficulties due to the features of sample locations. Sampling at locations with extreme characteristics, like hot spring locations, calls for particular methods and caution (Saxena et al., 2017).

Because extracting DNA from environmental samples is more challenging than extracting DNA from a single genome, this choice of DNA extraction methods presents another obstacle. DNA from an environment with more toxins, such as humic acid, protein, and carbohydrates, causes problems. Additionally, since the DNA is bonded to soil particles when a clay sample is used, DNA extraction will be more challenging (Zaveri et al., 2017). Therefore, samples with distinctive traits need to be given specific care by researchers.

3.7 The patented enzymes from metagenomic source for industry

In the past 5 years, several patented enzymes beneficial to the industry are published. Previously, the patented enzymes for commercialization are reported by Berini et al. (2017). This review presents novel patented enzymes published in the past 5 years (Table 3.1) (Prayogo et al., 2020). Table 3.1 lists patented enzymes from metagenomic source, including cellulase, protease, lipase, α-amylase, chitinase, β-glucosidase, and endoglucanase.

3.8 Standard methods for metagenomic analysis

Method selection is an important strategy in metagenomic research. In summary, the method is divided into two parts: molecular informatics and bioinformatics (Pucker et al., 2019).

3.8.1 Molecular method

Metagenomics is the study of the genome (metagenomics) of an ecological community as a research object. This study differs slightly from human-centric (single genome) genomic studies.

Table 3.1: Examples of patented enzymes from metagenomic source in the past 5 years (Prayogo et al., 2020).

Patent number	Country	Assignee	Enzyme	Source	Year	Application
CN108463551A	China	Scientific and Industrial Research Council	Cellulase	Soil	2016	Food and feed industry, detergent, weaving, and biofuel industry
BR102016000771A2	Brazil	Universidade Estadualhed De Santa Cruz	Protease	Mangrove sediment	2016	Antitumor, antifungal, antiviral, and antiparasitic treatment
KR102026836B1	South Korea	Korea Research Institute of Chemical Technology	Lipase	Soil	2018	Industrial mass production, biopharmaceuticals, and biodiesel
KR101816615B1	South Korea	Republic of Korea	α-Amylase	Black goat rumen	2015	Feed additive, detergents, and biofuel
CN107475273B	China	Chengdu Institute of Biology, Chinese Academy of Sciences	Chitinase	Wetland environment	2017	Food, medicine, agriculture, and cosmetic industries
CN107828806A	China	Guangdong Pharmaceutical University	β-Glucosidase	Soil	2017	Pharmaceutical, food, bioethanol, and medical industries
JP6552098B2	Japan	Honda Motor Co., Ltd., Kazusa DNA Research Institute	Endoglucanase	Hot spring soil	2016	Bioethanol

3.8.2 Metagenomic DNA extraction

Direct DNA metagenome extraction from ambient samples is performed. Accessing the DNA metagenome begins with this procedure. Based on the kind of research sample used, some researchers employ various techniques (Nazir, 2016). The metagenome's DNA was extracted by Tanveer et al. (2016) using commercial kits and accepted procedures.

The simplest way is to extract metagenomic DNA using commercial kits, which simply require reagents that the producer has provided. Lear et al. (2018) found that some researchers use dedicated kits depending on the type of sample to be extracted. Researchers most often use Power Soil and DNeasy PowerMax (Qiagen) kits to analyze soil samples, while researchers most often use DNeasy Blood and Tissue Kits (Qiagen) to analyze seawater and groundwater samples.

Using standard protocols takes more time than using commercial kits (Gutiérrez-López et al., 2015). Researchers prefer kits because they are more time efficient. In some cases, conventional methods provide better results when compared to experiments with kits. Tanveer et al. (2016) attempted to compare the HiPurA Soil DNA extraction kit (Himedia) with conventional metagenomic DNA extraction methodologies from soil. The results showed that the conventional approach produced the most DNA at the highest concentration. Standard methods result in higher concentrations than water DNA extraction kits, as shown (Hassan et al., 2018a, 2018b) (Epicenter).

The success of the subsequent stage will depend on the outcome of the metagenomic DNA extraction process, which is a critical step. Felczykowska et al. (2015) assert that the ideal DNA size must be produced during metagenomic extraction. Fragments utilized for metagenomic analysis typically range in size from 600 bp to 25 kbp. The extracted sample would not be suitable for additional metagenomic analysis due to poor results. As a result, the following needs to be taken into consideration: Do not physically tamper with the genetic material, and stay away from metals, humic acid, and protein contamination. pH, soil mineral content, and soil types are other variables that may have an impact on the outcomes of DNA extraction (Young et al., 2014).

3.9 Calculation of concentration and purity of metagenomic DNA extract

Three techniques, such as UV absorbance, fluorescence staining, and diphenylamine reaction, can be used to determine DNA concentrations and purity values (Li et al., 2014). The most common technique used by researchers to determine the DNA's content and purity is the UV absorbance method because of its simplicity, applicability, and low cost (Sun, 2010). A spectrophotometer is a tool used to determine DNA concentration and purity (Sun, 2010). Utilizing particular wavelengths that DNA molecules may absorb is the basis of the UV absorption ray technique (Li et al., 2014). Proteins absorb UV light at a wavelength of 280 nm, while DNA absorbs UV light at a wavelength of 260 nm. As a result, A260/A280 is the wavelength ratio used to determine the purity of nucleic acids. The purity ratio of DNA samples ranges from 1.8 to 2.0 (Sun, 2010). Proteins, guanidine HCL, EDTA, lipids, phenols, and salt compounds can all be detected using the ratio value of 260/230. The number of pollutants increases as the value decreases (Lucena-Aguilar et al., 2016). Results for DNA purity can be harmed by contaminants. Humic acid and protein are the most frequent contaminants in metagenomic samples (Felczykowska et al., 2015). Protein and phenol pollutants typically have absorption values that are less than 1.6, at 260/280. Meanwhile, if the 260/280 absorption ratio value is greater than 2.0, then RNA contamination of DNA is present (Lucena-Aguilar et al., 2016).

3.10 Gel electrophoresis

A common qualitative technique for separating, determining the size, and purifying nucleic acids is gel electrophoresis. This technique makes use of gel media with pores that nucleic acids can travel through (Barril & Nates, 2012). When activated, the negatively charged phosphate groups on the molecules of nucleic acids cause these molecules to migrate in the direction of the anode (positive electrode). The molecular weight, gel concentration, and electrical voltage have an impact on how quickly this transfer occurs (Yılmaz et al., 2012). The most common gels used in gel electrophoresis are agarose gels. Galactose and 3,6-anhydrogalactose are the two disaccharide units that make up agarose gels, which are polymers made of them. This gel has a lot of holes and is manufactured from seaweed extract (Barril & Nates, 2012). Gel concentration can affect the pore size. Each gel concentration profile represents the ideal length of a nucleic acid fragment to be used as a sample for gel electrophoresis. For samples from mud, Gutiérrez-Lucas et al. (2014) used a gel concentration of 0.8%. Due to the large size of environmental DNA fragments (eDNA), an agarose gel concentration of 0.8% is an approach for the electrophoresis of metagenomic materials. Depending on the length of the nucleic acid fragments used in the sample, Table 3.2 suggests recommended gel concentrations (Yılmaz et al., 2012).

3.11 Amplification of 16S rRNA gene

Protein production depends on ribosomes. It is fairly conservative and is often used as a reference when creating taxonomies. Prokaryotic bacteria typically contain 35% protein and 65% rRNA (ribosome-ribonucleic acid). Each bacterial ribosome consists of a large subunit (50S) containing two rRNA molecules (5S and 23S) and a small subunit (30S) containing one rRNA molecule (16S) (Ramazzotti & Bacci, 2018).

Prokaryotic organisms frequently employ the 16S rRNA region as a reference for taxonomy profiling study (Singer et al., 2016). The length of the nine hypervariable regions (V1–V9) of this gene is approximately 1,500 bp. The variety of prokaryotic organisms can be distinguished by these nine zones (Ramazzotti & Bacci, 2018; Roume et al., 2015). There are three factors that make 16 rRNAs as suitable markers for taxonomic profiling: (1) all prokaryotes have 16 rRNA genes; (2) lateral gene transfer is virtually unknown; and (3) the conserved structure of ribosomal proteins makes the sequence very strong (Ramazzotti & Bacci, 2018).

By employing sequence homology to read 16S rRNA sequences, it is possible to determine the identification and frequency of microorganisms. General and specific identification readings typically require a minimum level of 95% for genus and 97% for species, but a minimum level of 99% is required to read variant levels (Chierico et al., 2015). In metagenomic research, the V2–V3 region is typically a great location to employ a gene marker. However, a number of researchers have examined the diversity of microorganisms by using several target locations in the 16S rRNA gene's V region. It is stated that different target V regions will result in different compositions of the bacterial community (Zhang et al., 2018). The V1–V2 and V1–V3 zones produced the best outcomes.

3.12 Bioinformatics method

The analysis of metagenomics relies heavily on bioinformatics. The use of bioinformatics in metagenomic analysis, such as analysis of 16S rRNA data, **has been** described by Niu et al. (2018). The diversity of samples can be assessed using 16S rRNA data analysis, and the metabolic pathways of the bacteria in the sample can be predicted. MOTHUR is an illustration of a tool used for diversity analysis. Mallick et al. (2019) used the PICRUSt software to predict a community's metabolic pathway from a sample using 16S rRNA sequencing data. Utilizing bioinformatics tools depends on exploration goals. Environmental sample analyses are aided by bioinformatics technologies. Following are a few examples of bioinformatics analysis used in metagenomics approaches:

1. **Assembly:**

The process of recreating a lengthy sequence from a collection of short metagenome readings is called assembly. Contigs are the names for the lengthy sequence (Papudeshi et al., 2017). Assembly uses one of two commonly used methods: OLC or de Bruijn plot (Roumpeka et al., 2017). Assembling techniques like hybrid and iterative joining have also been developed by other researchers (Gregor et al., 2016; Ruby). The de Bruijn graph, however, is the most widely used approach. The de Bruijn graph has the benefit that it can be constructed more affordably than OLC (Roumpeka et al.,

2017). BBAP, Genovo, MegaGT, and MEGAHIT are examples of bioinformatic tools that can be applied in assembly (Ayling et al., 2020).

2. **Binning:**
The technique of clustering sequences created during the assembly process is known as binning. Binning categorizes sequences known as contigs into groupings that represent biological taxa (Roumpeka et al., 2017). Followed by the creation of contigs from raw sequence reads, this approach is used (Strous et al., 2012). MetaWatt (Strous et al., 2012) and CONCOCT (Alneberg et al., 2014) are two examples of binning analysis tools. Higher accuracy and ease of use than the previous techniques are two benefits of MetaWatt (Strous et al., 2012), although the author has claimed that CONCOCT has great precision and can group complex microbial communities (Alneberg et al., 2014).

3. **Sequence analysis:**
Finding pieces of the same biological sequence can be done using sequence analysis (Raza, 2012). There are two approaches of sequence analysis: simple alignments and multiple alignments. Multiple alignments are those that involve more than two sequences, whereas simple alignments only involve two sequences (Diniz & Canduri, 2017). BLAST (Basic Local Alignment Tool) is one of the alignment tools. A tool for comparing the sequences of distinct organism kinds is called BLAST. An expectation value (E value), a gauge of statistical significance, is assigned to the score of each alignment (Khandelwal et al., 2019).

4. **Pfam analysis:**
The protein family database is called Pfam. The double alignment generated by the hidden Markov model is the subject of Pfam's analysis. The goal of Pfam's investigation is to examine how protein sequences relate to one another at the family level (Khandelwal et al., 2019).

5. **Analysis of protein structure prediction:**
The primary structure is the amino acid sequence, which is determined by the sequence of the gene that encodes it. Proteins are divided into secondary, tertiary, and quaternary structures. To understand a protein's function, it is necessary to understand its structure. Bioinformatic predictive analysis of protein structure can help understand the physical properties of proteins and their functions (Raza, 2012).

6. **Phylogenetic analysis:**
Functional metagenomic phylogenetic analysis describes techniques for predicting specific properties of a molecule and reconstructing the evolutionary links between collections of protein molecules. The likelihood, parsimony, and distance methods are used to create phylogenetic trees. There is no perfect approach, and each has its advantages and disadvantages. Examples of phylogenetic analysis tools are MOLPHY, PHYLIP, and MEGA (Molecular Evolutionary Genetics Analysis) (Mehmood et al., 2014).

3.13 Industrial value of enzymes

The majority of the enzymes identified by metagenomic research have found a variety of uses in a wide range of fields, including food and pharmaceutical industries (Datta et al., 2020). Amylases are a diverse group of enzymes produced by numerous bacteria, archaea, and fungi that have the ability to catalyze the breakdown of starch into glucose monomers or short glucose chains. They are among the key industrial enzymes that are commercially exploited. These enzymes have been used to produce gold nanoparticles, alkane biodegradable, paper industry, food industry, textile industry, detergent, sugar, and food biodegradable (Mehta & Satyanarayana, 2016; Souza & Magalhães, 2010).

These enzymes have a wide range of applications, and metagenomics has facilitated the rapid discovery of new and useful variations of these enzymes. Table 3.2 lists that some of these enzymes have unique properties, are effective, and are more suitable for different industries than the enzymes currently on the market. Unique properties such as activity, heat resistance, and pH tolerance allow it to meet the needs of specific sectors (Datta et al., 2020). An example is the bakery industry, which benefits from the stability and activity of thermostable enzymes at higher temperatures. As a result, identifying gene clusters that produce new enzymes can help improve industrial processes and enhance the region's overall economic status.

Table 3.2: Enzymes identified using metagenomic techniques, including their isolation site, unique characteristics, and practical uses.

Enzyme	Source	Special feature	Application	References
β-Glucosidase	Topsoil of winter wheat field	Tolerant toward harsh detergents, glucose tolerant, and hydrogen peroxide tolerant	Dairy industry and biofuel	Biver et al. (2014)
Cellulase	Anaerobic beer	Halotolerant and biofuel	Textile industry and food processing	Yang et al. (2016)
Hemicellulase	Degrading wheat straw	Thermotolerant and halotolerant	Biofuel wood	Maruthamuthu et al. (2016)
Chitinase	Soil characterized as suppressive to club root disease of cabbage	Functional at a wide pH range (3–9)	Agriculture	Hjort et al. (2014)
α-Amylase	Arabian Sea sediments	Halotolerant	Food processing, fermentation, textile, detergent, and paper	Nair et al. (2017)

Table 3.2 (continued)

Enzyme	Source	Special feature	Application	References
Asparaginase	Soil	Better substrate affinity than commercially available asparaginase	Pharmaceutical and food processing	Arjun et al. (2018)
α-Galactosidase	Gut of *Hermetia illucens*	Halotolerant and protease resistant	Food processing	Lee et al. (2018)
β-Galactosidase	Not stated	Halotolerant	Dairy industry	Liu et al. (2019)

3.14 Environmental monitoring and bioremediation by microbial metagenomics

There are traces of heavy metals necessary for the maintenance of normal life of plants and animals (Samuel et al., 2019). However, excessive accumulation of heavy metals is believed to be harmful to both human health and natural ecosystems (Samuel et al., 2020). Through a variety of methods, including the use of pesticides and fertilizers, indiscriminate mining of production, metal electroplating, polishing, bioindustrial, and fertilizing, heavy metals have accumulated over time in different habitats and ecosystems. One of the main causes of heavy metal accumulation is human activity. With densities above 5 g/cm^3, about 53 elements have so far been identified as heavy metals and classified as global pollutants. The essential and nonessential heavy metal modules have been separated. Co, Fe, Cr, Cu, Mn, Ni, and Zn are essential heavy metals that were once considered life-sustaining micronutrients, but are also lethal in high doses. Pb, Cd, and Hg are nonessential heavy metals that can be tolerated in very small amounts by living organisms (Samuel et al., 2015).

Currently, adsorption, chemical precipitation, photocatalytic decomposition, ion exchange, coagulation, and solvent extraction are the main techniques used to remove heavy metal ions and other contaminants from water and other environments. (Crini & Lichtfouse, 2019; He et al., 2020). Every year, environmental contamination rises to an alarming level, endangering biological cycles and human health. According to the data produced by the World Health Organization, around 4.2 million people die each year as a result of breathing in air that contains particularly dangerous air pollutants (Samuel et al., 2020). In addition, pollution of the land and water has had serious repercussions, harming both the ecosystem and human health. In many nations, poor sanitation and a lack of reliable water supplies have contributed to the spread of infectious diseases as well as numerous other noninfectious health-related issues (Ali & Khan, 2017). Various organic and inorganic substances such as manure, heavy metals, herbicides, dyes, and hydrocarbons lead to ground poisoning. Therefore, it is important to encourage and implement the monitoring of soil contaminants to minimize hazards to public health and food safety (Yang et al., 2014).

Monitoring of air, soil, and water contamination relies heavily on microbial metagenomics. Metagenomic analysis has the ability to identify specific genes and the diversity of microbial communities, which might be used as pollution biomarkers (Kisand et al., 2012). For example, in one study it was noted that the use of biomarkers of oil pollution was emphasized in predictive and comparative metagenomic analysis on datasets obtained from contaminated habitats of 255 species and 414 functional modules (Wang et al., 2015). By comparing metagenomic datasets obtained in contaminated areas, software such as MetaBoot can be used to identify biomarkers of contamination (Wang et al., 2015). Shotgun metagenomics was used to characterize all of the microbial populations in each polluted environment, and the microbial communities' makeup can be used to create monitoring models for environmental contamination. By using targeted metagenomics, specific genes can be found in a population. Potential biomarkers for particular situations could be provided by these unique genes. For instance, the expression of genes that degrade cycloalkanes may indicate that the particular habitat was contaminated by hexane compounds (Techtmann & Hazen, 2016). In general, detailed graphs that provide clear thresholds and help to divide into different categories based on the number of stored contaminants can be generated based on expressed genes or microbial community composition.

Other than these, metagenomics has been used to identify other significant environmental remediating bacteria, bioactive substances, and enzymes. Numerous genes that break down hydrocarbons have been found in both soil and aquatic environments. This includes gene products that aid in the breakdown of hydrocarbons such as alkanes, phenanthrenes, naphthalenes, methylphenanthrenes, and some other aromatic compounds (Guerra et al., 2018). The metagenomic techniques have so far revealed a number of enzymes with bioremediation capabilities. These enzymes are obtained from soil, bovine rumen, activated sludge, environmental wastewater, sewage, oil-contaminated water, and soil that has been intentionally or naturally contaminated with chlorinated biphenyls. Dioxygenase, laccase, esterase, carboxylesterase, monooxygenase, phenol-degrading enzyme, polynuclear hydrocarbon-degrading enzyme, trichlorophenol-degrading enzyme, hexadecane-degrading enzyme, and alkane-degrading enzyme are some of the enzymes with contaminant-degrading properties. Pesticides, dyes, insecticides, plastics, fumigants, and diesel fuel are all degraded by these enzymes (Ufarté et al., 2015). The breakdown of hydrocarbons is thought to be facilitated by biosurfactants, which are amphoteric glycolipid molecules that promote the emulsification of hydrocarbons in water.

The purpose of using biosurfactants and biosurfactant-producing strains is to reduce oil spills in the marine environment. For the purpose of screening for biosurfactants, various functional tests such as oil smear test, green agar plate test, and emulsification test are used. Metagenomic analysis helps identify novel pathways and techniques to enhance the generation of novel biosurfactants by revealing novel biosurfactant-producing genes in marine bacteria. The *N*-acylated amino acids and palmitoyl putrescine are thought to be two new biosurfactants discovered by metagenomics (Jackson et al., 2015). Radionucleotides and hazardous metals are hazardous to both the environment and humans. Therefore, it is very important to remove radioactive nucleotides and other hazardous metals from soil and water resources. By screening bacteria in areas contaminated with heavy metals and radionuclides, metagenomics paved the way for the discovery of new bacteria specific to heavy metals and radioactive nucleotides. Areas contaminated with heavy metals have been reported to have nitrate-reducing microorganisms. Based on metagenomic analyses of heavy-metal-contaminated areas, several heavy metal resistance genes were found (Datta et al., 2020).

According to the metagenomic research, *Geobacter* species have continued to predominate among bacteria in uranium-contaminated sites. In addition to this, a number of metal resistance and radioresistant genes were discovered by applying metagenomic techniques with a variety of environmental samples (Xavier et al., 2019). The bioeconomy places a significant emphasis on bioremediation. A stable global economy and good health are maintained by using bioremediation technologies to remove contaminants from significant natural resources since clean water and healthy soil are vital to the economy of all nations.

3.15 Conclusion

We have discussed about the importance of metagenomics in earlier sections because of all the different ways it can be used, from finding new enzymes to using them in bioremediation processes. Sustainable industrial development has been facilitated by the use of metagenomics in search of enzymes and other bioactive molecules with industrial relevance. Monitoring of environmental bacteria and contaminants helps stop the spread of many diseases and epidemics. The minimization and ultimately elimination of health risks can be achieved by early clinical disease detection through the architecture of the viral microbiome. Additionally, bioremediation and the development of innovative medications contribute to an improvement in the quality of life on a worldwide scale. Overall, every single metagenomic application has the potential to be a socioeconomic driver that promotes the overall socioeconomic growth. This could ultimately improve everyone's quality of life. In addition, there were still a number of metagenomic applications that remained unexplored. The omics technology known as metagenomics had the potential to significantly improve the standard of living in our daily lives.

References

Aggarwala, V., Liang, G., & Bushman, F. D. (2017). Viral communities of the human gut: Metagenomic analysis of composition and dynamics. *Mobile DNA*, *8*(1), 1–10.

Ali, H., & Khan, E. (2017). Environmental chemistry in the twenty-first century. *Environmental Chemistry Letters*, *15*(2), 329–346.

Alneberg, J., Bjarnason, B. S., De Bruijn, I., Schirmer, M., Quick, J., Ijaz, U. Z., Lahti, L., Loman, N. J., Andersson, A. F., & Quince, C. (2014). Binning metagenomic contigs by coverage and composition. *Nature Methods*, *11*(11), 1144–1146.

Alves, L. D. F., Westmann, C. A., Lovate, G. L., De Siqueira, G. M. V., Borelli, T. C., & Guazzaroni, M.-E. (2018). Metagenomic approaches for understanding new concepts in microbial science. *International Journal of Genomics*, *2018*.

Amroun, A., Priet, S., De Lamballerie, X., & Quérat, G. (2017). Bunyaviridae RdRps: Structure, motifs, and RNA synthesis machinery. *Critical Reviews in Microbiology*, *43*(6), 753–778.

Arjun, J. K., Aneesh, B. P., Kavitha, T., & Harikrishnan, K. (2018). Characterization of a novel asparaginase from soil metagenomic libraries generated from forest soil. *Biotechnology Letters*, *40*(2), 343–348.

Ayling, M., Clark, M. D., & Leggett, R. M. (2020). New approaches for metagenome assembly with short reads. *Briefings in Bioinformatics*, *21*(2), 584–594.

Barril, P., & Nates, S. (2012). Introduction to agarose and polyacrylamide gel electrophoresis matrices with respect to their detection sensitivities. *Gel Electrophoresis-Principles and Basics*, 3–14.

Bashir, S., Maqsood, S. R., Peerzada, G. M., Khan, B., & Rizvi, M. A. (2014). Complexation of oxovanadium (IV) and dioxouranium (VI) with synthesized 1, 2-(diimino-4′-antipyrinyl)-1, 2-diphenylethane Schiff base: A thermodynamic, kinetic, and bioactivity investigation. *Journal of Inorganic Chemistry*, *2014*.

Berini, F., Casartelli, M., Montali, A., Reguzzoni, M., Tettamanti, G., & Marinelli, F. (2019). Metagenome-sourced microbial chitinases as potential insecticide proteins. *Frontiers in Microbiology*, *10*, 1358.

Berini, F., Casciello, C., Marcone, G. L., & Marinelli, F. (2017). Metagenomics: Novel enzymes from non-culturable microbes. *FEMS Microbiology Letters*, *364*(21), fnx211.

Biver, S., Portetelle, D., & Vandenbol, M. (2013). Characterization of a new oxidant-stable serine protease isolated by functional metagenomics. *Springerplus*, *2*(1), 1–10.

Biver, S., Stroobants, A., Portetelle, D., & Vandenbol, M. (2014). Two promising alkaline β-glucosidases isolated by functional metagenomics from agricultural soil, including one showing high tolerance towards harsh detergents, oxidants and glucose. *Journal of Industrial Microbiology and Biotechnology*, *41*(3), 479–488.

Chierico, F. D., Ancora, M., Marcacci, M., Camma, C., Putignani, L., & Conti, S. (2015). Choice of Next-generation Sequencing Pipelines. In *Bacterial Pangenomics* (pp. 31–47). Springer.

Choi, J.-E., Nguyen, C. M., Lee, B., Park, J. H., Oh, J. Y., Choi, J. S., Kim, J.-C., & Song, J. K. (2018). Isolation and characterization of a novel metagenomic enzyme capable of degrading bacterial phytotoxin toxoflavin. *PloS One*, *13*(1), e0183893.

Crini, G., & Lichtfouse, E. (2019). Advantages and disadvantages of techniques used for wastewater treatment. *Environmental Chemistry Letters*, *17*(1), 145–155.

Cui, J., Mai, G., Wang, Z., Liu, Q., Zhou, Y., Ma, Y., & Liu, C. (2019). Metagenomic insights into a cellulose-rich niche reveal microbial cooperation in cellulose degradation. *Frontiers in Microbiology*, *10*, 618.

Datta, S., Rajnish, K. N., Samuel, M. S., Pugazlendhi, A., & Selvarajan, E. (2020). Metagenomic applications in microbial diversity, bioremediation, pollution monitoring, enzyme and drug discovery. A review. *Environmental Chemistry Letters*, *18*, 1229–1241.

Devi, S. G., Fathima, A. A., Sanitha, M., Iyappan, S., Curtis, W. R., & Ramya, M. (2016). Expression and characterization of alkaline protease from the metagenomic library of tannery activated sludge. *Journal of Bioscience and Bioengineering*, *122*(6), 694–700.

Diniz, W. J. D. S., & Canduri, F. (2017). Bioinformatics: An overview and its applications. *Genetics and Molecular Research*, *16*(1), 10.4238.

Escuder-Rodríguez, -J.-J., DeCastro, M.-E., Cerdán, M.-E., Rodríguez-Belmonte, E., Becerra, M., & González-Siso, M.-I. (2018). Cellulases from thermophiles found by metagenomics. *Microorganisms*, *6*(3), 66.

Fan, X., Liang, M., Wang, L., Chen, R., Li, H., & Liu, X. (2017). Aii810, a novel cold-adapted N-acylhomoserine lactonase discovered in a metagenome, can strongly attenuate Pseudomonas aeruginosa virulence factors and biofilm formation. *Frontiers in Microbiology*, *8*, 1950.

Felczykowska, A., Krajewska, A., Zielińska, S., & Łoś, J. M. (2015). Sampling, metadata and DNA extraction-important steps in metagenomic studies. *Acta Biochimica Polonica*, *62*, 1.

Gregor, I., Schönhuth, A., & McHardy, A. C. (2016). Snowball: Strain aware gene assembly of metagenomes. *Bioinformatics*, *32*(17), i649–i657.

Guerra, A. B., Oliveira, J. S., Silva-Portela, R. C., Araújo, W., Carlos, A. C., Vasconcelos, A. T. R., Freitas, A. T., Domingos, Y. S., De Farias, M. F., & Fernandes, G. J. T. (2018). Metagenome enrichment approach used for selection of oil-degrading bacteria consortia for drill cutting residue bioremediation. *Environmental Pollution*, *235*, 869–880.

Gutiérrez-López, R., Martínez-de la Puente, J., Gangoso, L., Soriguer, R. C., & Figuerola, J. (2015). Comparison of manual and semi-automatic DNA extraction protocols for the barcoding characterization of hematophagous louse flies (Diptera: Hippoboscidae). *Journal of Vector Ecology*, *40*(1), 11–15.

Gutiérrez-Lucas, L. R., Montor-Antonio, J. J., Cortés-López, N. G., & Del Moral, S. (2014). Strategies for the extraction, purification and amplification of metagenomic DNA from soil growing sugarcane. *Advances in Biological Chemistry*, *2014*.

Hårdeman, F., & Sjöling, S. (2007). Metagenomic approach for the isolation of a novel low-temperature-active lipase from uncultured bacteria of marine sediment. *FEMS Microbiology Ecology*, *59*(2), 524–534.

Hassan, M., Essam, T., & Megahed, S. (2018a). Illumina sequencing and assessment of new cost-efficient protocol for metagenomic-DNA extraction from environmental water samples. *Brazilian Journal of Microbiology*, *49*, 1–8.

Hassan, S. W., Abd El Latif, H. H., & Ali, S. M. (2018b). Production of cold-active lipase by free and immobilized marine Bacillus cereus HSS: Application in wastewater treatment. *Frontiers in Microbiology*, *9*, 2377.

He, C., Gu, L., Xu, Z., He, H., Fu, G., Han, F., Huang, B., & Pan, X. (2020). Cleaning chromium pollution in aquatic environments by bioremediation, photocatalytic remediation, electrochemical remediation and coupled remediation systems. *Environmental Chemistry Letters*, *18*(3), 561–576.

Healy, F., Ray, R., Aldrich, H., Wilkie, A., Ingram, L., & Shanmugam, K. (1995). Direct isolation of functional genes encoding cellulases from the microbial consortia in a thermophilic, anaerobic digester maintained on lignocellulose. *Applied Microbiology and Biotechnology*, *43*(4), 667–674.

Hjort, K., Presti, I., Elväng, A., Marinelli, F., & Sjöling, S. (2014). Bacterial chitinase with phytopathogen control capacity from suppressive soil revealed by functional metagenomics. *Applied Microbiology and Biotechnology*, *98*(6), 2819–2828.

Jackson, S. A., Borchert, E., O'Gara, F., & Dobson, A. D. (2015). Metagenomics for the discovery of novel biosurfactants of environmental interest from marine ecosystems. *Current Opinion in Biotechnology*, *33*, 176–182.

Jansson, J. K., & Hofmockel, K. S. (2018). The soil microbiome – From metagenomics to metaphenomics. *Current Opinion in Microbiology*, *43*, 162–168.

Jia, G., Rao, Z., Zhang, J., Li, Z., & Chen, F. (2013). Tetraether biomarker records from a loess-paleosol sequence in the western Chinese Loess Plateau. *Frontiers in Microbiology*, *4*, 199.

Kanokratana, P., Mhuantong, W., Laothanachareon, T., Tangphatsornruang, S., Eurwilaichitr, L., Pootanakit, K., & Champreda, V. (2013). Phylogenetic analysis and metabolic potential of microbial communities in an industrial bagasse collection site. *Microbial Ecology*, *66*(2), 322–334.

Khandelwal, I., Sharma, A., Agrawal, P. K., & Shrivastava, R. (2019). Bioinformatics Database Resources. In *Biotechnology: Concepts, Methodologies, Tools, and Applications* (pp. 84–119). IGI Global.
Kisand, V., Valente, A., Lahm, A., Tanet, G., & Lettieri, T. 2012. Phylogenetic and functional metagenomic profiling for assessing microbial biodiversity in environmental monitoring.
Lear, G., Dickie, I., Banks, J., Boyer, S., Buckley, H. L., Buckley, T. R., Cruickshank, R., Dopheide, A., Handley, K. M., & Hermans, S. (2018). Methods for the extraction, storage, amplification and sequencing of DNA from environmental samples. *New Zealand Journal of Ecology, 42*(1), 10–50A.
Lee, C.-M., Kim, S.-Y., Song, J., Lee, Y.-S., Sim, J.-S., & Hahn, B.-S. (2018). Isolation and characterization of a halotolerant and protease-resistant α-galactosidase from the gut metagenome of Hermetia illucens. *Journal of Biotechnology, 279*, 47–54.
Leipold, L., Dobrijevic, D., Jeffries, J. W., Bawn, M., Moody, T. S., Ward, J. M., & Hailes, H. C. (2019). The identification and use of robust transaminases from a domestic drain metagenome. *Green Chemistry, 21*(1), 75–86.
Li, X., Wu, Y., Zhang, L., Cao, Y., Li, Y., Li, J., Zhu, L., & Wu, G. (2014). Comparison of three common DNA concentration measurement methods. *Analytical Biochemistry, 451*, 18–24.
Liu, P., Wang, W., Zhao, J., & Wei, D. (2019). Screening novel β-galactosidases from a sequence-based metagenome and characterization of an alkaline β-galactosidase for the enzymatic synthesis of galactooligosaccharides. *Protein Expression and Purification, 155*, 104–111.
Liu, X., & Kokare, C. (2017). Microbial enzymes of use in industry. Biotechnology of microbial enzymes: Production, biocatalysis and industrial applications. *Elsevier, 2017*, 267–298.
López-López, O., E Cerdan, M., & I Gonzalez Siso, M. (2014). New extremophilic lipases and esterases from metagenomics. *Current Protein and Peptide Science, 15*(5), 445–455.
Lucena-Aguilar, G., Sánchez-López, A. M., Barberán-Aceituno, C., Carrillo-Avila, J. A., López-Guerrero, J. A., & Aguilar-Quesada, R. (2016). DNA source selection for downstream applications based on DNA quality indicators analysis. *Biopreservation and Biobanking, 14*(4), 264–270.
Mallick, H., Franzosa, E. A., Mclver, L. J., Banerjee, S., Sirota-Madi, A., Kostic, A. D., Clish, C. B., Vlamakis, H., Xavier, R. J., & Huttenhower, C. (2019). Predictive metabolomic profiling of microbial communities using amplicon or metagenomic sequences. *Nature Communications, 10*(1), 1–11.
Maruthamuthu, M., Jiménez, D. J., Stevens, P., & Van Elsas, J. D. (2016). A multi-substrate approach for functional metagenomics-based screening for (hemi) cellulases in two wheat straw-degrading microbial consortia unveils novel thermoalkaliphilic enzymes. *BMC Genomics, 17*(1), 1–16.
Mehmood, M. A., Sehar, U., & Ahmad, N. (2014). Use of bioinformatics tools in different spheres of life sciences. *Journal of Data Mining in Genomics & Proteomics, 5*(2), 1.
Mehta, D., & Satyanarayana, T. (2016). Bacterial and archaeal α-amylases: Diversity and amelioration of the desirable characteristics for industrial applications. *Frontiers in Microbiology, 7*, 1129.
Mhuantong, W., Charoensawan, V., Kanokratana, P., Tangphatsornruang, S., & Champreda, V. (2015). Comparative analysis of sugarcane bagasse metagenome reveals unique and conserved biomass-degrading enzymes among lignocellulolytic microbial communities. *Biotechnology for Biofuels, 8*(1), 1–17.
Nair, H. P., Vincent, H., Puthusseri, R. M., & Bhat, S. G. (2017). Molecular cloning and characterization of a halotolerant α-amylase from marine metagenomic library derived from Arabian Sea sediments. *3 Biotech, 7*(1), 1–9.
Nazir, A. (2016). Review on metagenomics and its applications. *Imperial Journal of Intersdisciplinary Research, 2*, 10.
Nie, Y., Zhao, J.-Y., Tang, Y.-Q., Guo, P., Yang, Y., Wu, X.-L., & Zhao, F. (2016). Species divergence vs. functional convergence characterizes crude oil microbial community assembly. *Frontiers in Microbiology, 7*, 1254.
Niu, S.-Y., Yang, J., McDermaid, A., Zhao, J., Kang, Y., & Ma, Q. (2018). Bioinformatics tools for quantitative and functional metagenome and metatranscriptome data analysis in microbes. *Briefings in Bioinformatics, 19*(6), 1415–1429.

Papudeshi, B., Haggerty, J. M., Doane, M., Morris, M. M., Walsh, K., Beattie, D. T., Pande, D., Zaeri, P., Silva, G. G., & Thompson, F. (2017). Optimizing and evaluating the reconstruction of Metagenome-assembled microbial genomes. *BMC Genomics, 18*(1), 1–13.

Paula, C. C. D., Montoya, Q. V., Meirelles, L. A., Farinas, C. S., Rodrigues, A., & Seleghim, M. H. (2019). High cellulolytic activities in filamentous fungi isolated from an extreme oligotrophic subterranean environment (Catão cave) in Brazil. *Anais da Academia Brasileira de Ciências, 91*.

Pereira, M. R., Mercaldi, G. F., Maester, T. C., Balan, A., & De Macedo Lemos, E. G. (2015). Est16, a new esterase isolated from a metagenomic library of a microbial consortium specializing in diesel oil degradation. *PLoS One, 10*(7), e0133723.

Pessoa, T. B., Rezende, R. P., Marques, E. D. L. S., Pirovani, C. P., Dos Santos, T. F., Dos Santos Gonçalves, A. C., Romano, C. C., Dotivo, N. C., Freitas, A. C., & Salay, L. C. (2017). Metagenomic alkaline protease from mangrove sediment. *Journal of Basic Microbiology, 57*(11), 962–973.

Prayogo, F. A., Budiharjo, A., Kusumaningrum, H. P., Wijanarka, W., Suprihadi, A., & Nurhayati, N. (2020). Metagenomic applications in exploration and development of novel enzymes from nature: A review. *Journal of Genetic Engineering and Biotechnology, 18*(1), 1–10.

Pucker, B., Schilbert, H., & Schumacher, S. (2019). Integrating molecular biology and bioinformatics education. *Journal of Integrative Bioinformatics, 16*(3), 1–7.

Ramazzotti, M., & Bacci, G. (2018). 16S rRNA-based Taxonomy Profiling in the Metagenomics Era. In *Metagenomics* (pp. 103–119). Elsevier.

Raza, K. (2012). Application of data mining in bioinformatics. *arXiv Preprint arXiv:1205.1125*.

Razzaq, A., Shamsi, S., Ali, A., Ali, Q., Sajjad, M., Malik, A., & Ashraf, M. (2019). Microbial proteases applications. *Frontiers in Bioengineering and Biotechnology, 7*, 110.

Robinson, P. K. (2015). Enzymes: Principles and biotechnological applications. *Essays in Biochemistry, 59*, 1.

Rondon, M. R., August, P. R., Bettermann, A. D., Brady, S. F., Grossman, T. H., Liles, M. R., Loiacono, K. A., Lynch, B. A., MacNeil, I. A., & Minor, C. (2000). Cloning the soil metagenome: A strategy for accessing the genetic and functional diversity of uncultured microorganisms. *Applied and Environmental Microbiology, 66*(6), 2541–2547.

Roume, H., Heintz-Buschart, A., Muller, E. E., May, P., Satagopam, V. P., Laczny, C. C., Narayanasamy, S., Lebrun, L. A., Hoopmann, M. R., & Schupp, J. M. (2015). Comparative integrated omics: Identification of key functionalities in microbial community-wide metabolic networks. *npj Biofilms and Microbiomes, 1*(1), 1–11.

Roumpeka, D. D., Wallace, R. J., Escalettes, F., Fotheringham, I., & Watson, M. (2017). A review of bioinformatics tools for bio-prospecting from metagenomic sequence data. *Frontiers in Genetics, 8*, 23.

Ruby, P. Software for the targeted assembly of components of (Meta) genomic sequence data, G3.

Samuel, M. S., Bhattacharya, J., Raj, S., Santhanam, N., Singh, H., & Singh, N. P. (2019). Efficient removal of Chromium (VI) from aqueous solution using chitosan grafted graphene oxide (CS-GO) nanocomposite. *International Journal of Biological Macromolecules, 121*, 285–292.

Samuel, M. S., EA Abigail, M., & Ramalingam, C. (2015). Biosorption of Cr (VI) by Ceratocystis paradoxa MSR2 using isotherm modelling, kinetic study and optimization of batch parameters using response surface methodology. *PLoS One, 10*(3), e0118999.

Samuel, M. S., Jose, S., Selvarajan, E., Mathimani, T., & Pugazhendhi, A. (2020). Biosynthesized silver nanoparticles using Bacillus amyloliquefaciens; Application for cytotoxicity effect on A549 cell line and photocatalytic degradation of p-nitrophenol. *Journal of Photochemistry and Photobiology B: Biology, 202*, 111642.

Saxena, R., Dhakan, D. B., Mittal, P., Waiker, P., Chowdhury, A., Ghatak, A., & Sharma, V. K. (2017). Metagenomic analysis of hot springs in Central India reveals hydrocarbon degrading thermophiles and pathways essential for survival in extreme environments. *Frontiers in Microbiology, 7*, 2123.

Schröder, C., Elleuche, S., Blank, S., & Antranikian, G. (2014). Characterization of a heat-active archaeal β-glucosidase from a hydrothermal spring metagenome. *Enzyme and Microbial Technology, 57*, 48–54.

Sharma, V., Kumar, R., Sharma, V. K., Tiirola, M., & Sharma, P. K. (2020). Expression, purification, characterization and in silico analysis of newly isolated hydrocarbon degrading bleomycin resistance dioxygenase. *Molecular Biology Reports*, *47*(1), 533–544.

Simon, C., & Daniel, R. (2011). Metagenomic analyses: Past and future trends. *Applied and Environmental Microbiology*, *77*(4), 1153–1161.

Singer, E., Bushnell, B., Coleman-Derr, D., Bowman, B., Bowers, R. M., Levy, A., Gies, E. A., Cheng, J.-F., Copeland, A., & Klenk, H.-P. (2016). High-resolution phylogenetic microbial community profiling. *The ISME Journal*, *10*(8), 2020–2032.

Souza, P. M. D., & Magalhães, P. D. O. (2010). Application of microbial α-amylase in industry-A review. *Brazilian Journal of Microbiology*, *41*, 850–861.

Staley, C., & Sadowsky, M. J. (2018). Practical considerations for sampling and data analysis in contemporary metagenomics-based environmental studies. *Journal of Microbiological Methods*, *154*, 14–18.

Stewart, E. J. (2012). Growing unculturable bacteria. *Journal of Bacteriology*, *194*(16), 4151–4160.

Strous, M., Kraft, B., Bisdorf, R., & Tegetmeyer, H. (2012). The binning of metagenomic contigs for microbial physiology of mixed cultures. *Frontiers in Microbiology*, *2012*, 3.

Sun, W. (2010). General Procedures. In *Molecular Diagnostics: Techniques and Applications for the Clinical Laboratory* (pp. 92675). London: Academic Press. https://doi.org/10.1016/C2009-0-00182-2

Tanveer, A., Yadav, S., & Yadav, D. (2016). Comparative assessment of methods for metagenomic DNA isolation from soils of different crop growing fields. *3 Biotech*, *6*(2), 1–5.

Tavano, O. A., Berenguer-Murcia Secundo, F., & Fernandez-Lafuente, R. (2018). Biotechnological applications of proteases in food technology. *Comprehensive Reviews in Food Science and Food Safety*, *17*, 412–436.

Techtmann, S. M., & Hazen, T. C. (2016). Metagenomic applications in environmental monitoring and bioremediation. *Journal of Industrial Microbiology and Biotechnology*, *43*(10), 1345–1354.

Ufarté, L., Laville, É., Duquesne, S., & Potocki-Veronese, G. (2015). Metagenomics for the discovery of pollutant degrading enzymes. *Biotechnology Advances*, *33*(8), 1845–1854.

UOE, O. 2017. An overview of cultural, molecular and metagenomic techniques in description of microbial diversity.

van Wageningen-kessels, F., Van Lint, H., Vuik, K., & Hoogendoorn, S. (2015). Genealogy of traffic flow models. *EURO Journal on Transportation and Logistics*, *4*(4), 445–473.

Wang, X., Su, X., Cui, X., & Ning, K. (2015). MetaBoot: A machine learning framework of taxonomical biomarker discovery for different microbial communities based on metagenomic data. *PeerJ*, *3*, e993.

Xavier, J., Costa, P., Hissa, D., Melo, V., Falcão, R., Balbino, V., Mendonça, L., Lima, M., Coutinho, H., & Verde, L. (2019). Evaluation of the microbial diversity and heavy metal resistance genes of a microbial community on contaminated environment. *Applied Geochemistry*, *105*, 1–6.

Yang, C., Xia, Y., Qu, H., Li, A.-D., Liu, R., Wang, Y., & Zhang, T. (2016). Discovery of new cellulases from the metagenome by a metagenomics-guided strategy. *Biotechnology for Biofuels*, *9*(1), 1–12.

Yang, H., Huang, X., Thompson, J. R., & Flower, R. J. (2014). Soil pollution: Urban brownfields. *Science*, *344*(6185), 691–692.

Yılmaz, M., Ozic, C., & Gok, İ. (2012). Principles of nucleic acid separation by agarose gel electrophoresis. *Gel Electrophoresis – Principles and Basics*, *4*, 33.

Young, J. M., Rawlence, N. J., Weyrich, L. S., & Cooper, A. (2014). Limitations and recommendations for successful DNA extraction from forensic soil samples: A review. *Science & Justice*, *54*(3), 238–244.

Zaveri, P., Patel, R., Patel, M., Sarodia, D., & Munshi, N. S. (2017). Modification of extraction method for community DNA isolation from salt affected compact wasteland soil samples. *MethodsX*, *4*, 63–67.

Zhang, J., Ding, X., Guan, R., Zhu, C., Xu, C., Zhu, B., Zhang, H., Xiong, Z., Xue, Y., & Tu, J. (2018). Evaluation of different 16S rRNA gene V regions for exploring bacterial diversity in a eutrophic freshwater lake. *Science of the Total Environment*, *618*, 1254–1267.

Gaurav Yadav, Surati Kumari, Aneesha Polisety, Priyanka Prajapati, Devendra Singh, Sam Joy and Iffat Zareen Ahmad*

Chapter 4
Integrated omics platforms for biobutanol fermentation employing extremophiles

Abstract: Considering the over exhaustion and ecological effects of fossil fuels, there is a need to build an eco-friendly alternative. Biobutanol offers a suitable substitute for the future. However, only a few microorganisms have the ability to produce biobutanol by exploiting the acetone-biobutanol-ethanol (ABE) fermentation pathway, and lignocellulosic biomasses are the most feasible asset. However, it possesses limitations because of the necessity for pretreatment, hydrolysis, and resistance to degradation. To overcome these challenges, microorganisms require reprogramming to maintain industrial productivity standards, which serves as a constraint in production costs. Providentially, there are extremophiles that flourish in nature and offer an amazing substitution for enzymes and metabolites. Extremophiles can give momentum to biobutanol fermentation and lignocellulose degradation under extreme conditions. MS-based proteomics and metabolomics approaches can elucidate the extremoenzymes and extremometabolites for biobutanol fermentation because of their accuracy and productivity. Proteomics uncovers proteins needed for catalyzing biobutanol with increased activity, and metabolomics assists in selecting strains for ABE fermentation.

Acknowledgment: The authors are thankful to Integral University Lucknow for providing necessary facilities for the completion of this chapter.

***Corresponding author: Iffat Zareen Ahmad**, Natural Product Laboratory, Department of Bioengineering, Integral University, Lucknow, Uttar Pradesh, India, e-mail: iffat@iul.ac.in
Gaurav Yadav, Natural Product Laboratory, Department of Bioengineering, Integral University, Lucknow, Uttar Pradesh, India
Surati Kumari, School of Bio Sciences and Technology, Vellore Institute of Technology, Vellore, Tamil Nadu
Aneesha Polisety, Department of Industrial Engineering, University of Rome Tor Vergata, Rome, Italy
Priyanka Prajapati, Centre for Advanced Studies in Botany, Department of Botany, Institute of Science, Banaras Hindu University, Varanasi, Uttar Pradesh, India
Devendra Singh, Department of Biotechnology, Motilal Nehru National Institute of Technology Allahabad, Allahabad, Uttar Pradesh, India
Sam Joy, Indian Council of Medical Research-National Institute of Malaria Research, New Delhi, India

https://doi.org/10.1515/9783110793635-004

4.1 Introduction

Ever since the Industrial Revolution that started in the mid-eighteenth century, there has been a dramatic increase in energy demands worldwide, leading to increased exploration, production, and utilization of fossil fuels. Fossil fuels form an important and versatile substance, used in different forms across several applications, and even play a dominant role in driving global economies. Over the decades, with the current increasing energy consumption rates and exploding population, the Organization of the Petroleum Exporting Countries (OPEC) has predicted that global oil product demands will touch a mark of 109 million barrels per day (MB/D) by 2045 to meet a population projected to be 9.5 billion by 2045.

However, contemporary issues such as the volatile petroleum prices, increasing energy crisis, depletion of oil reserves, and growing awareness toward environmental safeguards and global climate policies and agreements have necessitated that global economies switch toward greener, abundant, clean, and affordable renewable resources. The sustainable production of biofuels has provided an attractive alternative to fossil fuels.

In the last two decades, among different biofuels, biobutanol has been gaining attention and is emerging as a feasible and sustainable biofuel due to its higher energy density (37 MJ/kg in comparison to only 30 MJ/kg for ethanol), fewer emissions, lower Reid vapor pressure, better anticorrosive nature, better blending compatibility with gasoline, and better transport options. Although the production of biobutanol through anaerobic bacterial fermentation dates back to as early as 1861 by Pasteur, the industrial production of biobutanol by Chaim Weizmann started only at the beginning of the twentieth century. Furthermore, the growth of the petroleum chemical industry and the cheaper cost of producing biobutanol from petroleum products in comparison to organic feedstock made the bio-based biobutanol plant obsolete. However, the increasing crude oil prices and environmental concerns have revived interest, research, and development into biobutanol.

According to the market survey carried out by 360 Market Updates, the global biobutanol market was found to be valued at 11,640 million US dollars in 2019, and it is projected to reach 18,510 million US dollars by 2027, growing at a compound annual growth rate of 6.8% from 2021 to 2027. The major factors driving the biobutanol market are the global demands for the reduction of carbon emissions and the emergence of biobutanol as an alternative biofuel and raw material for various safe and green chemical manufacturing.

Biobutanol is a versatile compound and finds applications in several industrial sectors. Biobutanol can be utilized as an alternative liquid fuel in internal combustion engines or in the blending of conventionally used fossil fuels. Since biobutanol is more chemically comparable to petroleum in comparison to ethanol, it can be incorporated into conventional internal combustion engines more easily than ethanol. It is also majorly used in several industries as an industrial solvent. It is used as an intermediate material, especially in the manufacturing of butyl acrylates, acetates, glycol ethers, and plasticizers, which are further used in paint, coating, adhesives, and textile manufacturing. It also finds application in the pharmaceutical industry as an extracting agent in the manufacture of drugs and cosmetics.

The major production strategy for biobutanol is through the fermentation of sugars using organic feedstock via the anaerobic acetone–biobutanol-ethanol (ABE) fermentative pathway observed in various microorganisms. In the ABE fermentation, biobutanol is produced as a by-product along with ethanol and acetone, mostly in the ratio of 6:3:1. However, a limited number of strains of clostridia such as *Clostridium aurantibutyricum* and *Clostridium beijerinckii* have also been reported to produce isopropanol instead of acetone. The most commonly reported biobutanol-producing bacteria are members of the genus *Clostridium.* Several species of *Clostridium* that have been reported in the literature to produce biobutanol include *C. acetobutylicum, C. aurantibutyricum, C. beijerinckii, C. cadaveris, C. pasteurianum, C. saccharoacetobutylicum, C. saccharoperbutylacetonicum, C. sporogenes,* and *C. tetanomorphum.* The ABE fermentation process is more complex and sensitive in comparison to other fermentative processes, involving two intricate phases of fermentation. The initial phase consists of an acidogenic phase in which major metabolic products such as butyrate, acetone, carbon dioxide, and hydrogen are produced, which lowers the pH of the fermentative growth medium. The lowering of the pH plays a very important role in switching the fermentation toward the next phase of solventogenesis, in which the metabolites produced in acidogenesis are converted to generate biobutanol, acetone, ethanol, H_2, and CO_2. In the acidogenesis phase, butyrate is supplied from butyryl-CoA and acetone from acetyl-CoA. However, during solventogenesis, both butyryl-CoA and acetyl-CoA act as intermediates for the production of biobutanol and ethanol, while in the case of acetone production, acetyl-CoA serves as the intermediate.

In spite of all the advantages and superior potential of biobutanol, the commercialization process still faces major hurdles due to high feedstock costs, low biobutanol concentrations, biobutanol inhibition, and high downstream processing costs. The low biobutanol product yield, due to the low tolerance of the producing bacterium, is one of the most important bottlenecks affecting the economics of production. The biobutanol production ranges as low as 10 to 20 g/L concentration in batch reactors with low productivity rates (<0.50 g/L/h) due to the toxicity of the biobutanol produced in the fermentative broth. The low biobutanol yield also affects the economics of downstream processing. Several methods have been studied to improve downstream efficiency with the aim of lowering processing costs. Some of the most commonly reported methods include the continuous or in situ removal of biobutanol from the fermentation broth using methods such as liquid-liquid extraction, adsorption, pervaporation, and gas stripping.

To overcome these bottlenecks and move toward the realization of the commercialization and industrial production of biobutanol, several extensive efforts have been exploited. Some of the efforts include the selection of waste/low-cost/zero-cost feedstock, improving productivity and tolerance, adopting various continuous or sequential fed-batch fermentation strategies along with various integrated downstream recovery techniques, strain improvement, and metabolic engineering of the ABE pathways.

Over the last few years, thermophilic microorganisms have been of industrial interest because of their ability to catalyze reactions of industrial significance at elevated temperatures. These organisms thrive in a variety of marine and terrestrial habitats and have enhanced rigidity compared to their mesophilic counterparts. Some of the advantages of using these organisms include lower contamination risk and production cost due to the elimination of the need for sterile environments and sterilization techniques. In addition, in biobutanol production, the thermophilic conditions provide the opportunity for continuous removal of volatile compounds. However, only a very few microorganisms have been reported over the years and characterized in detail with respect to physiological and genomic perspectives for biobutanol due to the limitation of conventional techniques. This thermophilic group of microorganisms serves as an abundant reservoir pool to isolate biobutanol producers, which can be isolated using integrated omics platforms. Thus, integrated omics platforms can serve as a tool to study the complex biological and metabolic processes involved in biobutanol production. Omics provides an integrative approach that can combine multi-omics data to highlight the interrelationships of the various biomolecules, enzymes, and their functions in the metabolic pathway of biobutanol fermentation. With the advent of present-day high-throughput rapid techniques and the availability of multi-omics data generated from a large set of samples, several promising tools and methods can serve in the collection, integration, and interpretation, paving new insights into biobutanol production as depicted in Figure 4.1. Thus, this chapter focuses on the use of integrated omics platforms with an emphasis on their salient features relevant to the utilization of genomics, proteomics, and metabolomics to improve their potential for biobutanol production employing thermophiles.

4.2 Role of extremophiles in industry

Increasing pollution and the greenhouse effect on the climate are major concerns in the modern scenario, resulting in significant environmental circumstances. As a result, researchers' primary goal is to resolve the environmental challenges in a budget-friendly and long-term manner (Zhu et al., 2020). The bulk of enzymes utilized in industries, on the other hand, come from mesophilic microorganisms, with only a few from extremophiles (Karim et al., 2020). In industrial operations, mesophilic enzymes are unable to function in severe conditions. Extremophiles are a safer choice for scientists and industry to tackle harsh environmental conditions such as pH, temperature, pressure, and so on. As a result, to deal with the cost and damage, the demand

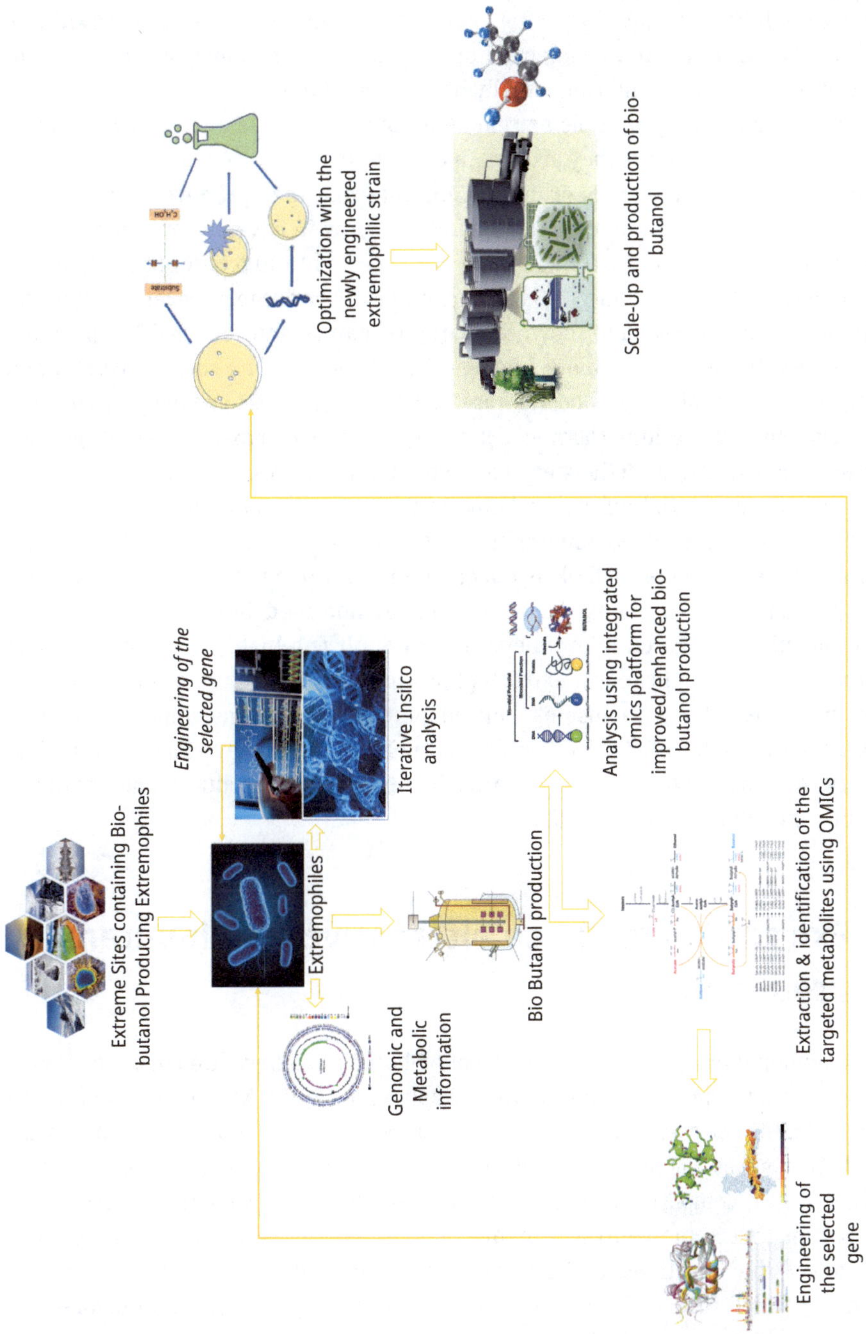

Figure 4.1: Extremophiles have the ability to catalyze reactions of industrial significance at elevated temperatures. Multi-omics tools and methods can serve in the collection, integration, and interpretation of data, paving new insights into biobutanol production using extremophiles.

for extremophilic enzymes is growing rapidly to suit industrial demands. Extremophilic microorganisms can be employed directly as a cell or as enzymes to carry out chemical reactions in an environmentally benevolent manner.

Microbes harboring multiple extreme environments are known as polyextremophiles. The adaptations that they achieve in order to flourish in a hostile environment are the result of diverse kinds of unique enzymes that they produce (Kumar et al., 2018). Considering the fact that extremophiles can grow in adverse environments, enzymes produced by them (extremozymes) are expected to function optimally under such conditions (van den Burg, 2003). Over the past few decades, several studies focused on extremozymes have shown their robust nature and unveiled the potential they hold for industrial and biotechnological applications. One such successful and widely known example is "Taq polymerase," a DNA polymerase enzyme purified from the thermophilic bacterium *Thermus aquaticus*, which is extensively used in PCR for amplification owing to its activity even at higher temperatures.

Given the substantial difference between the optimal conditions of enzymes obtained from extremophilic and mesophilic organisms, it can be observed that extremophiles and their enzymes will likely play an important role in many kinds of bioprocessing mechanisms, such as the conversion of non-food biomass into biofuels, biomining, bioremediation, biorefinery, and so on. It is a highly acknowledged fact that for industrial applications, extremophiles present themselves as better options rather than mesophiles because of their adaptability to extreme conditions (Coker, 2016). They can be beneficial for the biofuel industry as they have lower contamination risk, minimal environmental hazards, higher yields of products, and enhanced microbial growth (Bibra et al., 2015).

4.3 Extremophiles as an alternative for biobutanol production

The conventional mode for the production of biobutanol is either through the microbial route or the petrochemical route (Gottumukkala et al., 2019). Keeping in mind the (i) higher yield, (ii) low cost and easy recovery, (iii) faster process, and (iv) no requirement of land for biomass, the petrochemical route is more promising than microbial fermentation. However, compared to chemical catalytic synthesis, enzymes offer better chemical precision, that is, more efficient production of single stereoisomers, substrate specificity, fewer side reactions, and environmentally friendly. Besides, the continuous depletion of fossil fuels and growing concern over the environment have favored the utilization of biocatalysts. To date, mesophilic microbes are the main source for most industrial enzymes; though they offer certain benefits, they cannot withstand the extreme conditions of industrial processes. But these constraints can be abated by employing extremophiles.

In biobutanol production using renewable lignocellulosic, feedstock, depending on its type, the process undergoes four major steps: (i) pretreatment of the feedstock, (ii) hydrolysis of the pretreated feedstock, (iii) fermentation using the lignocellulosic hydrolysate, and (iv) downstream processing and product recovery. Currently, crops like rice, wheat, sugarcane, coffee beans, or cereal grains are used as raw materials (Ni et al., 2012). A few experiments reported that syngas produced from biomass can also be used as a substrate for the production of biobutanol using thermophiles and halophiles, thereby expanding the feedstock options (Bengelsdorf et al., 2013). However, over the years, the major emphasis is on utilizing lignocellulosic waste biomass to reduce the production cost and also to reduce the dependence on edible sources (Khandeparker & Sani, 2018). Pretreatment of lignocellulosic feedstock is a process in which complex recalcitrant lignocellulose is converted into its simple components, such as cellulose, hemicellulose, and lignin (Kumari & Singh, 2018), which can enhance cellulose access to enzymes during enzymatic hydrolysis (Vasić et al., 2021). An ideal industrial process is one that reduces overall production costs without sacrificing quality, resulting in a more cost-effective commodity. The pretreatment and hydrolysis unit accounts for 50% of the fixed capital investment. Thus, there is a need for a cost-effective pretreatment method.

"Extremophile" was coined by R.D. MacElroy in 1974 in his paper *"Some comments on the evolution of extremophiles"*. Extremophiles (extremus from Latin – extreme; philiâ from Greek – love) stand true to their name; not only do they live but thrive in extreme environments that are hostile to the majority of life forms. These extremes can be physical, for instance, temperature, pressure, or radiation, and geological, which include pH, salinity, and anoxia, and often coincide with each other, like the deep lake of Antarctica (hypersaline and extremely cold) or the black smoker chimney near Papua New Guinea (hypersaline and extremely hot) (Takai et al., 2001). It is important to distinguish extremophiles from those organisms that merely "tolerate" the extreme environment, the *extremotrophs,* but grow optimally under "normal" conditions. Unlike them, extremophiles are well "adapted" to extreme environments. The majority of extremophiles belong to the domain *Archaea*, thus mostly being unicellular and prokaryotic (Gupta et al., 2014).

In recent experiments, *Caldicellulosiruptor bescii*, a hyperthermophilic cellulolytic and hemicellulolytic bacterium, can successfully produce desirable bioproducts from untreated plant biomass as well as crystalline cellulose without pretreatment (Zhang et al., 2021), thus subsequently skipping the pretreatment cost. Typically, lignocellulosic biomass undergoes alkaline/acidic pretreatment, which is then neutralized before the hydrolysis phase (Sindhu et al., 2011). Nonetheless, using thermoacidophilic and thermoalkaliphilic enzymes for hydrolysis can eliminate the need to neutralize (Bhalla et al., 2013). It is also anticipated that by using extremely acidophilic and alkaliphilic enzymes at high temperatures, there is a possibility to avoid acid/alkali pretreatment (Bhalla et al., 2013).

Following the pretreatment, the residual of lignocellulose breakdown (mainly cellulose and hemicellulose) undergoes enzymatic hydrolysis, which converts them into monosaccharides before being used as the substrate in fermentation. However, present methods for enzymatic hydrolysis, which are carried out at or below 50 °C, exhibit various constraints (Taylor et al., 2008), some of which can be omitted by using thermophilic organisms. Another study stated that higher temperatures accelerate chemical reactions, resulting in higher specific activity of enzymes, which eventually decreases the amount of enzyme required for saccharification (Vaishnavi & Muralidhara, 2017). Also, fermentations involving thermozymes are at lower risk of microbial contamination, as there are fewer indigenous bacteria that are competitive at higher temperatures (Taylor et al., 2008) and require lower energy consumption in comparison to normal conventional fermentation, which results in the reduction of cooling steps required between the fermentation steps. The production of thermostable cellulase has been documented in a variety of bacteria such as *Acidothermus, Bacillus, Caldocellum, Caldibacillus, Clostridium*, and *Geobacillus* (Bhalla et al., 2013).

Hemicellulose is a heteropolymer composed of pentose and hexose sugars. Several thermostable, thermoalkalophilic, thermoacidophilic, and thermohalophilic organisms produce xylanases (Bhalla et al., 2013). Lignin is the most resistant among the three components of lignocellulose. A few extremophilic enzymes, such as laccase, manganese peroxidase, and lignin peroxidase, have been reported for lignin degradation and belong to the fungal family of basidiomycetes (Chandra et al., 2017).

Hydrolysates can undergo microbial fermentation to produce biobutanol. Most biobutanol-fermenting organisms are able to ferment hexose sugars only, while very few organisms can ferment pentose sugars. However, several thermophilic microorganisms (e.g., *Thermoanaerobacter*, *Clostridium*, and *Geobacillus*) have been reported to have the ability to ferment both hexose and pentose sugars, as well as many complex sugars (cellobiose, xylooligomers, glucose, fructose, and xylose) (Bibra et al., 2015). Furthermore, commercial biobutanol producers such as *Clostridium acetobutylicum* are anaerobes. Since the microbes used are anaerobes, anoxic conditions must be established before the onset of fermentation, and the bioreactor must remain closed during the process, which eventually increases the cost of downstream separation (Khandeparker & Sani, 2018) (Bibra et al., 2015). A similar study (Amiri et al., 2016) reported that a strain of moderately halophilic bacterium, *Nesterenkonia* sp., can grow and carry out ABE fermentation under both anaerobic and aerobic environments. Another work suggested that the technique of growing both aerobic and anaerobic organisms with a given set of improved fermentative conditions can eliminate the need for utilizing costly reducing agents, which are generally used in the anaerobic setup (Biswas et al., 2017). Some of the currently known biobutanol-producing extremophiles are listed in Table 4.1.

Table 4.1: Biobutanol production by some extremophilic microbes.

Environment condition	Extremophilic strains	Growth medium	References
High saline	*Nesterenkonia* sp.	Starch crops and sugars	Amiri et al. (2016)
	Vibrio furnissii	Starch crops and sugars	Park et al. (2007)
	Clostridium carboxidivorans	Lignocellulosic agricultural waste	Liou et al. (2005)
High alkaline	*Bacillus alcalophilus*	Starch crops and sugars	Meng et al. (2009)
	Clostridium cellulovorans	Starch crops and sugars	Wen et al. (2014)
High temperature	*Geobacillus thermoglucosidasius*	Culture media	Lin et al. (2014)
	Pyrococcus furiosus	Culture media	Keller and Schut (2014)
	Clostridium sp.	Corn stocks	Coker (2016)
	C. thermocellum + C. Saccharoper butylacetonicum N1-4	Crystalline Cellulose	Qureshi et al. (2007)

Microbial fermentation results not only in the production of biobutanol but also in other byproducts, such as acetone and ethanol, resulting in a lower biobutanol yield (Cao et al., 2016). In these circumstances, the discovery of *Thermoanaerobacterium thermosaccharolyticum* TG57, a thermophile capable of producing biobutanol as the sole product (without any acetone or ethanol produced) from crystalline cellulose, is of great importance (Li et al., 2018). Biobutanol toxicity or solvent toxicity of the producing organism is another major bottleneck in the cost-effective production of biobutanol. Host microbes producing biobutanol can tolerate less than 2% of biobutanol as it hampers their growth and cell wall composition (Bankar et al., 2013). Be that as it may, a few microbes, such as thermophilic *Bacillus* sp., may tolerate higher biobutanol concentrations of 2.5–7.0%. Furthermore, microbes belonging to the genus *Pseudomonas* exhibit higher tolerance owing to their inherent mechanism of physicochemical changes in their membrane lipid composition and also using solvent removal by efflux pumps (Segura et al., 2004). Rühl et al. (2009) reported in her paper that several strains of *Pseudomonas putida* can tolerate biobutanol concentrations as high as 6% and are desirable candidates for genomic studies for further improvement of biobutanol tolerance in microbes. Bibra et al. (2015) suggested that genes from extremophiles like *Bacillus* and *Pseudomonas* sp. should be studied for increased biobutanol tolerance and efficient production.

In recent years, several scientists have been working to bring improvements in the processes involved in biobutanol production. Several non-biobutanol-producing organisms such as *Escherichia coli*, *Saccharomyces cerevisiae*, *Pyrobaculum calidifontis*, *C. thermocellum*, and *Thermococcus kodakarensis* have been genetically modified with the help of various techniques for enhanced production of biobutanol (Mukhtar & Aslam, 2020). The discovery of new organisms, as well as a better understanding of their metabolic pathways, will provide new alternatives for improving biobutanol production. Further development of strains, through engineering of metabolic pathways, can provide modified strains that can more efficiently degrade raw materials, have higher biobutanol tolerance, favor one-step fermentation, and reduce the formation of by-products and inhibitory substances. Varied research using contemporary tools for genetic manipulation and study of various extremophilic organisms and their metabolic pathways, enzymes, and proteins using high-throughput techniques involving genomics, transcriptomics, and proteomics can pave the way to produce commercially feasible biobutanol (Mukhtar & Aslam, 2020).

4.4 Genetic alterations in microorganisms to improvise biobutanol production

The fermentation process of *C. acetobutylicum*, a biofuel generator, is divided in to two phases: acidogenesis and solventogenesis. Acidogenesis is characterized by the generation of acid in an exponential phase (acetate and butyrate). Solventogenesis is characterized by the formation of solvents (acetone, biobutanol, and ethanol/isopropanol) during the transition between exponential and stationary phases, along with a change in the pH during the transition between acidogenic and solventogenic phases in the kinetics of acetone, biobutanol, and ethanol synthesis. Metabolic remodeling demonstrates considerable variations in the concentration of all metabolites generated during the transition from acidogenesis to solventogenesis. By combining the mathematical modeling of the fermentation process with the induced gene, a mechanical representation of the pH-driven transition between the two phases is obtained. *Clostridium* has been metabolically modified to produce biobutanol through an *in vivo* methylation procedure using host strains lacking the very active restriction endonucleases *Cac824* (Lutke-Eversloh & Bahl, 2014). The disruption of metabolic pathways involving butyrate, lactate, acetone, and acetate is employed in biobutanol synthesis. The butyrate acetone lactate pathway is disrupted by: (1) gene-related butyrate, acetone, lactate, and acetate knockout/gene silencing, and (2) inserting genes or overexpressing genes linked with solvent production.

Integrational plasmid technology (IPT) is used, including replicating and non-replicating plasmids; however, TargeTron technology is widely employed due to low screening efficiency. Both approaches were employed to increase biobutanol concentrations in solvent fermentation and generate a 70% rise in biobutanol production ratio to 80–85%. Antisense RNA technology, on the other hand, is a powerful and versatile technique for microbial gene alteration (silencing) without altering the regulation of gene expression. This technology is used in the silencing of the *ctfB* gene for improving the biobutanol ratio in solvent fermentation. In two experiments, the fermentation process of biobutanol synthesis by *C. acetobutylicum* utilizing various lignocellulosic materials was validated using heterologous minicellulosomes. CRE (catabolic responsive element) deletion results in an increase in biobutanol production by 7.5-fold (Lutke-Eversloh & Bahl, 2014). The deletion of the gene *SpoA* in *Clostridium beijerinckii* via CRISPR study likewise resulted in ABE synthesis. CRISPR analysis is a method based on natural Clustered Regularly Interspaced Short Palindromic Repeats (CRISPR) and the CRISPR/Cas system (which provides the system by cleaving foreign DNA in archaea and bacteria groups). CRISPR analysis is a very effective and markerless genome engineering deletion technology that involves a simple cloning technique. In this method, a CRISPR array of target sequences is transcribed and processed to generate CRISPR-RNA (crRNA), which then guides Cas nuclease to cleave the target site using a Protospacer Neighboring Motif (PAM). Cas9 is only activated by Trans-Activating CRISPR in type-II CRISPR analysis. It is an RNA complex guiding system with two RNA complexes. The precision of CRISPR analysis is dependent on the microbes' homologous recombinant efficiency, and the precise prediction of the insertion site is difficult, particularly when the target site is short.

Geobacillus thermoglucosidasius, a thermoglucosidase-producing bacteria is a facultatively anaerobic, gram-positive, rod-shaped, endospore-forming bacterium. *Geobacillus* species may grow at temperatures ranging from 40 to 70 °C (Nazina et al., 2001). The *Geobacillus* was created with the intention of producing both ethanol and biobutanol solvents (Lin et al., 2014). The acetohydroxy acid synthase gene and 2-*ketoisovalerate dehydrogenase* gene from *B. subtilis* and *L. lactis*, respectively, were engineered in to *G. thermoglucosidasius*, as well as the promoter region of the lactate dehydrogenase gene from *Geobacillus thermodenitrificans*. From glucose as a substrate, 3.3 g/L biobutanol was generated. Lin et al., 2014 demonstrated that biobutanol could be generated even at a higher temperature of 50 °C.

Pyrococcus furiosus, an archaebacteria is heterophilic, and CO_2 and H_2 are the metabolic products of this flagellated, cocci-shaped bacterium (Ma et al., 1994). *Pyrococcus furiosus* was genetically modified to produce biobutanol at high temperatures. The gene for *lactate dehydrogenase* from *Caldicellulosiruptor bescii* was used to produce 3-hydroxypropionate (also known as electrofuel) from hydrogen (Basen et al., 2012). *Pyrococcus furiosus* has a mechanism for producing 1-biobutanol and 2-biobutanol. The genes for the enzymes involved in the first three processes, acetyl CoA to crotonyl CoA, *trans-2-enoyl-CoA reductase* (*ter*), *butyraldehyde dehydrogenase* (*Bad*), and *biobutanol dehydrogenase* (*Bdh*), were extracted from *Thermoanaerobacter tengcongensis* and *Spirochaeta thermophila*. After 48 h at 60 °C, the synthesis of 1-biobutanol and 2-biobutanol from genetically altered *Pyrococcus furiosus* was 70 mg/L and 15 mg/L, respectively (Keller et al., 2015).

Isolating additional biobutanol-producing extremophiles is a critical goal for increasing production levels. Biobutanol production utilizing cheap raw materials such as corn has been reported by Green Biologics from thermophilic *Clostridium* strains (Coker, 2016). In addition to this, *Thermoanaerobacterium*, *Pyrococcus*, and *Aeropyrum* were discovered, indicating that biobutanol has a high potential for industrial production (de Vrije, 2002). Metabolic engineering technologies have previously been utilized to produce biobutanol in recent years. To overcome *Clostridium*'s intractable gene operation, a number of approaches for improving electroporation efficiency were devised, including enhancing cell membrane modifications by adding solvent, weakening and disrupting the cell membrane with glycine or lysozyme, and adjusting electroporation operation parameters. *Clostridium* genetic transformation has also been effectively achieved using CRISPR-Cas technologies. For example, in *Clostridium acetobutylicum* DSM792T, the use of the *Streptococcus pyogenes* II CRISPR-Cas 9 system for genome editing has enhanced its ability to utilize both pentose and hexose sugars. The gene-editing technique of TargeTron, which generally depends on the mobility of group II introns, was also reported to be suitable for genetic modifications in *Clostridium*. In addition, metabolic engineering and functional gene identification were carried out using mesophilic-TargeTron and Thermo TargeTron technology.

In literature, the fermentation of biobutanol from halophiles utilizing sugar has been reported. Amiri et al. (2016), for example, reported that *Nesterenkonia* sp. strain F, a moderately halophilic bacterium isolated from the Aran-Bidgol hypersaline lake (Iran), under both aerobic and anaerobic conditions, can produce biobutanol, ethanol, and acetone. A biobutanol concentration of 105 mg/L was achieved from 50 g/L glucose in 72 h by *Nesterenkonia* sp. strain F grown under anaerobic conditions, whereas under aerobic circumstances, 66 mg/L biobutanol and 291 mg/L ethanol were produced utilizing the same feedstock. The natural synthesis of biobutanol, previously thought to be unique to the *Clostridia* class, was detected in *Nesterenkonia* sp. strain F, a halophilic bacterium of the family *Micrococcaceae* of the order Actinomycetales, under both aerobic and anaerobic conditions (Amoozegar et al., 2019).

Microorganisms have been used in biobutanol production for more than half a century. The *Clostridium* genus has been reported the most for the production of acetone, biobutanol, isopropanol, and ethanol in various yields and proportions. With solvent concentrations of 18–23 g/L and a yield of 30–33%, these strains were able to use 7.5% fermentable sugar. Hongo (1960) patented a biobutanol-producing organism that could produce up to 70% biobutanol. A study comparing the synthesis of acetone, biobutanol, isopropanol, and ethanol by 34 *Clostridia* strains representing 15 species. *C. acetobutylicum*, *C. aurantibutyricum*, *C. butylicum*, *C. cadaveris*, *C. beijerinckii*, *C. pasteurianum*, and *C. sporogenes* were the main species that produced solvent in reasonable proportions (>1 mmol/L) in a peptone, yeast extract, and dextrose medium. *C. aurantibutyricum* makes acetone, biobutanol, and isopropanol, but *C. tetanomorphum* makes ethanol and biobutanol in equal amounts. *Clostridia acetobutylicum*, *C. beijerinckii*,

C. saccharoperbutylacetonicum, and *C. saccharobutylicum* are the four species of industrial clostridia strains (Khandeparker & Sani, 2018).

These adaptation strategies primarily include improving the rate of cell membrane-repairing enzymes, producing distinct solvent-inactivating enzymes, having dynamic solvent efflux mechanisms via solvent efflux pumps (*tol C/mar/rob/sox S/acr AB* genes), and membrane vesicles releasing solvent molecules adhered to them. However, there is a significant gap in the information available on similar mechanisms in gram-positive bacteria. Organic solvent-emulsifying, deactivating, and solubilizing enzymes/substances are thought to have a critical role in reducing solvent toxicity in gram-positive bacteria. According to Segura et al. (2004), *Pseudomonas* achieves great solvent tolerance by removing solvent via efflux pumps and physicochemical changes in membrane lipids. De Carvalho et al. (2004) found that the *P. putida* S12 strain has a moderate tolerance to biobutanol, but other *P. putida* strains have been developed to tolerate 6% w/v biobutanol (Ruhl et al., 2009). This finding together with the recent engineering of *P. putida* makes solvent-tolerant organisms a new ushered era in biobutanol production.

Several genetically engineered bacteria from the genera *Zymomonas*, *Enterococcus*, *Rhodococcus*, *Pichia*, *Candida*, *Saccharomyces*, *Pseudomonas*, and *Klebsiella* have been described as possible biobutanol makers, albeit their potential has not been thoroughly investigated. In order to optimize biobutanol production, it is critical to choose the right microbial strain(s) and fermentation process, as well as to understand the function of genetic manipulation in strain and process improvement.

4.5 Enhancing biobutanol-resistant genes in extremophiles

One of the most difficult aspects of creating an efficient bioprocess is reducing the toxicity of the host strains. By entering and accumulating in the membrane and cytoplasm, 1% (v/v) biobutanol is sufficient to limit cellular development. This constraint makes it difficult to build an effective bioprocess that requires titers of 100 g/L. To enhance biobutanol tolerance, several techniques have been used. Overexpressing heat-shock proteins is one way to improve the fluidity and integrity of the membrane. Chaperones respond to stress by folding and transporting proteins and disaggregating denatured proteins, resulting in the highest tolerance to biobutanol in cells expressing *SecB* (Ferreira et al., 2020).

Several molecular chaperones, including as *groES*, *dnaKJ*, *hsp18*, and *hsp90*, have been identified and characterized as possible target genes to promote cellular tolerance as an alternate method. The recombinant strains (designated 824 (*dnaK* R12) and 824 (*groESL* R12)) had higher tolerances against various stresses (such as biobutanol) and produced 13.0 g/L and 11.2 g/L of biobutanol with increases of 49.4% and 28.7%, respectively, compared to the parent strain, derived from the extremely radio-resistant bacterium *Deinococcus wulumuqiensis*. Furthermore, a genomic library was developed to enrich genes involved in biobutanol tolerance under enhanced biobutanol stress conditions, and the most enriched gene (*CAC1869*) contributed to a variety of tolerance pathways, with overexpression increasing biobutanol tolerance by more than 80% (Ferreira et al., 2020).

The *gshAB* genes, which code for *glutamate-cysteine ligase* and *glutathione synthase*, could be investigated as a viable technique for improving cellular robustness. Overexpression of the *gshAB* genes resulted in a more robust *C. acetobutylicum* phenotype with biobutanol resistance and aerotolerance (Ferreira et al., 2020).

4.6 Metabolomics approach to optimize biobutanol production

Metabolomics is the study of the interaction of small molecules known as metabolomes, present in the cells and tissues of different organisms. It allows structured characterization and computation of the metabolomes. It even enables one to review the genetic mechanisms and the metabolism involved in an organism.

Due to the various chemical compositions of metabolites, the metabolome study is extremely difficult. Metabolites are the outcome of the system's genome interacting with its environment, and they are not only the end product of gene expression but also an integral part of the regulatory system (Ohtake et al., 2017). Although metabolomics had its origin in the profiling investigations of early metabolites, it now holds its unique position in the area of scientific research.

The metabolome is considered a contemplation of the phenotype. The metabolome is interrelated to transcriptomics, genomics, and proteomics. This omics is therefore observed as the best omics approach to review any type of health and disease, on the basis of the phenotype. In biology, food science, and pathology, metabolomics provides particular advantages in enabling the knowledge of pathological and physiological processes as one of the omics domains (Ohtake et al., 2017).

4.6.1 Techniques involved in metabolomics

The complete coverage of the metabolome is mainly possible by the two widely used techniques – nuclear magnetic resonance and mass spectrometry – as shown in Figure 4.2. They are the most primarily used analytical metabolomics platforms. Targeted and untargeted metabolomics investigations are the common methodologies used in metabolomics. A set of metabolites is selected for examination and then measured using standards in targeted metabolomics research. In untargeted metabolomics,

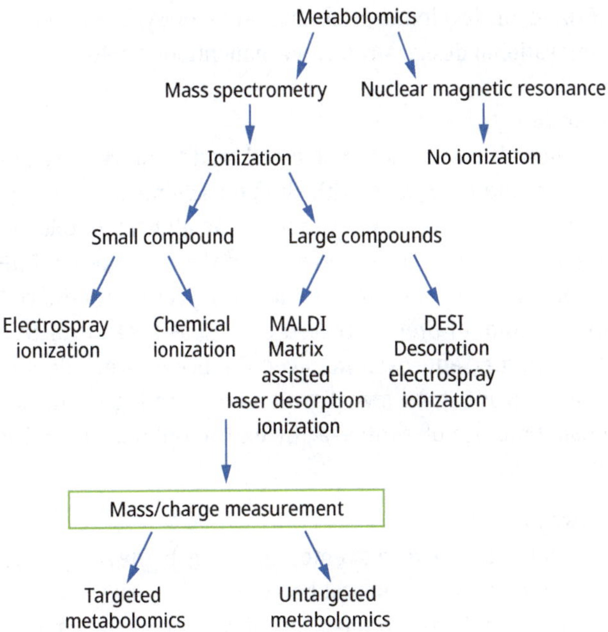

Figure 4.2: An overview of metabolomics methods.

researchers analyze as many metabolites as possible from materials without having any prior preconceptions about which molecules might be present.

4.6.2 Extremophilic metabolomics for biobutanol production

Several microorganisms have the ability to generate biobutanol through their metabolism. Similarly, the diversity of ecological niches inhabited by microbes allows for the improved conversion of residual biomass. In the biorefinery industry, genomics, proteomics, and transcriptomic technologies can be used to define new targets and metabolic engineering for the development of strains (Liu et al., 2019). In this way, metabolomics technologies can be used to investigate metabolic regulations and pathways in order to boost strain efficiency and performance (Vaishnav et al., 2020).

The conversion of inexpensive lignocellulosic substrates to biobutanol is necessary for achieving a cost-effective method. *Clostridial* strains are more vulnerable to the inhibitors produced by the pretreatment methods than other bacteria. The complex regulatory and metabolic mechanisms of *Clostridial* strains hinder the development of better cell factories. To manufacture biobutanol in extremophilic microorganisms, a variety of techniques are used, including overexpression of enzymes responsible for metabolite conversion, conversion of other substrates, testing different enzymes for specific steps in the *Clostridial* route, or development of alternative biosynthetic routes (Ohtake et al., 2017). The different rational design strategies are mentioned below:

i. **Sustainable substrate conversion**
Biofuels made from low-cost, renewable raw materials could lead to more environmentally friendly procedures that could compete with the petrochemical industry. Glycerol, cellulose, hemicellulose, switchgrass, and fatty acids have all been employed as potential sources for biobutanol synthesis. Overexpression of the transporter *GlpF* boosted biobutanol synthesis from glycerol. The consumption of glycerol into the cell was increased by 25%, and the biobutanol titer increased by 23% as a result of this alteration (Barnard et al., 2010). Another approach involved the use of a cellulolytic strain similar to *E. coli*, capable of growing on switchgrass and secreting cellulases and hemicellulases to establish biobutanol synthesis in extremophiles (Ferreira et al., 2020).

ii. **Deletion of competing pathways**
Internal modification of biochemical fluxes of extremophiles can be developed to meet their particular needs, resulting in environmental benefits, which usually necessitate genetic interventions to adjust the fluxes to meet industrial purposes. Gene interruption and then determining the required metabolome is a typical approach for enforcing overproduction of a specific biochemical molecule by removing competing pathways (Ferreira et al., 2020). This method was successfully used in *E. coli* to produce a wide range of chemicals, including succinate, ethanol, and L-alanine. Numerous genes were damaged in three different pathways – ketoacids, reversal oxidation cycle routes, and the ACP-dependent FA-pathway – mostly to divert the flux into biobutanol. Gene substitutions were used to boost the primary precursor (2-ketobutyrate) pool in biobutanol synthesis as part of the ketoacids pathway research. The genes *acetolactate synthase* (*ilvB* and *ilvI*) were selectively eliminated using genetic engineering, resulting in a twofold rise in the biobutanol titer (Ferreira et al., 2020).

iii. **Balancing cofactors to enhance the NADH pool**
Growing cells in low-oxygen environments is a simple technique to boost *NADH* availability. The molecules of *NADH* generated in glycolysis should be recycled to *NAD* + by creating limited compounds, such as alcohols, because there is no oxygen to act as an energy sink (Ferreira et al., 2020). The *NADH* recycling systems help maintain the redox equilibrium inside the cells, which is necessary for living cells to continue to expand.

Although *NADH* regeneration via oxidative phosphorylation creates *ATP* in aerobic conditions, in the oxygen-free environment, *NADH* recycling results in an increase of *NADH*-consuming intermediates. The native *formate dehydrogenase* in *E. coli* involves the conversion of formate to CO_2 and H_2. Moreover, that very same protein can hydrolyze formate into CO_2 and *NADH,* and similar technology can be employed in the case of extremophilic microorganisms. To expand the *NADH* pool, the *formate de-*

hydrogenase from *Candida boidinii* can be produced more, which results in a greater biobutanol titer whenever the media is supplied with formate (Ferreira et al., 2020).

iv. **Refinement of heterologous pathway activity**
Six catalytic steps transform two molecules of acetyl-CoA into one molecule of biobutanol in the *Clostridial* biobutanol biosynthetic pathway. This pathway's seven genes (*thl, hbd, crt, bcdetfAB, adhE*) are adequate to support biobutanol synthesis in *E. coli*. Furthermore, identifying rate-limiting stages and expressing alternative enzymes that are more compatible with the host can result in a more efficient cell factory (Ferreira et al., 2020). A strain of *E. coli* was created to alternatively produce biobutanol and propanol by exploiting the ketoacid pathway, which was the first alternative method to create biobutanol. *E. coli* was modified to boost the pool of 2-ketobutyrate, a common keto acid intermediary in isoleucine biosynthesis. By using the similar heterologous decarboxylases and dehydrogenases pathway, this ketoacid can be transformed into 1-propanol or biobutanol via the norvaline biosynthetic route in extremophiles (Ohtake et al., 2017).

4.7 Proteomics approach to optimize biobutanol production

The major problems for biobutanol synthesis are its accumulation within the bioreactor, which becomes lethal for the microorganisms involved in the production. Several recent research studies have reported an in-depth analysis of the proteomics profile of extremophiles, for example, *C. beijerinckii*, accounting for biobutanol shock. It was found that the most resistant strains of *Clostridium acetobutylicum*, T64 and NT642, were tolerant to biobutanol at approximately 2.4% and 3.2%, respectively (Arsov et al., 2021). The biobutanol synthetic production route was also introduced into several other strains, such as *B. subtilis, E. coli, P. putida*, and *S. cerevisiae*, but these microbial species produce much lower biobutanol tolerance than the above *clostridial* strains. In recent times, another method was used to increase biobutanol production and tolerance in model organisms, e.g., the library enhancement approach was applied to a gram-negative strain of *E. coli*, and a total of 55 proteins involved in the biobutanol challenge were tested by removal or overexpression, out of which eleven overexpressing proteins provided improved biobutanol tolerance while three of them provided increased biobutanol resistance during removal (Reyes et al., 2011).

The proteins that are associated with biobutanol stress play a role in iron metabolism and transportation (Reyes et al., 2011). The reports of the over-expressed proteins for the synthesis are fatty acid, the absorption of iron, and the efflux pump, which leads to an increase in biobutanol tolerance of ±2% (Lee et al., 2015). Among the developed or engineered *E. coli* strains, *S. cerevisiae* and *S. elongatus*, which produce biobutanol, showed a low tolerance range to biobutanol (0.5–2%) (Liu et al., 2017). Hence, it is very important to develop a higher biobutanol tolerance range of extremophiles for the biobutanol fermentation industry. Therefore, there is a requirement for identifying more tolerant species and improving the engineering strategy to achieve a biobutanol fermentation titer greater than 20 g/L, so that feedback inhibition can be avoided.

If it were possible to achieve a higher titer, then the cost of the recovery and separation of biobutanol might be reduced. Researchers in different countries (including both developing and developed) are continuously working in this direction, and it has recently been shown that chemically induced strains of *Clostridium beijerinckii* NRRL B-598 are tolerant to a maximum of 25g/L of biobutanol (Vasylkivska et al., 2020). The modification for more vigorous biobutanol production from the host is based on identifying the biobutanol-resistant proteins with optimized codons and a better understanding of the mechanisms involved in biobutanol resistance. However, the current knowledge about the proteomics of the gram-positive species involved in biobutanol resistance is very limited. Several research studies have shown that gram-positive LAB (lactic acid bacteria) are unsurprisingly more tolerant and adaptive to higher concentrations of biobutanol. Different *Lactococcus* strains, such as *L. diolivorans, L. brevis,* and *L. delbrueckii,* can grow even in 2.5% and 3% biobutanol concentrations (Russmayer et al., 2019).

A review based on the proteomics studies of biobutanol tolerance demonstrates that extremophiles and LAB can be utilized as possible hosts or as a gene pool for studying biobutanol tolerance (Arsov et al., 2021). Because of the continuous adaptive research laboratory evolution of the *L. lactis* strain against increasing biobutanol concentration, it will result in a species-tolerant to 28 g/L of biobutanol (Gupta et al., 2020). Researchers have utilized these adaptive strategies in order to select biobutanol-resistant strains of LAB. Ten strains of LAB from different species, like *L. amylovorus, L. mucosae, L. crispatus, W. confusa,* and *P. parvulus,* showed increased tolerance to biobutanol after a long period of adjustment, and several selected strains were able to grow in anaerobic media containing 3–4% of biobutanol (Liu et al., 2012). At the same time, the stress assay of biobutanol (4%) exposure demonstrates that the strains *L. mucosae* BR0713–33 and *L. mucosae* BR0713–30 were the two most biobutanol-resistant strains (Liu et al., 2012).

4.7.1 Extremophilic proteins and their roles in biobutanol production

In an effort to increase the Earth's declining storage of fossil fuels, a combined effort was made to create similar fuels by utilizing biomass like sugar cane, beets, corn, and wheat (Coker, 2016). Depending on the material, these biofuels were divided into two generations: first generation and second generation. Those biofuels that are easily derived from the hydrolyzed sugars, oils, and starches offered by crops fall under the first-generation category, whereas those biofuels obtained from lignocellulosic raw material and resistant to hydrolysis fall under the second-generation category (Coker, 2016).

On the basis of the final end products, biofuels may also be characterized as hydrogen, biobutanol, biodiesel, methane, and ethanol. At the same time, the traditional techniques of producing bioethanol and biobutanol involve chemical processes along with mesophilic microorganisms like *Clostridium* species and *Saccharomyces cerevisiae* strains (Lee et al., 2008). In comparison with the other products produced, methane was constantly produced by using a microorganism consortium, including methanogens. Methanogens (extremophiles) are the only known biological methane producers (Barnard et al., 2010). Extremophiles can easily thrive and survive in intolerably lethal conditions like extremely salty lakes, arctic frost, and hot places. A few extremophiles can even grow in environments that are harmful to other organisms, such as in the presence of organic waste, heavy metals, and toxic contaminants. Scientists have found extremophiles in volcanoes, hydrothermal vents, and even in the deep sea (Rampelotto, 2013).

Multiple steps involved in the production of biofuel require extreme pH environments and higher temperatures; thus, extremophiles can be an ideal substitute for the mesophilic microbes commonly used in traditional methods. For example, a thermophilic anaerobe, *Thermoanaerobacterium saccharolyticum*, utilizes pentose sugars (xylose) and hemicellulose as preliminary materials for producing high-yield ethanol (Demain et al., 2005). However, these engineered thermophiles have been found to be more promising in producing biobutanol in bulk quantities and also minimize other side products/reactions (Basen et al., 2014). In hydrogen production, via hydrogenases and anaerobic fermentation, there are multiple applications of extremophiles. The use of different strains, including *Aeropyrum*, *Caldicellulosiruptor*, *Pyrococcus*, and *Thermoanaerobacterium*, shows enormous potential (Ren et al., 2008).

On the other side, the two most important products which were produced by microbes and were also commercially very successful are biobutanol and biodiesel. While the biodiesel yields depend upon the lipid content (i.e., greater than 75% of dry weight) of algae, a maximum of which contains long-chain hydrocarbons similar to those found in petroleum. Presently, there are some extremophilic algae-like *Galdieria sulphuraria* and *Cyanidium caldarium* that can fulfill these necessities (Pulz et al., 2004). Similarly, modified halophilic algae may also seem promising because they can be easily grown in wide-open vessels due to the higher salinity required for their growth, which will also inhibit the growth of other microbes. Thus, these algae can be easily grown in the oceans and arid environments (DasSarma et al., 2009).

When compared with ethanol, biobutanol was found to be quite effective in hindering the microbial growth of different microbes; even most microbes cannot survive in more than 2% of biobutanol. Specifically, in order to control the inhibition of the product and also withstand the enormous biobutanol quantities, microbes must be modified. Presently, organizations named Joule Unlimited, Solazyme, and Gevo are capable of producing huge volumes of biodiesel as well as bioethanol and also jet fuel for military and civilian use. Moreover, Sapphire Energy Company has taken one step back in the production process and produces Green Crude that will replace the requirement of crude oil in the present petroleum set-up (Coker, 2016).

4.8 Discussion

Despite the beneficial role of heterotrophic microorganisms in improving biobutanol production, the extensively acknowledged yield and tolerant properties of extremophiles have showcased their potential. Extremophiles play an important role in increasing nutrient conversion systems at elevated temperatures, adjusting tolerance against feedback toxicity, and nurturing biobutanol production with no environmental concerns. Among omics sciences, metabolomics and proteomics are emerging applications primarily concerned with high-throughput profiling of molecules at a given point and under specific physiological conditions. Metabolomics is of special relevance for understanding the extremophiles' response to environmental signals. As metabolites are the final products of cellular regulatory processes, their study is potentially a more direct indicator of functional phenotype. Proteomics analyses can be classified as either targeted or untargeted, where untargeted analysis focuses on pattern-based classification of as many proteins as possible in the system with unbiased and global screening. Genetic engineering of extremophiles is more commonly used for research-based questions such as the characterization of metabolic changes followed by pretreatment, genetic perturbation, or increased tolerance. A future prospect of extremophiles omics study is to enhance the production, yield, and tolerance of biobutanol for advanced biodiesel scale-up.

References

Amiri, T. D., Azarbaijani, R., Yeganeh, L. P., fAZELI, A. S., Tabatabaei, M., Salekdeh, G. H., et al. (2016). *Nesterenkonia sp.* strain F, a halophilic bacterium producing acetone, butanol and ethanol under aerobic conditions. *Scientific Reports*.

Amoozegar, M. A., Safarpour, A., Noghabi, K. A., Bakhtiary, T., & Ventosa, A. (2019). Halophiles and their vast potential in biofuel production. *Frontiers in Microbiology*, *10*, 1895.

Arsov, A., Petrov, K., & Petrova, P. (2021). How to outwit nature: Omics insight into butanol tolerance. *Biotechnology Advances*, *46*, 107658.

Bankar, S. B., Survase, S. A., Ojamo, H., & Granstrom, T. (2013). Biobutanol: An outlook of an academist and industrialist. *RSC Advances*.

Basen, M., Schut, G. J., Nguyen, D. M., Lipscomb, G. L., Benn, R. A., Prybol, C. J., . . . & Adams, M. W. (2014). Single gene insertion drives bioalcohol production by a thermophilic archaeon. *Proceedings of the National Academy of Sciences*, *111*(49), 17618–17623.

Basen, M., Sun, J., & Adams, M. W. (2012). Engineering a hyperthermophilic archaeon for temperature-dependent product formation. *MBio*, *3*(2), e00053-12.
Bengelsdorf, F., Straub, M., & Durre, P. (2013). Bacterial synthesis gas (syngas) fermentation. *Environmental Technology*.
Bhalla, A., Bansal, N., Kumar, S., Bischoff, K. M., & Sani, R. K. (2013). Improved lignocellulose conversion to biofuels with thermophilic bacteria and thermostable enzymes. *Bioresource Technology*, 751–759.
Bibra, M., Wang, J., Squillace, P., & Pinkelman, R. J. (2015). Biofuels and Value-added Products from Extremophiles. In *Advances in Biotechnology* (pp. 17–51). I.K. International Publishing House New Delhi.
Biswas, S., Katiyar, R., Gurjar, B. R., & Purthi, V. (2017). Role of different feedstocks on the butanol production through microbial and catalytic routes. *International Journal of Chemical Reactor Engineering*.
Cao, G., Sheng, Y., Zhang, L., Song, J., Cong, H., & Zhang, J. (2016). Biobutanol production from lignocellulosic biomass: Prospective and challenges. *Journal of Bioremediation & Biodegradation*.
Chandra, R., Kumar, V., & Yadav, S. (2017). Extremophilic ligninolytic enzymes. *Extremophilic Enzymatic Processing of Lignocellulosic Feedstocks to Bioenergy*.
Coker, J. A. (2016). Extremophiles and biotechnology: current uses and prospects. F1000Research.
DasSarma, P., Coker, J. A., Huse, V., & DasSarma, S. (2009). Halophiles, industrial applications. *Encyclopedia of Industrial Biotechnology: Bioprocess, Bioseparation, and Cell Technology*, 1–43.
De Carvalho, C. C., Da Cruz, A. A., Pons, M. N., Pinheiro, H. M., Cabral, J. M., Da Fonseca, M. M. R., . . . & Fernandes, P. (2004). *Mycobacterium sp., Rhodococcus erythropolis*, and *Pseudomonas putida* behavior in the presence of organic solvents. *Microscopy Research and Technique*, *64*(3), 215–222.
De Vrije, T. D., De Haas, G. G., Tan, G. B., Keijsers, E. R. P., & Claassen, P. A. M. (2002). Pretreatment of Miscanthus for hydrogen production by *Thermotoga elfii*. *International Journal of Hydrogen Energy*, *27* (11–12), 1381–1390.
Demain, A. L., Newcomb, M., & Wu, J. D. (2005). Cellulase, *clostridia*, and ethanol. *Microbiology and Molecular Biology Reviews*, *69*(1), 124–154.
Ferreira, S., Pereira, R., Wahl, S. A., & Rocha, I. (2020). Metabolic engineering strategies for butanol production in *Escherichia coli*. *Biotechnology and Bioengineering*, *117*(8), 2571–2587.
Gottumukkala, L. D., Mathew, A., Abraham, A., & Sukumaran, R. K. (2019). Biobutanol Production: Microbes, Feedstock, and Strategies. In *Biofuels: Alternative Feedstocks and Conversion Processes for the Production of Liquid and Gaseous Biofuels* (pp. 355–377).
Gupta, G. N., Srivastava, S., Khare, S., & Prakash, V. (2014). Extremophiles: An overview of microorganism from extreme environment. *International Journal of Agriculture, Environment & Biotechnology*.
Gupta, J. A., Thapa, S., Verma, M., Som, R., & Mukherjee, K. J. (2020). Genomics and transcriptomics analysis reveals the mechanism of isobutanol tolerance of a laboratory evolved *Lactococcus lactis* strain. *Scientific Reports*, *10*(1), 1–18.
Hongo, M. (1960). Butanol by fermentation.
Karim, A. S., Dudley, Q. M., Juminaga, A., Yuan, Y., Crowe, S. A., Heggestad, J. T., . . . & Jewett, M. C. (2020). In vitro prototyping and rapid optimization of biosynthetic enzymes for cell design. *Nature Chemical Biology*, *16*(8), 912–919.
Keller, M., & Schut, G. J. (2014). A hybrid synthetic pathway for butanol production by a hyperthermophilic microbe. *Metabolic Engineering*.
Keller, M. W., Lipscomb, G. L., Loder, A. J., Schut, G. J., Kelly, R. M., & Adams, M. W. (2015). A hybrid synthetic pathway for butanol production by a hyperthermophilic microbe. *Metabolic Engineering*, *27*, 101–106.
Khandeparker, R., & Sani, R. K. (2018). Biobutanol production using recombinant microorganisms. *Extremophilic Microbial Processing of Lignocellulosic Feedstocks to Biofuels, Value-Added Products, and Usable Power*, 47–62.

Kumar, A., Alam, A., Tripathi, D., Rani, M., Khatoon, H., Pandey, S., et al. (2018). Protein adaptations in extremophiles: An insight into extremophilic connection of mycobacterial proteome. *Seminars in Cell and Developmental Biology*.

Kumari, D., & Singh, R. (2018). Pretreatment of lignocellulosic wastes for biofuel production: A critical review. *Renewable and Sustainable Energy Reviews*.

Lee, J. Y., Geraldi, A., Rahman, Z., Lee, J. H., & Kim, S. C. (2015). Improved n-butanol tolerance in Escherichia coli by controlling membrane related functions. *Journal of Biotechnology, 204*, 33–44.

Lee, S. Y., Park, J. H., Jang, S. H., Nielsen, L. K., Kim, J., & Jung, K. S. (2008). Fermentative butanol production by *Clostridia. Biotechnology and Bioengineering, 101*(2), 209–228.

Li, T., Zhang, C., Yang, K.-L., & He, J. (2018). Unique genetic cassettes in a *Thermoanaerobacteroium* contribute to simultaneous conversion of cellulose and monosugars into butanol. *Science Advances*.

Lin, P. P., Rabe, K. S., Takasumi, J. L., Kadisch, M., Arnold, F. H., & Liao, J. C. (2014). Isobutanol production at elevated temperatures in thermophilic *Geobacillus thermoglucosidasius. Metabolic Engineering, 24*, 1–8.

Liou, J. S., Balkwill, D. L., Drake, G. R., & Tanner, R. S. (2005). *Clostridium carboxidivorans sp. nov*, a solvent-producing clostridium isolated from an agricultural settling lagoon, and reclassification of the acetogen *Clostridium scatologenes strain SL1* as *Clostridium drakei sp. nov. International Journal of Systematic and Evolutionary Microbiology*.

Liu, H., Zhang, J., Yuan, J., Jiang, X., Jiang, L., Zhao, G., . . . & Liu, B. (2019). Omics-based analyses revealed metabolic responses of *Clostridium acetobutylicum* to lignocellulose-derived inhibitors furfural, formic acid and phenol stress for butanol fermentation. *Biotechnology for Biofuels, 12*(1), 1–20.

Liu, S., Bischoff, K. M., Leathers, T. D., Qureshi, N., Rich, J. O., & Hughes, S. R. (2012). Adaptation of lactic acid bacteria to butanol. *Biocatalysis and Agricultural Biotechnology, 1*(1), 57–61.

Liu, S., Qureshi, N., & Hughes, S. R. (2017). Progress and perspectives on improving butanol tolerance. *World Journal of Microbiology and Biotechnology, 33*(3), 51.

Lütke-Eversloh, T. (2014). Application of new metabolic engineering tools for *Clostridium acetobutylicum. Applied Microbiology and Biotechnology, 98*(13), 5823–5837.

Ma, K., Zhou, H. Z., & Adams, M. W. (1994). Hydrogen production from pyruvate by enzymes purified from the hyperthermophilic archaeon, *Pyrococcus furiosus*: A key role for *NADPH. FEMS Microbiology Letters, 122*(3), 245–250.

Meng, X., Yang, J., Xu, X., Zhang, L., Nie, Q., & Xian, M. (2009). Biodiesel production from oleaginous microrganisms. *Renewable Energy*.

Mukhtar, S., & Aslam, M. (2020). Biofuel synthesis by extremophilic microorganisms. *Biofuels Production – Sustainability and Advanves in Microbial Bioresources*, 115–138.

Nazina, T. N., Tourova, T. P., Poltaraus, A. B., Novikova, E. V., Grigoryan, A. A., Ivanova, A. E., . . . & Ivanov, M. V. (2001). Taxonomic study of aerobic thermophilic bacilli: Descriptions of *Geobacillus subterraneus gen. nov., sp. nov.* and *Geobacillus uzenensis sp. nov.* from petroleum reservoirs and transfer of *Bacillus stearothermophilus, Bacillus thermocatenulatus, Bacillus thermoleovorans, Bacillus kaustophilus, Bacillus thermodenitrificans* to *Geobacillus* as the new combinations *G. stearothermophilus, G. th. International Journal of Systematic and Evolutionary Microbiology, 51*(2), 433–446.

Ni, Y., Wang, Y., & Sun, Z. (2012). Butanol production from cane molasses by *Clostridium saccharobutylicum DSM 13864*: Batch and semicontinuous fermentation. *Applied Biochemistry and Biotechnology*.

Ohtake, T., Pontrelli, S., Laviña, W. A., Liao, J. C., Putri, S. P., & Fukusaki, E. (2017). Metabolomics-driven approach to solving a CoA imbalance for improved 1-butanol production in Escherichia coli. *Metabolic Engineering, 41*, 135–143.

Park, H. J., Jeon, J. H., Kang, S. G., Lee, J. H., Lee, S. A., & Kim, H. K. (2007). Functional expression and refolding of new alkaline esterase, *EM2L8* from deep-sea sediment metagenome. *Protein Expression and Purification*.

Pulz, O., & Gross, W. (2004). Valuable products from biotechnology of microalgae. *Applied Microbiology and Biotechnology, 65*(6), 635–648.

Qureshi, N., Ezeji, T., Ebener, J., & Dien, B. (2007). Butanol production by *Clostridium beijerinckii*. Part I: Use of acid and enzyme hydrolyzed corn fiber. *Bioresource Technology*.

Rampelotto, P. H. (2013). Extremophiles and extreme environments. *Life*, *3*(3), 482–485.

Ren, N., Cao, G., Wang, A., Lee, D. J., Guo, W., & Zhu, Y. (2008). Dark fermentation of xylose and glucose mix using isolated *Thermoanaerobacterium thermosaccharolyticum W16*. *International Journal of Hydrogen Energy*, *33*(21), 6124–6132.

Reyes, L. H., Almario, M. P., & Kao, K. C. (2011). Genomic library screens for genes involved in n-butanol tolerance in *Escherichia coli*. *PloS One*, *6*(3), e17678.

Rühl, J., Schmid, A., & Blank, L. M. (2009). Selected *Pseudomonas putida* strains able to grow in the presence of high butanol concentrations. *Applied and Environmental Microbiology*, *75*(13), 4653–4656.

Russmayer, H., Marx, H., & Sauer, M. (2019). Microbial 2-butanol production with *Lactobacillus diolivorans*. *Biotechnology for Biofuels*, *12*(1), 1–11.

Segura, A., Duque, E., Rojas, A., Godoy, P., Delgado, A., Hurtado, A., . . . & Ramos, J. L. (2004). Fatty acid biosynthesis is involved in solvent tolerance in *Pseudomonas putida DOT-T1E*. *Environmental Microbiology*, *6*(4), 416–423.

Sindhu, R., Kuttiraja, M., Binod, P., Janu, K. U., Sukumaran, R. K., & Pandey, A. (2011). Dilute acid pretreatment and enzymatic saccharification of sugarcane tops for bioethanol production. *Bioresource Technology*.

Takai, K., Komatsu, T., Inagaki, F., & Horikoshi, K. (2001). Distribution of archaea in a black smoker chimney structure. *Applied and Environmental Microbiology*.

Taylor, M. P., Eley, K. L., Martin, S., Tuffin, M. I., Burton, S. G., & Cowan, D. A. (2008). Thermophilic ethanologenesis:future prospects fpr second-generation bioethanol production. *Trends in Biotechnology*.

Vaishnav, A., Sahu, J., & Singh, H. B. (2020). Genomics of extremophiles for sustainable agriculture and biotechnological applications (Part II). *Current Genomics*, *21*(4), 238.

Vaishnavi, G. S., & Muralidhara, R. (2017). Biological chemistry of extreme thermophile "*Caldicellulosiruptor Saccharolyticus*"- A future biofuel producer. *Biosciences Biotechnology Research Asia*.

van den Burg, B. (2003). Extremophiles as a source for novel enzymes. *Current Opinion in Microbiology*.

Vasic, K., Knez, Z., & Leitgeb, M. (2021). Bioethanol production by enzymatic hydrolysis from different lignocellulosic sources. *Molecules*.

Vasylkivska, M., Branska, B., Sedlar, K., Jureckova, K., Provaznik, I., & Patakova, P. (2020). Phenotypic and genomic analysis of clostridium beijerinckii NRRL B-598 mutants with increased butanol tolerance. *Frontiers in Bioengineering and Biotechnology*, *8*, 1307.

Wen, Z., Wu, M., Lin, Y., Yang, L., Lin, J., & Cen, P. (2014). Artificial symbiosis for acetone-butanol-ethanol (ABE) fermentation from alkali extracted deshelled corn cobs by co-culture of *Clostridium beijerinckii* and *Clostridium cellulovorans*. *Microbial Cell Factories*.

Zhang, K., Zhao, W., Rodionov, D. A., Rubinstein, G. M., Nguyen, D. N., Tanwee, T. N., et al. (2021). Genome-scale metabolic model of *Caldicellulosiruptor bescii* reveals optimal metabolic engineering strategies for bio-based chemical production. *mSystems*.

Zhu, D., Adebisi, W. A., Ahmad, F., Sethupathy, S., Danso, B., & Sun, J. (2020). Recent development of extremophilic bacteria and their application in biorefinery. *Frontiers in Bioengineering and Biotechnology, 483*.

Shreya Anand, Abhimanyu Kumar, Vishakha Chaturvedi
and Padmini Padmanabhan*

Chapter 5
Metagenomics and its diverse application in microbial bioremediation processes

Abstract: An ecotoxicological imbalance in the environment is brought on by rapid industrialization and pollution from industrial discharges. Bioremediation is a great, environmentally friendly, and cost-effective solution to this environmental imbalance. The procedure of bioremediation, which uses a variety of bacteria, is effective, economical, straightforward, and environmentally acceptable for treating contaminants. However, a solid understanding of microbial metabolism, microbial colonies, their important enzymes, and their fundamental traits is required to carry out the bioremediation method. The genomes of numerous different microorganisms that are present in the environment are combined to form a metagenome. Understanding the metabolic capacity of the local microbial populations and how positive reactions are managed in situ are essential for the effective development of bioremediation techniques. A clever method for analyzing microbial communities at the genetic level is called metagenomics. It is among the best technological developments in the bioremediation industry. With the aid of short read and next generation sequencing, it gives culture independent information on environmental microorganisms. Through screening technologies, it is possible to see the variety of non-cultivable organisms present in a given environment. These instruments facilitate the discovery of new bacteria and enhance pathogen surveillance. The concept and use of metagenomics in examining microbial diversity, dynamics, and function in a metal/radionuclide-contaminated environment will be discussed in this chapter.

5.1 Introduction

Pollutants such polychlorinated biphenyl compounds, hydrocarbons, dyes, pesticides, esters, heavy metals, petroleum products, and chemicals containing nitrogen continue to be released into the environment through a variety of industrial and agricultural sources. These pollutants are very carcinogenic and deadly by nature, and their buildup threatens the ecosystem as well as our flora and wildlife. The most economi-

*Corresponding author: Padmini Padmanabhan, Department of Biotechnology and Bioengineering, Birla Institute of Technology, Mesra 835215, India, e-mail: padmini@bitmesra.ac.in
Shreya Anand, Abhimanyu Kumar, Vishakha Chaturvedi, Department of Biotechnology and Bioengineering, Birla Institute of Technology, Mesra 835215, India

https://doi.org/10.1515/9783110793635-005

cal and environmentally benign way to solve this problem is through bioremediation. Generally speaking, bioremediation involves using microorganisms and their enzymes to transform pollutants into ones that are less harmful to the environment. Archaea, bacteria, algae, and fungi are capable of bioremediation. The genetic makeup of plants and microorganisms can be changed to enhance bioremediation thanks to recent developments in molecular and recombinant DNA technologies. Oxidoreductases, laccases, hydrolases, and peroxidases are among the enzyme types actively participating in the bioremediation process. Temperature, moisture content, and pH are just a few of the environmental variables that have an impact on microorganism development. The rate of degradation can be considerably accelerated through manipulation and modification of these parameters. Only a small number of bacteria (*Pseudomonas aeruginosa, Rhodococcus erythropolis*, and others), fungi (*Tramates versicolor, Pleurotuseryngii, Phanerochaete chrysosporium*, and others), and algae (*Monoraphidium braunii, Chlamydomonas reinhardtii*, and others) have been found to have a catabolic pathway for the degradation of pollutants in the environment.

Similar to **genomics**, **metagenomics** is a scientific field with a wide range of related concepts and procedures. Greek word **meta** means "transcendent." Thanks to its methodology and procedures, metagenomics bypasses the fundamental obstacles to clinical and environmental microbiology breakthroughs – the noncultivability and genetic diversity of most bacteria. In order to fully understand the genetic makeup and actions of communities that are so complicated that they can only be partially characterized, Meta acknowledges the need to build computational tools. By focusing on the community of genes and how they interact to carry out collective functions rather than just the individual genes of an organism, the second definition of the term "meta" refers to a branch of science that seeks to understand biology at the aggregate level. Of course, individual organisms will remain the center of community activities, and we anticipate that metagenomics will support and guide research on individuals and their genomes. In the ensuing decades, we anticipate convergence between the top-down metagenomics technique, the bottom-up classical microbiology method, and organism-level genomics. Communities and the group of communities that make up the biosphere will be seen as nested systems of systems in which humans participate and whose survival is dependent. In some circumstances, it will be possible to apply the new understanding to pressing and significant issues.

However, the term encompasses cultivation-independent genome-level characterization of communities or their members, high-throughput gene-level studies of communities using genomics methods, and other "**omics**" studies. **Metagenomics** in either sense will almost certainly never be tightly defined by a definition, and it would be foolish to try now. They seek to comprehend transorganismal behavior and the environment from a genetic perspective. Even though the current emphasis of metagenomics is non-eukaryotic bacteria, there is no doubt that its ideas and techniques will ultimately transform all life.

In this chapter, the overview of **bioremediation**, description of omics approaches involved in bioremediation, systems biology in environmental science, what is metagenomics, how metagenomics work, what is the role of metagenomics in bioremediation, the tools and techniques used in metagenomics, limitations of metagenomics are discussed supported with the case studies.

5.2 Bioremediation

Three broad groups of bioremediation techniques can be distinguished according to the degree of intervention. The least intrusive method is called "natural attenuation," in which local organisms are used to cleanse pollutants through organic means. This strategy is appealing because it does not call for expensive or potentially ecodamaging additives. The rates of natural attenuation, however, may be too slow in many systems and unresponsive to the threats to the environment and human health. While using native organisms, **biostimulation** aims to accelerate biodegradation by removing some environmental restrictions. Limiting nutrients are frequently added to achieve this. Biostimulation still causes slow rates of biodegradation in some environments. These delayed biodegradation rates may be a result of the native microbial community's incapacity to break down the harmful pollutant. In an effort to improve the rates of biodegradation, non-native organisms or enzymes might be introduced to a system during **bioaugmentation** to address this problem. This strategy adds a non-native organism to a habitat, making it the most invasive. The most effective method of cleanup, however, has occasionally been bioaugmentation (Payne et al., 2011). The possibility that non-native organisms won't be able to live in the contaminated system is a common worry with bioaugmentation. The possibility that these non-native creatures continue to exist long after the pollutant has been eliminated and affect the environment is another cause for concern.

To best activate the bioremediation processes, it's critical to comprehend the microbial populations involved in bioremediation rather than merely the outcome and rates of pollutant breakdown. Changes in the makeup and activity of a microbial population may have an effect on how a pollutant behaves in the environment because **microorganisms** are the primary agents in bioremediation (Lovely, 2003). The microbial communities participating in various bioremediation strategies have been studied using next-generation sequencing techniques (Chakraborty et al., 2012). Our knowledge of the microbiological processes involved in bioremediation as well as the effects of different response techniques for contamination cleanup have both been considerably enhanced by these approaches.

5.3 Systems biology in environmental science

A microbial community is used during the **bioremediation** process to remove an environmental contamination. The composition of the native microbial population, the environmental circumstances, and the nature of the contamination are some of the variables that affect the rates of contaminant detoxification. In order to comprehend and forecast the destiny of environmental toxins, optimization of bioremediation hence necessitates the integration of complicated variables. **Systems biology** has been used to comprehend complex biological systems and how they would react to different perturbations (Alivisatos et al., 2015). It is the study of the systematic qualities and dynamic interactions in a biological system (Hazen & Sayler, 2016). Investigating complex environmental microbial communities and the environmental limitations on pollutant degradation can be done using a systems biology approach to understanding environmental systems and bioremediation (Hazen & Sayler, 2016). Environmental systems biology is a method that aims to comprehend biological systems in the environment at all informational levels, from molecules to ecosystems and every level in between. The application of **computational methods** to create a predictive understanding of the systems reaction to a disturbance is a key aspect of environmental systems biology. Because it integrates multiple levels of a system to forecast the destiny of environmental contaminants, environmental systems biology is intriguing for understanding contamination remediation.

Getting a thorough characterization of the various layers of the biological system of interest is one of the systems biology's limiting phases (Shade & Handelsman, 2012). In order to get over this restriction, next-generation sequencing is appealing since it offers comprehensive insights into a microbe or microbial community. In order to define the relevant environmental population, environmental systems biology frequently uses several omics methods (such as metagenomics, metatranscriptomics, and metaproteomics) (Smith et al., 2015; Hazen et al., 2013; Hazen et al., 2010). This method allows for the interrogation of the system at many levels. The cost of sequencing has significantly decreased, enabling more thorough characterization of microbial communities and hypothesis-driven experiments into how environmental ecosystems react to environmental pollutants. **Predictive models** can be built using these sizable sequencing data sets to learn how various system components would behave in various scenarios. These forecasting models have a great deal of potential as diagnostic tools to track microbial communities in the environment and forecast how they will react to different **contaminants** and environmental situations (Smith et al., 2015).

The lengthy process of microorganisms degrading contaminants limits the practical practicality of bioremediation. In recent years, microbial enzymes isolated from their cells have been used for bioremediation instead of whole microorganisms to get over the aforementioned restriction. In recent years, **microbial enzymes** isolated from their cells have been used for bioremediation instead of whole microorganisms to get over the aforementioned restriction. Enzymes can quicken a reaction by lowering the energy required for molecules to activate.

5.4 Omics approaches involved in bioremediation

Environmental systems biology relies heavily on 'omics methods. Understanding microbial communities in environmental systems has been greatly aided by the application of metagenomics, which examines a microbial community's entire genomic content (Techtmann & Hazen, 2016). Environmental **microbial communities** have more recently been studied using other omics techniques, such as metatranscriptomics and metaproteomics.

5.5 Metagenomics

By extracting, sequencing, or cloning the DNA of microbial communities, metagenomics allows scientists to study populations of organisms in their natural habitat. The diversity of catabolic genes, entire operons, and taxonomic and phylogenetically significant genes is thus exposed. The metagenomics process includes the extraction of the metagenome for fragmentation, cloning, transformation, and later screening of the generated metagenomics library. The main objective of **metagenomics** was to screen environmental communities for a specific biological activity and identify the genes/gene clusters associated with it because not all microbes can be grown in a lab. This strategy is referred to as "function-based screening."

Environmental systems are increasingly using metatranscriptomics and metaproteomics. These methods offer important new insights into the genes that are being actively expressed in a microbial community, and they are therefore reliable predictors of the microbial functions that are being expressed under the circumstances at the time of sampling. RNA is isolated from an environmental sample for metatranscriptomics. Similar to metagenomics, the RNA is transformed into cDNA and sequenced. This method gives a list of the genes that are being actively expressed in a sample. High-resolution mass spectrometry paired with enzymatic protein digestions and liquid chromatography are used in metaproteomics instead of nucleic acid sequencing (Hettich et al., 2013). Metaproteomics offers information on the range of proteins present in an environmental sample, as well as any post-translational alterations that may affect the function of the proteins. The advantages and disadvantages of these approaches have been outlined in a number of reviews (Hettich et al., 2013; Moran et al., 2013; Xiong et al., 2015).

5.5.1 Flow of metagenomics

Two common metagenomic techniques are **targeted metagenomics** and **shotgun metagenomics**. It is most frequently used to explore both the phylogenetic diversity and relative abundance of a certain gene in a sample. The diversity of a single gene is investigated in targeted metagenomics or micro-biomics to locate the whole complement sequences of a specific gene in an environment. Small subunit rRNA (16S/18S rRNA) diversity in a sample is routinely assessed using this technique. Microbial ecologists typically employ small subunit **rRNA sequencing** to assess the taxonomic diversity of an environment. It can also be used to investigate how contaminants in the environment impact the composition of microbial populations. The gene of interest is amplified by PCR using primers created to amplify the widest variety of sequences for the gene of interest. The ambient DNA is removed. Additionally, next-generation sequencing (NGS), which generates thousands of small subunit rRNA reads per sample and may simultaneously examine hundreds of samples, is used to sequence these amplified genes. Targeted metagenomics, which captures the diversity of a particular gene of interest, is limited by the universality of the PCR primers used in the analysis (Shakya et al., 2013; Prosser, 2012; Klindworth et al., 2013; Parada et al., 2016).

Additionally, the calculation of total diversity may be distorted by numerous bioinformatics analysis (Rinke et al., 2013). **Targeted metagenomics** creates an exhaustive list of the microbial species present in a group of samples and allows for detailed comparisons of changes in microbial diversity before and after a disturbance. Shotgun metagenomics uses genomic sequencing to analyze the whole genetic complement of an environmental population. This technique involves the collection and fragmentation of environmental DNA to produce sequencing libraries, which are subsequently analyzed to determine the genetic makeup of the sample as a whole. It's a strong tool for figuring out the functional potential of a microbial community, but its scope is constrained by the depth of sequencing. A thorough investigation of every creature's genetic makeup is necessary to fully assess the functioning capacity of a community, though. **Shotgun metagenomics** typically favors the bacterial species that are dominant in a community and merely reads the genomes of the members that are less prevalent. Furthermore, it is challenging to analyze metagenomic sequencing data since doing so requires correctly annotating a variety of gene sequences, many of which lack homologs in the sequence databases used today (Delmont et al., 2012). A functional gene and a disease are often connected through research. When using metagenomics sequencing, it can be difficult to apply a phylogenetic anchor to connect a functional gene to a taxonomic classification unless sufficient **sequencing** depth is attained and the reads can be consistently assembled into long enough contigs. To gain a deeper knowledge of the functional potential of particular species within a community to address the issues, numerous computer tools have sought to assemble metagenomic sequences into entire genomes (Sharon & Banfield, 2013; Imelfort et al., 2014). Recent evaluations described both potential risks and the main phases in metagenomics (Faust et al., 2015; Mason et al., 2014; Vuono et al., 2015).

5.5.2 Role of metagenomics in bioremediation

Monitoring the level of pollution in the air, soil, and water requires the use of microbial metagenomics. The variety of **microbial communities** and particular genes that could be exploited as pollution biomarkers can be revealed through metagenomics analysis. Software like **Meta Boot** can be used to find pollution biomarkers by comparing metagenomics datasets obtained from contaminated locations. The composition of the microbial communities can be utilized to develop pollution monitoring models. Shotgun metagenomics was employed to characterize the full spectrum of microbial populations in each polluted area. Using targeted **metagenomics**, it is possible to find specific genes in a population. These genes might be employed as biomarkers in particular circumstances. For instance, the expression of genes that break down cycloalkanes may be a sign of hexane contamination in a particular ecosystem. In general, specific charts may be produced based on the expression genes or the makeup of microbial communities, which offer defined cutoffs and support classification into particular classes based on the amount of pollutants present.

Important environmental remediating bacteria, bioactive substances, and enzymes have been discovered via metagenomics. Genes that break down hydrocarbons have been discovered in great numbers in both soil and aquatic habitats. Alkanes, phenanthrene, naphthalene, methyphenanthrene, and a number of other aromatic compounds are hydrocarbons that can be broken down with the help of certain genes. Some of the enzymes that break down pollutants include dioxygenases, laccases, esterases, carboxylesterases, monooxygenases, phenol-degrading enzymes, polyaromatic hydrocarbon-degrading enzymes, trichlorophenol-degrading enzymes, hexadecane hydrolyzing enzymes, and **alkane-degrading enzymes**. These enzymes decompose gasoline, plastics, fumigants, pesticides, dyes, and insecticides. A socially and environmentally viable alternative to conventional remediation techniques is micro bioremediation. Even though numerous microorganisms with bioremediation potential have been isolated and characterized, many of them are either unable to completely degrade the target contaminant or perform poorly when exposed to mixed waste.

5.6 Tools used in metagenomics

5.6.1 Short gene gun

Shotgun metagenomics uses genomic sequencing to analyze the whole genetic complement of an environmental population. In this technique, environmental DNA is extracted, fragmented, and then assembled into sequencing libraries. The libraries are then sequenced to determine the complete genetic makeup of the sample. A powerful tool for assessing a microbial community's functional potential is shotgun metagenomics. To obtain a complete inventory of the genes in an environmental sample, **deep sequencing** is typically necessary. A thorough analysis of the entire genetic makeup of each organism in the community is necessary to fully assess its functional potential. Shotgun metagenomics commonly oversamples the bacterial species that predominate in a community while sparingly sampling the genetic diversity of the minority species.

Furthermore, it is challenging to analyze **metagenomic sequencing** data because doing so requires accurately annotating a variety of gene sequences, many of which lack homologs in current sequence databases. A phylogenetic anchor is often used in research to connect a functional gene to a taxonomic classification. This can be difficult with metagenomics sequencing until appropriate sequencing depth is attained and the reads can be consistently assembled into adequately lengthy contigs. To gain a better knowledge of the functional potential of certain species within a community, many computer techniques have sought to assemble metagenomics sequences into entire genomes. Recent reviews have made an effort to draw attention to both the many potential risks and the main stages of metagenomics. Additionally, a number of microarray-based methods have been created.

The two microarray technologies that are most frequently used are **PhyloChip** and **GeoChip** (Hazen et al., 2013). A microarray based on 16S rRNA called PhyloChip can be used to examine the diversity of 10,993 subfamilies within 147 phyla (Hazen et al., 2010). In order to examine 152,414 genes from 410 distinct gene categories, GeoChip is a functional gene microarray (Zhou et al., 2013). Without relying on the depth of sequencing, microarray techniques provide significant insights into the microbial population. Additionally, they offer complete annotation for the numerous taxa or genes on the chip, which eliminates the need for reliable homologs in the database for appropriate categorization. The ability to find novel genes or pathways in a sample is constrained by the fact that microarray-based approaches can only detect the genes on the chip.

5.6.2 Next-generation sequencing technologies

In order to understand the structure and makeup of microbial communities, **NGS approaches** are useful. A full understanding of the ecology of microbial mediated processes, such as wastewater quality, environmental pollutant breakdown and detoxification, and disease transmission, is made possible by NGS technologies, which are currently recognized as the most powerful **gene sequencing techniques**. Applications for these NGS technologies include shotgun metagenome sequencing, whole-genome sequencing, and single-gene targeted sequencing. NGS, also referred to as high-throughput sequencing, is a term that describes a number of contemporary sequencing techniques that have transformed the fields of environmental genomics and microbial ecology research, including Illumina (Solexa) sequencing, Roche 454 sequencing, Ion torrent: Proton/PGM sequencing, SOLiD sequencing, and others.

5.7 Techniques involved in metagenomic sequencing

Different microbial species' metagenomes are sequenced using a number of next-generation genomic sequencing techniques. The Sangers sequencing method is frequently employed during the process of sequencing the human genome. **Pyrosequencing**, single-molecule sequencing, ligation sequencing, and reverse terminator sequencing are examples of next-generation sequencing techniques that have been made possible by technical drift and enable high throughput readings in a shorter amount of time. The majority of metagenomics researchers favor pyrosequencing as a method for sequencing the variety of microorganisms from various sites. Furthermore, metagenome analyses of *Burkholderia, Rhodanobacter, and* Proteobacteria from the environment with high metal, nutritional, and other pollutant contents were reported in genome sequencing of metal-polluted water. Bioremediation is a common method for applying new scientific and technological developments to the creation of better ecosystems, and metagenomics can be considered the latest approach in the field of metagenomics bioremediation (Sharma et al., 2021).

5.7.1 Smash community

A **stand-alone metagenomics** inquiry pipeline called SHell for microbial communities shares architecture principles and practices with **SmashCell** (Arumugam et al., 2010). This is ideal for data compiled using Sanger's sequencing method. This makes it easier to do crucial metagenomic tasks like gene prediction and assembly. Additionally, it provides strategies for calculating the quantitative phylogenetic and functional compositions of metagenome, contrasting the compositions of different metagenome, and producing useful graphical representations of these tests. This provides customized parameter sets for Arachne and Celera as well as **GeneMark** and **MetaGene** for the discovery of protein-coding genes on the proteome.

5.7.2 Metagenomic analysis technique

The most popular and significant strategy is analysis for broad bioinformatics segments (El Hadidi et al., 2013). This approach is excellent for assessing metaproteomics and metagenomics from a functional and taxonomical perspective. The program displays information on the National Center for Biotechnology Information (NCBI) taxonomy, **Kyoto Encyclopedia of Genes and Genomes (KEGG),** Clusters of Orthologous Groups of proteins (COG), and Science for Equity Empowerment and Development (SEED) categories are assigned readings for analysis, and functional evaluation. Moreover, samples may differ functionally and taxonomically be compared by utilizing a wide range of mapping and visualization. Its program uses principal coordinate analysis (PCoA) and clustering algorithms to compare a huge number of samples at a high level. Additionally, MEGAN supports a number of data input formats and can provide study results in various image and text-based applications.

5.7.3 Metagenome subsystems technology rapid annotation

A sophisticated metagenome technique called Metagenome Subsystems Technology Rapid Annotation (MG-RAST) provides microbial communities with quantitative insights based on sequencing analysis. This pipeline performed protein prediction, nucleic acid sequence quality control, and grouping datasets using a variety of bioinformatics tools. Users are encouraged to access sequence input data in **FASTA format**, and databases will be designed to generate data in the file promptly. For many forms of information, including metabolic reconstructions, phylogeny, one or more metagenome, genomes with metabolism, and annotations, the **MG-RAST** database provides multiple access methods. The pipeline is implemented in Perl using a variety of open-source modules, including the Sun Grid Engine, NCBI, SQLite, BLAST, and SEED system (Markowitz et al., 2007).

5.8 Case studies on metagenomics involved in bioremediation

5.8.1 Biostimulation to increase uranium oxidation in a contaminate aquifer

A substantial portion of the population relies on groundwater for drinking water. On the other hand, groundwater is susceptible to contamination. **Bioremediation** is one of the best ways to deal with groundwater contamination. Groundwater habitats can become contaminated by heavy metals as well as organic contaminants. Due to discharges from mine, processing, and natural sources, **uranium-contaminated groundwater** is a significant risk (Brugge & Buchner, 2011). As a result of their ability to interact with uranium, microbes can help lessen the impacts of uranium pollution. Both uranium's speciation and redox state affect how mobile it is (Newsome et al., 2014). U(VI) is soluble in aqueous solutions; however, U(IV) is not. By using electron donors such hydrogen, lactate, acetate, and ethanol, several metal-reducing bacteria can convert U(VI) to U(IV) (Lovely et al., 1991). As a result, the use of metal-reducing microbes as a method of limiting the spread of soluble uranium in uranium-contaminated environments has been suggested. Uranium reduction can be accelerated by adding an electron donor to the groundwater (Istok et al., 2004; Wu et al., 2006; William et al., 2011). The impact of **biostimulation** on uranium reduction and the main microorganisms involved in uranium reduction in groundwater have been studied using metagenomics in many studies in two uranium-contaminated aquifers. This study investigated the stimulation by ethanol, acetate, and emulsified vegetable oil. Numerous investigations were carried out at the US DOE Rifle facility in Colorado. At the Rifle site, mill tailings from a mine that processed uranium seeped into the groundwater. The microbial population at the Rifle site has been examined using metagenomic methods. The microbial life in background groundwater was very diverse. The use of acetate amendment for bioacceleration of uranium reduction was studied in 2002 (Anderson et al., 2003). During **uranium bioreduction**, *Geobacter* sp. dominated the microbial population (Chang et al., 2005). To identify which bacteria in the community were metabolizing the modified acetate, stable isotope probing and analysis of the 16S rRNA gene were combined with 13C-labeled acetate. This demonstrated that the primary organisms eating and integrating acetate into the system were *Geobacter* sp. biomass. According to a whole genome microarray study, the expression of rpsC is a highly accurate predictor of growth rates in *Geobacter uraniireducens*. This was confirmed in situ at the Rifle site during acetate stimulation, proving that the rpsC expression data accurately reflects the growth and metabolic rates of the ***Geobacter* species**. Proteomic analysis revealed that the planktonic microbial community in the acetate-stimulated groundwater was dominated by *Geobacter* proteins related to acetate metabolism and energy production. 16S rRNA analysis revealed that biostimulation reduced the overall diversity of organisms in the groundwater. According to **PhyloChip analysis**, the number of microbial species involved in the iron and sulfur cycle has increased. During stimulation, iron reducers lost their dominance over the population and were replaced by sulfate reducers.

5.8.2 Metagenomic analysis of biochemical passive reactors during acid mine drainage bioremediation

Biochemical passive reactors (BPRs) are a promising sustainable and safer choice among remediation technologies to address this problem (Villegas-Plazas et al., 2021; Johnson & Hallberg, 2005; Fetch, 2015). They are made up of columns loaded with a reactive mixture of organic and inorganic substrates and their microbial load, which neutralizes the running Advanced molecular detection (AMD) and removes or changes metals to less hazardous states (Simate & Ndlovu, 2014). Sulfate reduction, which is carried out by **sulfate-reducing bacteria** while oxidizing low-molecular carbon sources, is the main process in these passive treatments. In order to precipitate metals in the form of compounds with extremely poor solubility, this process produces hydrogen sulfides (Jong & Parry, 2003).

This work outlined the attempts to characterize the functional potential of the microbial communities from these laboratory-scale BPRs during the treatment of particular synthetic AMD, which is characterized by high sulfate concentration and low metal loading. This kind of AMD is prone to H_2S buildup, which is harmful to the microbial population and may decrease BPR effectiveness (Vasquez et al., 2018).

Shotgun metagenomics analysis of samples from the AMD-input section (bottom) of the 2-day HRT bioreactors, including sampling points of 8, 17, and 36 operating weeks, was done to carry out this characterization. These BPRs that met the following requirements were chosen as model microbiomes. The 2-day HRT samples had the optimal rate and removal efficiency balances (Vasquez et al., 2016), the reactor-input part was the most taxonomically diverse section among the bioreactors and included those microbial communities that were directly exposed to the AMD. Based on this justification, we examined microbial communities that had grown more diverse over the course of their operations, identifying microbiomes shifts in terms of gene abundance, in order to provide evidence for the potential connections between the inferred functions and the metabolic pathways that maintain the BPRs' constant efficacy and (ii) reveal the co-selection of crucial functions during efficient AMD bioremediation. This is the first metagenomics examination of BPR microbiomes, as far as we are aware. Our research offers fresh perspectives on the functional characteristics of the microbiome's activities in a model BPR system that is used to treat synthetic AMD with a high sulfate content (Villegas-Plazas et al., 2019).

5.8.3 Biofilms on whiteleg shrimp (*Litopenaeus vannamei*) effluents: metagenomic analysis

Invertebrates play a major role in aquaculture, which is considered to be the fastest growing food source, in particular, whiteleg shrimp (*Litopenaeus vannamei*) (Mendoza et al., 2021). Although they may have some unfavorable environmental repercussions, they still contribute to human welfare. Indeed, **aquaculture** has less of an impact on the environment than employing wild harvests. So, from an ecological standpoint, shrimp farming is examined in relation to bioremediation and sustainability. Given the high potential of metagenomics for aquaculture, an experimental study using biofilms on the effluents of whiteleg shrimp is also included as a proof of concept. This exposed a wide range of microorganisms, including bacteria, archaea, viridian plants, fungi, and protozoa. They comprised 972 species, divided into 221 families and 638 genera.

Additionally, **metagenomics** has been suggested as a viable technique for in-depth assessments of aquaculture, which includes information from open-access databases (Cornejo-Granados et al., 2018). According to studies (Andreotti et al., 2011; Follows & Dutkiewicz, 2011; Shange et al., 2013), marine habitats exhibit a very high richness of **microorganisms**. As a result, different microbial communities can be compared in one or more locations based on their diversity and abundance (Schloss & Handelsman 2008; Shange et al., 2013; Bai et al., 2014; Koo et al., 2015).

5.9 Limitations of metagenomics

The field of functional **metagenomics** has various difficulties. (1) As the microbial community's life conditions are complex, it may not be possible to harvest the genes of the species with low abundance. (2) The **gene cloning** procedure results in the loss of DNA fragments. (3) Heterologous expression of foreign genes occurs infrequently. (4) The currently available screening techniques are constrained and unable to satisfy all enzyme screening needs. (5) Only a few numbers of positive clones can be chosen from tens of thousands of clones due to low screening efficiency. Therefore, in the future, we must develop screening techniques that offer higher throughput, greater sensitivity, and are quick and straightforward. (6) Only a small portion of the recently discovered enzymes may be employed for industrial purposes due to restrictions on temperature, pH, and other factors. (7) Geographic diversity in recent functional metagenomic research is poor. It is primarily concentrated in industrialized nations. Underutilized are the vast microbial resources of developing nations (Zhang et al., 2021).

5.10 Future of metagenomics in bioremediation

Complex bioremediation techniques have proved effective in recovering contaminated environments. **Metagenomic** bioremediation is seen by environmental scientists as a viable method of getting rid of pollutants. Metagenomic bioremediation offers the highest results with higher degradation percentages when compared to the other bioremediation techniques. Microorganisms are crucial to bioremediation, therefore their variety, abundance, and community structure in the contaminated area might shed light on the likelihood of bioremediation. Additionally, the possible microbial strains, putative genes, enzymes, and metabolites that degrade/biotransform these pollutants from the environment can be examined using metagenomics, **metatranscriptomics**, **metaproteomics**, and metabolomics techniques. Utilizing microbial approaches such as genomics, proteomics, metabolomics, transcriptomics, and advanced molecular techniques such as "**Omics**" have enabled microbial recognition of roles, mechanisms of catabolism, and metabolism. The efficacy of bioremediation may be hampered by the availability of nutrients, a deficient microbiome with bioavailability, and the degradation capacity of contaminants. Although microbiological techniques include ways to accelerating biostimulation and bioaccumulation in contaminated wastewater are used in the treatment process. Biostimulation promotes the development of microorganisms by supplying nutrients to a contaminated sample. Microbes are widely distributed in a variety of environmental settings. This indicates that microorganisms are present in contaminated areas, and that the degree of wastewater contamination influences their ability to utilize nutrients such as nitrogen, phosphorus, and potassium for growth and metabolic processes. The microbial consortium has been shown to digest pollutants more efficiently than pure isolates.

Different types of environmental contaminants are being naturally remedied thanks to the metabolic profile of the microbial community. Numerous **bioremediation** techniques could be implemented to improve remediation effectiveness (Cassidy et al., 2015). When it comes to biomedicine, agriculture, environmental stewardship, uncounted microbial diversity of contaminated areas, and other technical issues, metagenomics offers a solution. One of the most important big potential improvements is this one:

- **Biomedical sciences**: a global assessment of the microbiome's role in biomedical waste and the development of new detoxification treatment approaches based on the knowledge gained.
- **Earth sciences**: switch, conservation, frameworks for deciphering and predicting global environmental processes, and the design of bacteria' environmental genetic sequence.
- **Life sciences**: use of novel theories and statistical methods, community-based microbial biology, and ecosystems.
- Bioenergy: it would be quicker and more environmentally responsible to produce new bioenergy resources for microbial systems and processes.

- Biotechnology: the discovery and application of the highly intricate and flexible biosynthetic capabilities of microbial populations to the production of valuable industrial, food, and health additives.
- Agriculture: to develop management strategies that maximizes the positive traits of microbial communities in and around domestic plants and animals as well as to establish more effective and systematic procedures for the early detection of hazards to food production and food safety.
- Biodefense and microbial forensics: the development of much more effective vaccines and treatments against potential bioterror agents, the use of genomic biosensors for the detection of known and potential pathogens in microbial environments, and the capability to accurately identify and categorize microbes that have contributed to conflict, terrorism, and crime.

5.11 Conclusion

To remove pollutants quickly and with the least amount of disruption, remediation solutions can be adopted with the help of metagenomics. Numerous case studies demonstrate the widespread use of metagenomic applications in microbial bioremediation processes for the detection and remediation of contaminants and pollutants in various sources of water bodies. The environment will recover more organically if these new methods for creating healthier surroundings are used in a more cost-conscious and environmentally conscious way. Due to their adaptability and immense potential, bioremediation techniques can therefore be employed to reduce the ecotoxicological consequences of wastewater.

References

Alivisatos, A. P., Blaser, M. J., Brodie, E. L., Chun, M., Dangl, J. L., Donohue, T. J., . . . & Unified Microbiome Initiative Consortium (2015). A unified initiative to harness Earth's microbiomes. *Science*, *350*(6260), 507–508.

Anderson, R. T., Vrionis, H. A., Ortiz-Bernad, I., Resch, C. T., Long, P. E., Dayvault, R., . . . & Lovley, D. R. (2003). Stimulating the in situ activity of Geobacter species to remove uranium from the groundwater of a uranium-contaminated aquifer. *Applied and Environmental Microbiology*, *69*(10), 5884–5891.

Andreotti, R., Pérez de León, A. A., Dowd, S. E., Guerrero, F. D., Bendele, K. G., & Scoles, G. A. (2011). Assessment of bacterial diversity in the cattle tick Rhipicephalus (Boophilus) microplusthrough tag-encoded pyrosequencing. *BMC Microbiology*, *11*(1), 1–11.

Arumugam, M., Harrington, E. D., Foerstner, K. U., Raes, J., & Bork, P. (2010). SmashCommunity: A metagenomic annotation and analysis tool. *Bioinformatics*, *26*(23), 2977–2978.

Bai, Y., Liang, J., Liu, R., Hu, C., & Qu, J. (2014). Metagenomic analysis reveals microbial diversity and function in the rhizosphere soil of a constructed wetland. *Environmental Technology*, *35*(20), 2521–2527.

Brugge, D., & Buchner, V. (2011). Health effects of uranium: new research findings.
Cassidy, D. P., Srivastava, V. J., Dombrowski, F. J., & Lingle, J. W. (2015). Combining in situ chemical oxidation, stabilization, and anaerobic bioremediation in a single application to reduce contaminant mass and leachability in soil. *Journal of Hazardous Materials*, *297*, 347–355.
Chakraborty, R., Wu, C. H., & Hazen, T. C. (2012). Systems biology approach to bioremediation. *Current Opinion in Biotechnology*, *23*(3), 483–490.
Chang, Y. J., Long, P. E., Geyer, R., Peacock, A. D., Resch, C. T., Sublette, K., . . . & White, D. C. (2005). Microbial incorporation of 13C-labeled acetate at the field scale: Detection of microbes responsible for reduction of U (VI). *Environmental Science & Technology*, *39*(23), 9039–9048.
Cornejo-Granados, F., Gallardo-Becerra, L., Leonardo-Reza, M., Ochoa-Romo, J. P., & Ochoa-Leyva, A. (2018). A meta-analysis reveals the environmental and host factors shaping the structure and function of the shrimp microbiota. *PeerJ*, *6*, e5382.
Delmont, T. O., Simonet, P., & Vogel, T. M. (2012). Describing microbial communities and performing global comparisons in the 'omic era. *The ISME Journal*, *6*(9), 1625–1628.
El Hadidi, M., Ruscheweyh, H. J., & Huson, D. (2013, July). Improved metagenome analysis using MEGAN5. *Joint 21st Annual International Conference on Intelligent Sys-tems for Molecular Biology (ISMB) and 12th European Conference on Computational Biology (ECCB)*.
Faust, K., Lahti, L., Gonze, D., De Vos, W. M., & Raes, J. (2015). Metagenomics meets time series analysis: Unraveling microbial community dynamics. *Current Opinion in Microbiology*, *25*, 56–66.
Fitch, M. (2015). Mine-impacted Water and Biochemical Reactors. In *Food, Energy, and Water* (pp. 129–159). Elsevier.
Follows, M. J., & Dutkiewicz, S. (2011). Modeling diverse communities of marine microbes. *Annual Review of Marine Science*, *3*(1), 427–451.
Hazen, T. C., Dubinsky, E. A., DeSantis, T. Z., Andersen, G. L., Piceno, Y. M., Singh, N., . . . & Mason, O. U. (2010). Deep-sea oil plume enriches indigenous oil-degrading bacteria. *Science*, *330*(6001), 204–208.
Hazen, T. C., Rocha, A. M., & Techtmann, S. M. (2013). Advances in monitoring environmental microbes. *Current Opinion in Biotechnology*, *24*(3), 526–533.
Hazen, T. C., & Sayler, G. S. (2016). Environmental systems microbiology of contaminated environments. *Manual of Environmental Microbiology*, 5–1.
Hettich, R. L., Pan, C., Chourey, K., & Giannone, R. J. (2013). Metaproteomics: Harnessing the power of high performance mass spectrometry to identify the suite of proteins that control metabolic activities in microbial communities. *Analytical Chemistry*, *85*(9), 4203–4214.
Imelfort, M., Parks, D., Woodcroft, B. J., Dennis, P., Hugenholtz, P., & Tyson, G. W. (2014). GroopM: An automated tool for the recovery of population genomes from related metagenomes. *PeerJ*, *2*, e603.
Istok, J. D., Senko, J. M., Krumholz, L. R., Watson, D., Bogle, M. A., Peacock, A., . . . & White, D. C. (2004). In situ bioreduction of technetium and uranium in a nitrate-contaminated aquifer. *Environmental Science & Technology*, *38*(2), 468–475.
Johnson, D. B., & Hallberg, K. B. (2005). Acid mine drainage remediation options: A review. *Science of the Total Environment*, *338*(1–2), 3–14.
Jong, T., & Parry, D. L. (2003). Removal of sulfate and heavy metals by sulfate reducing bacteria in short-term bench scale upflow anaerobic packed bed reactor runs. *Water Research*, *37*(14), 3379–3389.
Klindworth, A., Pruesse, E., Schweer, T., Peplies, J., Quast, C., Horn, M., & Glöckner, F. O. (2013). Evaluation of general 16S ribosomal RNA gene PCR primers for classical and next-generation sequencing-based diversity studies. *Nucleic Acids Research*, *41*(1), e1–e1.
Koo, H., Mojib, N., Huang, J. P., Donahoe, R. J., & Bej, A. K. (2015). Bacterial community shift in the coastal Gulf of Mexico salt-marsh sediment microcosm in vitro following exposure to the Mississippi Canyon Block 252 oil (MC252). *3 Biotech*, *5*(4), 379–392.
Lovley, D. R. (2003). Cleaning up with genomics: Applying molecular biology to bioremediation. *Nature Reviews Microbiology*, *1*(1), 35–44.

Lovley, D. R., Phillips, E. J., Gorby, Y. A., & Landa, E. R. (1991). Microbial reduction of uranium. *Nature, 350*(6317), 413–416.

Markowitz, V. M., Ivanova, N. N., Szeto, E., Palaniappan, K., Chu, K., Dalevi, D., . . . & Kyrpides, N. C. (2007). IMG/M: A data management and analysis system for metagenomes. *Nucleic Acids Research, 36* (suppl_1), D534–D538.

Martínez-Porchas, M., & Vargas-Albores, F. (2017). Microbial metagenomics in aquaculture: A potential tool for a deeper insight into the activity. *Reviews in Aquaculture, 9*, 42–56.

Mason, O. U., Scott, N. M., Gonzalez, A., Robbins-Pianka, A., Bælum, J., Kimbrel, J., . . . & Jansson, J. K. (2014). Metagenomics reveals sediment microbial community response to Deepwater Horizon oil spill. *The ISME Journal, 8*(7), 1464–1475.

Mendoza, O., Pretell, K., Ramirez, B., Sandoval, J., Caballero, J. L., & Dorado, G. (2021). Metagenomic analyses of biofilms on whiteleg shrimp (Litopenaeus vannamei) effluents: Implications for worldwide aquaculture bioremediation and environmental sustainability in the current trend of climate change and global warming–state of the art and experimental proof of concept. *Reviews in Fisheries Science & Aquaculture, 29*(4), 431–444.

Moran, M. A., Satinsky, B., Gifford, S. M., Luo, H., Rivers, A., Chan, L. K., . . . & Hopkinson, B. M. (2013). Sizing up metatranscriptomics. *The ISME Journal, 7*(2), 237–243.

Newsome, L., Morris, K., & Lloyd, J. R. (2014). The biogeochemistry and bioremediation of uranium and other priority radionuclides. *Chemical Geology, 363*, 164–184.

Parada, A. E., Needham, D. M., & Fuhrman, J. A. (2016). Every base matters: Assessing small subunit rRNA primers for marine microbiomes with mock communities, time series and global field samples. *Environmental Microbiology, 18*(5), 1403–1414.

Payne, R. B., May, H. D., & Sowers, K. R. (2011). Enhanced reductive dechlorination of polychlorinated biphenyl impacted sediment by bioaugmentation with a dehalorespiring bacterium. *Environmental Science & Technology, 45*(20), 8772–8779.

Prosser, J. I. (2012). Ecosystem processes and interactions in a morass of diversity. *FEMS Microbiology Ecology, 81*(3), 507–519.

Rinke, C., Schwientek, P., Sczyrba, A., Ivanova, N. N., Anderson, I. J., Cheng, J. F., . . . & Woyke, T. (2013). Insights into the phylogeny and coding potential of microbial dark matter. *Nature, 499*(7459), 431–437.

Schloss, P. D., & Handelsman, J. (2008). A statistical toolbox for metagenomics: Assessing functional diversity in microbial communities. *BMC Bioinformatics, 9*(1), 1–15.

Shade, A., & Handelsman, J. (2012). Beyond the Venn diagram: The hunt for a core microbiome. *Environmental Microbiology, 14*(1), 4–12.

Shakya, M., Quince, C., Campbell, J. H., Yang, Z. K., Schadt, C. W., & Podar, M. (2013). Comparative metagenomic and rRNA microbial diversity characterization using archaeal and bacterial synthetic communities. *Environmental Microbiology, 15*(6), 1882–1899.

Shange, R., Haugabrooks, E., Ankumah, R., Ibekwe, A. M., Smith, R. C., & Dowd, S. (2013). Assessing the diversity and composition of bacterial communities across a wetland, transition, upland gradient in Macon County Alabama. *Diversity, 5*(3), 461–478.

Sharma, P., Kumar, S., & Pandey, A. (2021). Bioremediated techniques for remediation of metal pollutants using metagenomics approaches: A review. *Journal of Environmental Chemical Engineering, 9*(4), 105684.

Sharon, I., & Banfield, J. F. (2013). Genomes from metagenomics. *Science, 342*(6162), 1057–1058.

Simate, G. S., & Ndlovu, S. (2014). Acid mine drainage: Challenges and opportunities. *Journal of Environmental Chemical Engineering, 2*(3), 1785–1803.

Smith, M. B., Rocha, A. M., Smillie, C. S., Olesen, S. W., Paradis, C., Wu, L., . . . & Hazen, T. C. (2015). Natural bacterial communities serve as quantitative geochemical biosensors. *MBio, 6*(3), e00326–15.

Techtmann, S. M., & Hazen, T. C. (2016). Metagenomic applications in environmental monitoring and bioremediation. *Journal of Industrial Microbiology and Biotechnology*, *43*(10), 1345–1354.

Vasquez, Y., Escobar, M. C., Neculita, C. M., Arbeli, Z., & Roldan, F. (2016). Biochemical passive reactors for treatment of acid mine drainage: Effect of hydraulic retention time on changes in efficiency, composition of reactive mixture, and microbial activity. *Chemosphere*, *153*, 244–253.

Vasquez, Y., Escobar, M. C., Saenz, J. S., Quiceno-Vallejo, M. F., Neculita, C. M., Arbeli, Z., & Roldan, F. (2018). Effect of hydraulic retention time on microbial community in biochemical passive reactors during treatment of acid mine drainage. *Bioresource Technology*, *247*, 624–632.

Villegas-Plazas, M., Sanabria, J., Arbeli, Z., Vasquez, Y., Roldan, F., & Junca, H. (2021). Metagenomic analysis of biochemical passive reactors during acid mine drainage bioremediation reveals key co-selected metabolic functions. *Microbial Ecology*, 1–8.

Villegas-Plazas, M., Sanabria, J., & Junca, H. (2019). A composite taxonomical and functional framework of microbiomes under acid mine drainage bioremediation systems. *Journal of Environmental Management*, *251*, 109581.

Vuono, D. C., Benecke, J., Henkel, J., Navidi, W. C., Cath, T. Y., Munakata-Marr, J., . . . & Drewes, J. E. (2015). Disturbance and temporal partitioning of the activated sludge metacommunity. *The ISME Journal*, *9*(2), 425–435.

Williams, K. H., Long, P. E., Davis, J. A., Wilkins, M. J., N'Guessan, A. L., Steefel, C. I., . . . & Lovley, D. R. (2011). Acetate availability and its influence on sustainable bioremediation of uranium-contaminated groundwater. *Geomicrobiology Journal*, *28*(5–6), 519–539.

Wu, W. M., Carley, J., Fienen, M., Mehlhorn, T., Lowe, K., Nyman, J., . . . & Criddle, C. S. (2006). Pilot-scale in situ bioremediation of uranium in a highly contaminated aquifer. 1. Conditioning of a treatment zone. *Environmental Science & Technology*, *40*(12), 3978–3985.

Xiong, W., Abraham, P. E., Li, Z., Pan, C., & Hettich, R. L. (2015). Microbial metaproteomics for characterizing the range of metabolic functions and activities of human gut microbiota. *Proteomics*, *15*(20), 3424–3438.

Zhang, L., Chen, F., Zeng, Z., Xu, M., Sun, F., Yang, I., & Xie, Y. (2021). Advances in metagenomics and its application in environmental microorganisms. *Frontiers in Microbiology*, *12*.

Zhou, A., He, Z., Qin, Y., Lu, Z., Deng, Y., Tu, Q., . . . & Zhou, J. (2013). StressChip as a high-throughput tool for assessing microbial community responses to environmental stresses. *Environmental Science & Technology*, *47*(17), 9841–9849.

Shreya Anand and Padmini Padmanabhan*

Chapter 6
Study of microbial communities in degrading toxic pollutants in wastewater and solid waste treatment industries

Abstract: Industrial wastewater is one of the major sources releasing variety of pollutants into the environment. Release of wastes containing toxic substances, such as heavy metals and anions, without proper treatment into the environment can cause severe environmental and health hazards leading to deterioration of ecosystem. Removal of toxic pollutants from industrial wastewater is a major challenge. There are various physical, chemical, as well as biological methods of wastewater treatment. Traditional methods of removal of pollutants are undesirable from the environmental and economic aspects. Industrial wastewater treatment employs microbes such as bacteria, fungi, or protozoa for the degradation and removal of toxic pollutants. Biological treatment involving diverse range of microbes for bioremediation is a proficient, cost-efficient, simple, and environment-friendly process. Advantages of biological treatment processes include shorter process time, low energy requirements, no secondary pollution, and long time viability. Microbial diversity contributes to the removal of toxicants through enzymes and several other interactions that break down the complex toxicants compounds into simpler compounds and help in degradation. Bioremediation is emerging as a promising alternative to conventional methods of treatment. This chapter offers valuable insights into the fundamental role of microorganisms in the bioremediation process for pollutant(s) removal.

6.1 Introduction

Environmental hazards from industrialization and other anthropogenic human activities are significant (Saravanan et al., 2021). The varied living things in their respective ecosystems may contract harmful diseases as a result of hazardous pollution. Different industries produce a variety of wastes in the form of solid waste or wastewater (Zhang et al., 2015). If these wastes are discharged without being properly treated, there may be serious dangers and hazards. Cyanide, fluoride, heavy metals (HMs),

***Corresponding author: Padmini Padmanabhan,** Department of Biotechnology and Bioengineering, Birla Institute of Technology, Mesra 835215, Jharkhand, India, e-mail: padmini@bitmesra.ac.in
Shreya Anand, Department of Biotechnology and Bioengineering, Birla Institute of Technology, Mesra 835215, India

https://doi.org/10.1515/9783110793635-006

chemical discharge, and others are examples of harmful industrial pollutants (Vardhan et al., 2019). Therefore, these harmful pollutants must be treated before being released into the environment.

To lessen the hazardous contaminants in the environment, a variety of physical, chemical, and biological approaches have been used. Among these, **bioremediation** is one of the biological methods used to degrade harmful contaminants using microorganisms such as bacteria, fungus, and yeast to produce nontoxic metabolites. The most often obtained by-products of hazardous chemicals' biodegradation are carbon dioxide and water (Ojuederie & Babalola, 2017). In comparison to other approaches, bioremediation has a number of benefits, including being simple to use, less expensive, ecologically beneficial, and safer. Diverse categories of enzymes, including oxidoreductases, hydrolases, oxygenases, peroxidases, lipases, and phosphodiesterases, are released by the microorganisms involved in the bioremediation process, which aids in the process via catalytic action (Sharma et al., 2018). By reducing the activation energy, these enzymes make it easier for harmful environmental contaminants to be broken down. Since the interaction of the enzyme and substrate totally determines how something degrades, degradation of enzymes is more exact. The enzyme interacts with the substrates (pollutants) through a special binding region known as the active site to carry out the catalysis (Karthik et al., 2021; Bilal et al., 2019a). **Biodegradation** of enzymes is beneficial than bioremediation, including an extremely precise, quick, extra adaptable operating circumstances, actual outcomes even during unfavorable conditions, a lack of dependence on nutrient supplies, a reduction in by-product formation, and a lower threshold for mass transfer than bioremediation techniques (Kensa, 2011). Degradation through enzymes offers better outcomes than bioremediation, but choosing the right microorganism for the production of enzyme is crucial, and producing and purifying enzymes on an industrial scale are more expensive. Immobilization strategies have been used by researchers to address these issues now, increasing process effectiveness and lowering the price associated with manufacturing of enzyme and purification of enzyme (Antecka et al., 2018; Ba et al., 2013). Industrial-scale **enzymatic degradation** calls for a greater quantity of pure enzymes, which could be constrained by the economics of enzyme manufacturing and purification. This can be done with the use of genetic engineering, which produces the appropriate number of enzymes more specifically and more affordably than the standard enzyme production procedure (Huy et al., 2021). Various **immobilizations** and genetic engineering techniques have recently been developed to increase enzyme effectiveness and reduce the cost of the pollution removal process.

In this chapter, the toxic pollutants in wastewater, metagenomic analysis for profiling microbial communities, microbial communities involved in the process of bioremediation, role of microbial communities in degradation toxic wastes, and the cutting edge technologies involved in the treatment of toxic pollutants in wastewater and solid wastes are discussed.

6.2 Toxic pollutants in wastewater

Environmental **pollution** is caused by several **contaminants**. Among these are the principal harmful contaminants that pollute the air, water, and soil. The toxic pollutants are HMs, dyes, pesticides, PAH, plastic, and so on (Table 6.1).

Table 6.1: Toxic environmental pollutants.

Toxic pollutants				
Heavy metals	**Dyes**	**Pesticides**	**PAH**	**Plastic**
Chromium	Sulfur dye	Organochlorines	Naphthalene	Polyvinyl chloride
Arsenic	Reactive dye	Carbamates	Perylene	Low-density polyethylene
Lead	Acid dye	Pyrethroids	Pentacene	Polyethylene terephthalate
Cadmium	Basic dye	Organophosphates	Anthracene	Polystyrene

6.2.1 Heavy metals

A class of **metals** with a density greater than 4,000 kg/m^3 is referred to as HMs. Popular HMs include copper, cadmium, zinc, cobalt, lead, and chromium. The lighter metals are arsenic, selenium, and aluminum. The two categories of HMs are essential **HMs** that are less harmful and nonessential HMs that are more harmful. Some of the HMs that serve as cofactors in a range of biological processes include cobalt, copper, and zinc. However, once they exceed acceptable limits, they become toxic to environments. HMs were initially discovered in the Earth's crust (Briffa et al., 2020; Duan et al., 2020; Abdelhaleem et al., 2019). **Contamination** with HM inevitably has a number of negative effects on human health and the environment. To reduce the impact of emissions on the environment, methods for removing hazardous metals from various systems must be developed.

6.2.2 Dyes

Dyes, which are water-soluble colorants, are used in the garment, paint, paper, and cosmetic industries.. Depending on their composition, application, and color, dyes are divided into a number of categories. Real and artificial dyes are two types that have different sources. Synthetic dyes are frequently utilized in industry due to their chromophoric groups and good fixing ability (Tan et al., 2015; Zhuo et al., 2015). Despite the fact that these colors are important for day-to-day activities, the health of people and the **environment** is endangered when they are released as industrial effluents. Dyes are a complex mixture of nitro, aromatic, and HM compounds, some of which, when used in excess, can have harmful effects. Even at low quantities, dyes are harmful due to their carcinogenic nature. **Chlorine** also pollues some marine creatures (Katheresan et al., 2018).

6.2.3 Pesticides

Pesticides are a combination of chemical compounds designed to inhibit, lessen, or prevent microbial and insect growth. It is categorized based on where they are used and how they work. Pyrethroid, carbamate, organophosphate, and based on the source of pesticide, they are of four major pesticide classes. It is created to harm a certain class of chemicals or microbes, but they could also have important implications for other living beings (Souza et al., 2020; Hassaan & Nemr, 2020).

Persistent organic pollutants are unaffected to weakening and may persist in the atmosphere for years, and are present in the majority of pesticides. Pesticides that are transmitted from industrial sources are harmful to both marine and land life. Any type of herbicide results in the death of the dead plants, which uses up additional oxygen & interferes with the natural respiratory system of fish. The air quality worsens when pesticides are volatilized and released into the atmosphere. **Leaching** of pesticides affects the quality of groundwater, the fertility of the soil, and the normal growth of terrestrial vegetation. Pesticides can enter the body through the mouth, nose, or skin.

As pesticides are mixed and put to the field, the majority of cutaneous contamination occurs (Bose et al., 2021). The fluid form is significantly added engrossing and destructive to the covering when uncovered than the aerosol. **Aerosolized pesticides** are breathed by the system, which results in oxidative stress in people. **Reactive oxygen species** (ROS), which are present in pesticides such as rotenone, maneb, and paraquat, cause Parkinson's disease when there is an excess of ROS. They also encourage inflammation, which then results in immunosuppression and severe asthma. Organophosphorus pesticides also induce leukemia and genetic abnormalities (Bilal et al., 2019b; Sharma et al., 2020). Therefore, increased efforts are required to limit the usage of pesticides and to look for affordable alternatives.

6.2.4 Plastics

Plastics are lightweight, inexpensive, corrosion-resistant, and long-lasting polymers with synthetic or semisynthetic origins. In densely populated industrial settings, plastic bags are frequently present, and when they are dumped, they pollute the coastal habitat. **Macroplastic** and microplastic are the two categories used to describe plastic. With a size bigger than 25 mm, macroplastics are type of malleable debris found in oceans, coast, and shorelines. Despite their size, fish and marine animals consume them. There are two categories of microplastics: primary and secondary, the primary

microplastic is made on a microscopic size for industrial and home use, and secondary microplastics are produced when macroplastics break down (Tiwari et al., 2020).

Until they have gathered, they are pushed lower on the food chain. Polychlorinated biphenyls and dichlorodiphenyltrichloroethane are two destructive chemicals released by plastics, harms and deadly outcomes in marine species. Once plastic waste is not correctly disposed in the atmosphere, it can clog pipes and soil, which can disrupt regular operations.

Chemical buildup in the soil is the main source of crop contamination. Other harmful substances released by plastics alter the soil's makeup and taint groundwater. Plastics frequently cause harm to land animals through ingestion and entanglement. Chemicals **bioaccumulate** over time and go up the food chain. Humans become contaminated with plastics by ingesting micro and nanoplastics through their skin. Other negative consequences include necrosis, genotoxicity, apoptosis, and oxidative stress. Therefore, it must be remedied in order to lessen the impacts.

6.2.5 Polycyclic aromatic hydrocarbons (PAHs)

A complicated class of compounds known as **polycyclic aromatic hydrocarbons (PAHs)** have flat, clustered, or angularly joined aromatic carbon rings. Small PAHs have less aromatic rings, and large PAHs have more than six aromatic rings. Most incomplete burning of coal and oil, some wastewater discharges, release PAHs into the air. Low cell yields are caused by PAHs, which are mutagenic to microorganisms and limit their development. Oil leaks cause the discharge of PAH into aqueous habitats, which has an impact on the marine ecosystem. Fish that consume PAHs experience immunological reactions that resemble humoral and cell-mediated reactions. Lymphocytes are extremely sensitive to PAHs, and exposure to them can cause macrophages to swell (Kadri et al., 2017).

Particularly **petroleum-based PAHs** seem to glide on the surface of water and form a reedy coating that blocks sunlight and the photosynthesis. Industrial **wastewater** discharge can cause PAHs deposition in agricultural lands, changing the roughness of those areas. Plants raised on it eat PAH, which eventually builds up in the food chain. PAHs may inhibit the normal cellular membrane activities and the enzymes involved when inhaled or ingested. **Dihydrodiols** and epoxides in PAHs fix to DNA and other biological proteins, causing mutations, illness, and cell disintegration.

The metabolic changes of PAH lead to the production of electrophilic derivatives. Red blood cells break down when naphthalene is inhaled. Additional long-term effects of PAHs exposure include lung abnormalities, cataracts, kidney disorder, and liver damage. Potential therapeutic options must be used to prevent PAH toxicity's detrimental impacts on the environment and human health.

6.3 Metagenomic analysis for profiling of microbial communities

The **metagenomic** analysis is a great method for analyzing the quantitative traits of microbial communities, including those found in industrial wastewater treatment. Wastewater from the pulp and paper sector helps the diverse microbial populations break down organic matter and remove developing contaminants. One of the major energy and water consumers in the world is the pulp and paper sector. The pulp and **paper industry** in India produce over 100 million kg of harmful pollutants per year (Sharma et al., 2021).

Deoxyribonucleic acid and ribose nucleic acid, which are essential for cell growth, are just two examples of the genetic materials that make up the entire genetic material in an organism as defined by the **metagenomics** study.

Both total DNA collected from an ambient sample (shotgun MGA) and 16S rRNA amplified PCR fragments can be used for metagenomic analysis. In order to classify and define every species (or genes), if they are present in the still vast and expanding databases, the nucleotide sequences of millions of DNA fragments are determined. **Computational studies** are then utilized to do so. A powerful method for gaining insights into systems that are becoming fundamentally important to wastewater is emerging in science as MGA. Significant amounts of wastewater from the disposal site are enhanced with nutrients, pollutants, and bacteria. Studies that have linked HMs to mobile genetic elements like the class 1 integron integrase gene (*intI1*) and *sul* genes suggest that the presence of metals in wastewater contributes to the co-selection of **antibiotic resistance genes**. Because the environmental growing conditions are unknown or varied, more than 99% of microbial species are considered "uncultivated." The pulp and paper industry's processes offer a variety of chances to use microbial enzymes.

Microorganisms and the use of energy as a nutrition source can mineralize the organometallic compounds in wastewater. Because they have a range of mechanisms for resisting metal toxicity, microbes were crucial for the clearance of toxic metal-polluted locations. Microorganisms exist in a variety of interrelated ecosystems that affect both the natural environment and human activity. Using the metagenomics technique, classification of the microorganism function and distribution of coal-mining wastewater treatment plants has previously been published. Microorganisms in particular play a key role in the biodegradation of pollutants, making bioremediation an economical and environmentally benign method for remediating contaminated soils. **Microbiome** ecosystem assessments have been transformed by the development of **next-generation sequencing** (NGS) technology with an emphasis on 16S rRNA (Awasthi et al., 2020).

NGS analysis was used to decode how microbial communities participate in the detoxification, degradation, and transformation of dangerous chemicals at polluted areas in microbial phylogenetic research. For the microbial world to reach its full potential, metagenomics provides a wealth of information that must be used. Metagenomics analysis of wastewater at various stages of treatment was conducted in central Germany to determine the potential for in-situ bioremediation. Diverse wastewater treatment techniques have been used in numerous sectors to effectively purify wastewater and reduce end waste. It is important to keep track of how microbial diversity affects the treatment of biologically **contaminated wastewater**.

Characterization of natural bacterial populations' functions through surviving in wastewater will help with management and sustainable approaches for management and reuse. Conclusions show how vital it is for the understanding of local bacterial species rehabilitation of contaminated ecosystems through sustainable ecology. Research examining the wastewater's microbial community's makeup will not only can you study the concerns linked with pollution in the future wastewater and offer guidance on potential methods for cleaning up of pollution using local bacteria.

One of the newest tools for the sequences of various bacteria that can remove various toxins from contaminated wastewater by metagenomics analysis is the metagenomic approach. In this experiment, we investigated potential resistance or tolerance of **microbes** capable of surviving the contaminated ground by using metagenomics analysis to examine the structure and function of the microorganisms in metals contaminated wastewater and expose the variations of the microbial community (Lundberg et al., 2012).

6.4 Microbial community involved in bioremediation

Microorganisms that actively participate in **microbial consortiums** are those that perform biodegradation in a variety of conditions. Included in this group are the microbes *Acinetobacter, Actinobacter, Alcaligenes, Arthrobacter, Bacillus, Beijerinckia, Flavobacterium, Methylosinus, Mycobacterium, Mycococcus, Nitrosomonas, Nocardia, Penicillium, Phanerochaete, Pseudomonas, Rhizoctonia.* Most dangerous chemicals cannot be mineralized by microorganisms alone. Complete mineralization causes consortia of microorganisms to sequentially degrade the material; this process involves synergism and co-metabolism. A staggering amount of organic molecules can be metabolized and frequently mineralized by natural populations of microorganisms found in a variety of settings. Although the exact number of molecules that some communities of **bacteria** and fungus can break down, thousands are known to be destroyed as a result of microbial activity in one setting or another. The majority of bioremediation systems operate aerobically, although operating a system anaerobically may allow microbial organisms to break down normally resistant compounds.

6.4.1 Use of pure culture

Several experimental studies have shown that pure cultures of various algae, bacteria, and fungi can be used to treat commonly found pollutants such as iopromide, carbamazepine, ibuprofen, sulfamethoxazole, diclofenac, paracetamol, and triclosan. These samples include activated sludge, wastewater, or sediment. **Sulfamethoxazole** and other sulfonamides, for instance, can be mineralized by the bacterium *Achromobacter denitrificans* (Reis et al., 2018). The study found that *Stenotrophomonas* biosorption considerably contributed to the elimination of paracetamol, while *Delftia tsuruhatensis* and *Pseudomonas aeruginosa* contributed relatively little. Different enzymes involved in the degradation process may be the cause of the variations. The main prerequisite to enable the efficient biodegradation of dangerous recalcitrant Personal and pharmaceutical care products (PPCPs) is to activate the microorganisms to produce the degradative enzymes specific to the environmental concern (Singh et al., 2020).

6.4.2 Use of mixed cultures

As it can be difficult to isolate the pure culture in particular environmental conditions, mixed cultures are more effective than pure cultures for degrading PPCPs biologically. A little amount of research has been done to evaluate the effectiveness of mixed cultures for the elimination of PPCPs. **Mixed cultures** is effective at eliminating PPCPs. The role of a mixed culture of heterotrophic and ammonia-oxidizing bacteria in enhancing 17a-ethinylestradiol elimination was discussed. Actually, in order to eradicate PPCPs, the generally used biological therapy procedures rely on the synergistic activities of mixed cultures. However, occasionally activated sludge has a low rate of PPCPs removal. As a result, research has conducted to improve the **PPCP degradation** through the activated sludge method (Zhao et al., 2013).

6.4.3 Use of activated sludge

In conventionally used **wastewater treatment plants**, the biological treatment process has been mostly carried out using the activated sludge approach. The interaction of volatilization, surface binding, and microbial decomposition is blamed for the PPCPs deterioration during biological therapy. However, the volatilization and adsorption processes have a minor role in the degradation of PPCPs (Li et al., 2015). In most cases, aeration and volatilization occur concurrently. The **physicochemical** properties of the compounds that will be degraded have a substantial impact on the adsorption of PPCPs during biological therapy. Additionally, modifications in environmental factors such as pH, oxygen content, temperature, the makeup of the microbial population, and nutrient status can have a significant impact on the overall effectiveness of the activated sludge system used for the elimination of PPCPs.

6.4.4 Use of genetically engineered microorganism

Currently, a number of scientific organizations worldwide are working to eliminate man-made pollution by altering the genetic makeup of microorganisms. **Genetically modified organisms** provide the possibility of pollution degradation for a variety of substances. In managing damaged soil, groundwater systems, and activated sludge environments, they have demonstrated bioremediation potential and improved degrading capabilities for a variety of chemical and petrochemical contaminants.

However, their extensive **in situ applications** are constrained by regulatory requirements and environmental considerations. Although many in vitro experiments on **bioremediation** have been conducted, the participation of numerous environmental elements, such as biosafety, has always made it difficult to study the fate, behavior, and removal rate of toxins in the natural ecosystem (Singh et al., 2011).

6.5 Role of microbial communities in pollutant removal

The complicated **lipophilic** chemical compounds, formerly thought to be refractory, may now be cleaned and modified mostly by microorganisms to simple water soluble products. When exposed to these particular or structurally comparable substances, they first assault these organic chemicals using the enzymatic machinery gained during enrichment. The enzymatic function of microorganisms is either stimulated or suppressed by the presence of these pollutants in the environment. The selective **microbial community**, as well as the structural and functional classes of hazardous chemicals, plays a significant role in this ability. These water-soluble intermediates typically undergo complete biodegradation when they are attacked by primary or secondary groups of organisms, which produce inorganic end products. Bioremediation is the process of converting environmental pollutants into less harmful forms by using living organisms, primarily microbes. It uses plants, naturally existing bacteria, and fungi to break down or detoxify pollutants that are harmful to the environment or to human health. The microorganisms may be separated from another location and delivered to the contaminated site, or they may be local to the contaminated area. Living things spread contaminated substances through reactions that happen as part of their metabolic processes. Multiple species' activity frequently leads to the biodegradation of a chemical. The technique is referred to as "**bioaugmentation**" when microorganisms are imported to a contaminated location to speed up degradation. To take place in a bioremediation process, the microorganisms with the genetic ability to change chemicals of interest must be present in contaminant metabolism. In some circumstances, bioaugmentation – the addition of organisms adapted to particular contaminants – can shorten the time that lag periods last. The process employed determines how well a bioremediation system can be enhanced biologically. Utilizing **natural biological processes**, bioremediation is a method that provides the opportunity to eliminate or render harmless a variety of toxins.

6.6 Mechanism of reduction of toxic wastes

6.6.1 Biomethylation

The **biological reducers** use an alkylation reaction to transform harmful metals into forms that can be easily broken down. *S*-Adenosylmethionine, methylcobalamin, and methyltetrahydrofolate are important co-factors in carrying out the basic biorestorative process, which involves the production of methylated intermediates throughout the detoxification process. The occurrence of oxygen along with the metals involved in redox processes also regulates **biomethylation**. A coenzyme (methylcobalamin) is found in both bacterial cells and animal cells. It is the primary component in the mechanism. This can carry out the transformation in one of two methods and transmission of the methyl group. A process uses *S*-adenosyl methionine which releases positively charged carbon species (methyl) known as the carbonium ion, which is typically less reactive than non-methylated metal ions. Aquatic system, especially groundwater, contains a lot of biomethylating communities, such as **Proteobacteria**, *Nitrospira*, *Rhodopseudomonas*, and Firmicutes, which can be used to detoxify arsenic. In the Hetao plain (China), it was discovered that archaea methylating species predominated in wells with elevated arsenic concentration. The biomethylation of inorganic mercury was influenced by oxygen level in another investigation (Begum et al., 2022).

According to the findings, both facultative anaerobic and aerobic environments resulted in notable biomethylation and **demethylation**. The methylation and demethylation steps of the mercury transformation occurred and the overall methylation ratio determined the effectiveness of the entire process. The demethylation was carried out to protect the bacterial cells from the negative effects of an excessive buildup of mercury (Cao et al., 2021). The mobilization and metal ions volatilization are enhanced under reducing conditions, which favors biomethylation.

6.6.2 Bioprecipitation

The method of **precipitation** involving chemicals are regarded as the go-to approach for recovering dissolved useful byproducts or reducing the amount of harmful metal in the wastewater stream. The maintenance and operation costs for this approach, which uses precipitating agents like sulfate or hydroxide, can be extremely high depending on the amount of harmful effluent. Therefore, the rather slow but environmentally friendly and cost-effective biological techniques are used. One of these cutting-edge methods for recovering metal from leached waste or for sequestering hazardous metal waste is bioprecipitation.

Microorganisms aid in the **immobilization** of metals and cause the modification of chemical of metal complexes during bioprecipitation. When metals in the industrial water interact with the metabolites produced during precipitation, an insoluble species or precipitate is created. The microbial population, temperature, pH, and hazardous metal all have an impact on the process. The capacity of the *Citrobacter* N14 strain to accumulate the waste from uranium nuclear plants was examined. It was noted that the uranium precipitate was primarily found on one side of the cell and at the cell's edge.

The first stage of **biomineralization**, which is connected to the nucleation of harmful elements and mineral crystals, is called microbial-derived precipitation. Through the formation of intracellular nanoscale inclusions, the bacterial culture predominantly endures the metal. Numerous microbial cells have cell wall components or enzymes that they can use to start the precipitation of the nucleation site. Their capacity to thrive in severely contaminated environments and their capacity to withstand metal toxicity are the elements that determine the effectiveness of microbial precipitation (Guo et al., 2014).

6.6.3 Biotransformation

Microbial cells use biotransformation, a technique that transforms toxic metal waste into less toxic compounds, as a protective measure to survive in highly contaminated environments. Metal-binding proteins of microorganisms assist in the mobility of metals during the transformation, which leads to their solubilization or precipitation. **Biological agents** are employed to chemically change the fatal toxicants in a polluted environment by oxidizing, condensing, or hydrolyzing the bonds. The multiplication and survival of the bacterial population in the highly contaminated area can be attributed to the presence of genes for particular enzymes. *Pseudomonas chengduensis* MBR strain has specific active enzymes that may quickly biotransform cadmium(II) to Cd. The creation and deposition of a biofilm or nanoparticles on the surface of the cells is the initial step in the remediation process. The phosphate hydrolase enzyme supplies the energy required for the metal's transit, while membrane proteins like P-type ATPase help to chemically recast the metal (Wang et al., 2020). Another efflux pump called **resistant nodulation division** (RND) transports metals like cadmium, zinc, lead, or cobalt outside of the cells. For the transporters to work effectively, an aerobic environment is essential, which aids in the elimination of even very high cadmium concentrations. Redox reactions, which microorganisms use as a defense against the harmful effects of metals, are a major component of biotransformation. In photosynthetic bacteria and chemolithoautotrophs, arsenic(III) serves as an electron donor and the first step of oxidation is a process that generates energy (Rahman & Hassler, 2014).

6.6.4 Bioleaching

The term "**bioleaching**" describes the technique of removing metal ions from their ores using microorganisms. A broad population of microorganisms with various leaching capacities typically releases certain chemical intermediates that aid in their attachment to metal. The microbial cultures have the ability to bioleach has been successfully used to recover metals from an intricate blend of acid mine waste chemicals. Leaching of zinc, copper, uranium, cobalt, and so on from their sulfide or oxides has been taking place for the past 10 years.

The population of **bioleachers** is highly diverse, with the majority of them being acidophilic in nature. Additionally, thermophiles such as *Sulfolobus* bacterial group can be used to address the issue of acid mine drainage. In refineries, hazardous metals including nickel, aluminum, cobalt, and tungsten that are parts of the catalysts used to treat feed, are removed via a process called bioleaching. Based on their characteristics, the leached metals can also be used in other industrial applications. Acidophilic bacteria involved in bioleaching the catalyst include *Rhodotorula* species, *Acidithiobacillus ferrooxidans*, *Acidithiobacillus thiooxidans*, and *Acidomyces acidophilus* (Phyo et al., 2020; Yu et al., 2020).

Protons from the **sulfur** react with the metal sulfide during the bioleaching process, breaking the connection between the metal and sulfur group of compounds. As a result, the valence bond's electrons are released, and the sulfide groups are subsequently solubilized in an acidic growth media.

6.7 Cutting-edge technologies for solid waste remediation

6.7.1 Floating treatment wetland (FTW)

A cutting-edge device called a **floating treatment wetland** (FTW) uses a system of floating plants to treat wastewater in a way that is unlike any other (Jones et al., 2017; Headley & Tanner, 2008). In FTWs, plants are held up by a raft or buoyant mat that floats on the water's surface (Tanner & Headley, 2011). The plant's roots grow below the floating mat, extending down the water column, and creating a vast root network below the surface of the water (Shahid et al., 2018). For FTWs to function as intended, a wide-spread, dense root structure must be developed (Ladislas et al., 2015). In comparison to emergent roots, FTWs can move freely and cover a larger area of water. The dense root structure and rhizomes in an FTW system create a unique **hydraulic movement** in the water zone between the mat and the water's bottom, and the floating roots serve as filters (Chen et al., 2016). Due to the increased root surface area available for adsorption and **absorption**, this effectively removes contaminants from the water (Kumar & Tripathi, 2014). An environment for microbial formation and growth is provided by the roots and rhizomes. Different physical and biological processes are carried out by the roots and connected biofilms to remove pollutants from the polluted water (Li & Li, 2009; Zhang et al., 2016). Three basic procedures, including adsorption, **sedimentation**, and biodegradation, are used in FTWs to remove **pollutants** (Pavlineri et al., 2017).

6.7.2 Microbial fuel cells (MFCs)

Wild bacterial consortia have been extensively studied in **microbial fuel cells** (MFCs) and their reactor systems for energy production and **wastewater treatment** (Hassan et al., 2018). As a potential technology for wastewater treatment and sustainable energy generation, MFCs can reduce chemical oxygen demand (COD). The primary focus of current research is on improving the quantity of electricity produced by MFCs, which simultaneously biodegrade organic pollutants found in wastewater while using **exoelectrogenic** bacteria to produce energy. MFCs are seen as a vital and sustainable technology for wastewater treatment due to the growing environmental implications of wastewater from industrial activity. The MFC technology has a number of benefits, including direct energy and value-added product production, dependable wastewater treatment performance through an efficient combination of biological and **electrochemical processes**, and operating stability that can be attained through real-time monitoring and control (Li et al., 2014).

MFCs may generate up to 1.43 kWh/m^3 of primary sludge or 1.8 kWh/m^3 of secondary sludge treatment of wastewater (Ge et al., 2013). MFCs typically only use 0.024 kW or 0.076 kWh/kg-COD, It uses less energy than activated sludge bioprocesses by around ten times (0.3 kW or 0.6 kWh/kg-COD) (Zhang et al., 2013). This indicates that MFCs have very low energy consumption rates compared to traditional wastewater treatment methods, while simultaneously recapturing energy through **bioelectrochemical** techniques based on exo-electrogens. In addition, a far lesser proportion of activated sludge produces less sludge (0.07–0.16 gVSS/gCOD cell yield) than an MFC (0.35–0.45 gVSS/gCOD) procedure (Gude, 2016). This is significant since managing sludge is a treatment of **industrial wastewater** is a significant issue.

The most recent study findings show encouraging systematic outcomes in the production of energy and the treatment of wastewater utilizing MFCs. For instance, it was discovered that noble materials made ideal MFC electrodes because they improved COD reduction (by about 53%) and power generation (about 4.5 mW/m^2) (Papaharalabos et al., 2015). **Magnesium oxide catalytic** electrode with **graphene oxide** support had improved power density of 756 mW/m^2 and COD reduction of 80% (Li et al., 2017). A very high cell density retention and maximum power generation (372 mW/m^2) have been combined in an up-flow MFC system. A lab-scale MFC system using a flame-oxidized hematite anode catalyzed by anodic dominating *Geobacter* sp. was found to produce more power (1,063 mW/m^2). However, the actual usefulness of these MFC designs for large-scale processes is somewhat disputed because to the high-profile architecture and very expensive materials.

6.7.3 Enzyme immobilization

Microbial enzymes that have been immobilized are used in a wide range of scientific and environmental processes. To maintain high stability in terms of pH, temperature, packing, reuse, and separation, enzymes are immobilized into solid and stable supports. In a recent work, the degradation of synthetic reactive dyes by the cross-linking agent glutaraldehyde enabled the covalent **immobilization** of horseradish peroxidase (HRP) onto calcium-alginate beads. Reactive red, reactive orange, and reactive blue showed strong decolorization efficiencies with immobilized HRP, ranging from 70% to 80%. Additionally, it maintained 40% of its initial enzyme activity after seven cycles (Bilal et al., 2016).

To treat industrial wastewater containing dyes, laccase enzyme isolated from the fungus *Pleurotus nebrodensis* was immobilized on chitosan beads. To boost the yield of enzyme manufacturing, the researchers adopted a **genetic engineering** technique. Either genetic experimentation to choose natural forms or in vitro gene mutagenesis and expression in a host cell can be used to accomplish this. Most often, **laccase** enzymes are used for genetic engineering. In order to effectively degrade plastics in a batch system, the monohydroxy ethyl terephthalate hydrolase enzyme from *Ideonella sakaiensis* 201-F6 was genetically engineered and expressed in *E. coli BL21* (Janatunaim & Fibriani, 2020).

6.8 Conclusion

Microbial enzymes offer a promising method for removing pollutants from the environment because anthropogenic activities have greatly increased the discharge of toxins. Instead of using harsh chemicals, microbial enzymes have been employed to biodegrade a wide range of pollutants, including HMs, poisons, dyes, PAHs, and plastics. By using high levels of contaminants as substrates, these enzymes effectively break them down. However, several factors, including pH, **temperature**, microbes, **incubation period**, and the availability of pollutants, must be taken into consideration for successful degradation.

As a result of improvements in processing and modification techniques, enzymes with enhanced physiological properties can now be produced, enabling processes to be carried out at extremely high or low temperatures. Additionally, it is necessary to build enzyme systems that produce high yields and efficiencies while minimizing the use of nutrients, water, and energy. The economic challenges associated with genetic engineering techniques still need to be addressed, despite the fact that genetic engineering procedures improve the yield of the enzyme with higher efficiency. Additionally, use of FTW and MFC are the **cutting-edge technology** being used for the treatment of toxic pollutants in wastewater and solid waste.

References

Abdelhaleem, H. A., Zein, H. S., Azeiz, A., Sharaf, A. N., & Abdelhadi, A. A. (2019). Identification and characterization of novel bacterial polyaromatic hydrocarbon-degrading enzymes as potential tools for cleaning up hydrocarbon pollutants from different environmental sources. *Environmental Toxicology and Pharmacology*, *67*, 108–116.

Antecka, K., Zdarta, J., Siwińska-Stefańska, K., Sztuk, G., Jankowska, E., Oleskowicz-Popiel, P., & Jesionowski, T. (2018). Synergistic degradation of dye wastewaters using binary or ternary oxide systems with immobilized laccase. *Catalysts*, *8*(9), 402.

Ba, S., Arsenault, A., Hassani, T., Jones, J. P., & Cabana, H. (2013). Laccase immobilization and insolubilization: From fundamentals to applications for the elimination of emerging contaminants in wastewater treatment. *Critical Reviews in Biotechnology*, *33*(4), 404–418.

Begum, S., Rath, S. K., & Rath, C. C. (2022). Applications of microbial communities for the remediation of industrial and mining toxic metal waste: A review. *Geomicrobiology Journal*, *39*(3–5), 282–293.

Bilal, M., Adeel, M., Rasheed, T., Zhao, Y., & Iqbal, H. M. (2019a). Emerging contaminants of high concern and their enzyme-assisted biodegradation–a review. *Environment International*, *124*, 336–353.

Bilal, M., Iqbal, H. M., & Barceló, D. (2019b). Persistence of pesticides-based contaminants in the environment and their effective degradation using laccase-assisted biocatalytic systems. *Science of the Total Environment*, *695*, 133896.

Bilal, M., Iqbal, H. M., Shah, S. Z. H., Hu, H., Wang, W., & Zhang, X. (2016). Horseradish peroxidase-assisted approach to decolorize and detoxify dye pollutants in a packed bed bioreactor. *Journal of Environmental Management*, *183*, 836–842.

Bose, S., Kumar, P. S., Vo, D. V. N., Rajamohan, N., & Saravanan, R. (2021). Microbial degradation of recalcitrant pesticides: A review. *Environmental Chemistry Letters*, *19*(4), 3209–3228.

Briffa, J., Sinagra, E., & Blundell, R. (2020). Heavy metal pollution in the environment and their toxicological effects on humans. *Heliyon*, *6*(9), e04691.

Cao, D., Chen, W., Xiang, Y., Mi, Q., Liu, H., Feng, P., . . . & Wang, D. (2021). The efficiencies of inorganic mercury bio-methylation by aerobic bacteria under different oxygen concentrations. *Ecotoxicology and Environmental Safety*, *207*, 111538.

Chen, Z., Cuervo, D. P., Müller, J. A., Wiessner, A., Köser, H., Vymazal, J., & Kuschk, P. (2016). Hydroponic root mats for wastewater treatment – A review. *Environmental Science and Pollution Research, 23*(16), 15911–15928.

de Souza, R. M., Seibert, D., Quesada, H. B., De Jesus Bassetti, F., Fagundes-Klen, M. R., & Bergamasco, R. (2020). Occurrence, impacts and general aspects of pesticides in surface water: A review. *Process Safety and Environmental Protection, 135*, 22–37.

Duan, C., Ma, T., Wang, J., & Zhou, Y. (2020). Removal of heavy metals from aqueous solution using carbon-based adsorbents: A review. *Journal of Water Process Engineering, 37*, 101339.

Ge, Z., Zhang, F., Grimaud, J., Hurst, J., & He, Z. (2013). Long-term investigation of microbial fuel cells treating primary sludge or digested sludge. *Bioresource Technology, 136*, 509–514.

Gude, V. G. (2016). Wastewater treatment in microbial fuel cells–an overview. *Journal of Cleaner Production, 122*, 287–307.

Guo, J., Zhai, P., Wu, L., Cribb, M., Li, Z., Ma, Z., . . . & Zhang, J. (2014). Diurnal variation and the influential factors of precipitation from surface and satellite measurements in Tibet. *International Journal of Climatology, 34*(9), 2940–2956.

Hassaan, M. A., & El Nemr, A. (2020). Pesticides pollution: Classifications, human health impact, extraction and treatment techniques. *The Egyptian Journal of Aquatic Research, 46*(3), 207–220.

Hassan, H., Jin, B., Donner, E., Vasileiadis, S., Saint, C., & Dai, S. (2018). Microbial community and bioelectrochemical activities in MFC for degrading phenol and producing electricity: Microbial consortia could make differences. *Chemical Engineering Journal, 332*, 647–657.

Headley, T. R., & Tanner, C. C. (2008, November). Floating Treatment Wetlands: An Innovative Option for Stormwater Quality Applications. In *11th International Conference on Wetland Systems for Water Pollution Control* (pp. 1101–1106).

Huy, N. D., Le, N. T. M., Chew, K. W., Park, S. M., & Show, P. L. (2021). Characterization of a recombinant laccase from Fusarium oxysporum HUIB02 for biochemical application on dyes removal. *Biochemical Engineering Journal, 168*, 107958.

Janatunaim, R. Z., & Fibriani, A. (2020). Construction and cloning of plastic-degrading recombinant enzymes (MHETase). *Recent Patents on Biotechnology, 14*(3), 229–234.

Jones, T. G., Willis, N., Gough, R., & Freeman, C. (2017). An experimental use of floating treatment wetlands (FTWs) to reduce phytoplankton growth in freshwaters. *Ecological Engineering, 99*, 316–323.

Kadri, T., Rouissi, T., Brar, S. K., Cledon, M., Sarma, S., & Verma, M. (2017). Biodegradation of polycyclic aromatic hydrocarbons (PAHs) by fungal enzymes: A review. *Journal of Environmental Sciences, 51*, 52–74.

Karthik, V., Senthil Kumar, P., Vo, D. V. N., Selvakumar, P., Gokulakrishnan, M., Keerthana, P., & Jeyanthi, J. (2021). Enzyme-loaded nanoparticles for the degradation of wastewater contaminants: A review. *Environmental Chemistry Letters, 19*(3), 2331–2350.

Katheresan, V., Kansedo, J., & Lau, S. Y. (2018). Efficiency of various recent wastewater dye removal methods: A review. *Journal of Environmental Chemical Engineering, 6*(4), 4676–4697.

Kensa, V. M. (2011). Bioremediation-an overview. *I Control Pollution, 27*(2), 161–168.

Kumar Awasthi, M., Ravindran, B., Sarsaiya, S., Chen, H., Wainaina, S., Singh, E., . . . & Zhang, Z. (2020). Metagenomics for taxonomy profiling: Tools and approaches. *Bioengineered, 11*(1), 356–374.

Kumari, M., & Tripathi, B. D. (2014). Effect of aeration and mixed culture of Eichhornia crassipes and Salvinia natans on removal of wastewater pollutants. *Ecological Engineering, 62*, 48–53.

Ladislas, S., Gerente, C., Chazarenc, F., Brisson, J., & Andres, Y. (2015). Floating treatment wetlands for heavy metal removal in highway stormwater ponds. *Ecological Engineering, 80*, 85–91.

Li, M., Zhou, S., & Xu, M. (2017). Graphene oxide supported magnesium oxide as an efficient cathode catalyst for power generation and wastewater treatment in single chamber microbial fuel cells. *Chemical Engineering Journal, 328*, 106–116.

Li, W., & Li, Z. (2009). In situ nutrient removal from aquaculture wastewater by aquatic vegetable Ipomoea aquatica on floating beds. *Water Science and Technology*, *59*(10), 1937–1943.

Li, W., Shi, Y., Gao, L., Liu, J., & Cai, Y. (2015). Occurrence, fate and risk assessment of parabens and their chlorinated derivatives in an advanced wastewater treatment plant. *Journal of Hazardous Materials*, *300*, 29–38.

Li, W. W., Yu, H. Q., & He, Z. (2014). Towards sustainable wastewater treatment by using microbial fuel cells-centered technologies. *Energy & Environmental Science*, *7*(3), 911–924.

Lundberg, D. S., Lebeis, S. L., Paredes, S. H., Yourstone, S., Gehring, J., Malfatti, S., . . . & Dangl, J. L. (2012). Defining the core Arabidopsis thaliana root microbiome. *Nature*, *488*(7409), 86–90.

Ojuederie, O. B., & Babalola, O. O. (2017). Microbial and plant-assisted bioremediation of heavy metal polluted environments: A review. *International Journal of Environmental Research and Public Health*, *14*(12), 1504.

Papaharalabos, G., Greenman, J., Melhuish, C., & Ieropoulos, I. (2015). A novel small scale Microbial Fuel Cell design for increased electricity generation and waste water treatment. *International Journal of Hydrogen Energy*, *40*(11), 4263–4268.

Pavlineri, N., Skoulikidis, N. T., & Tsihrintzis, V. A. (2017). Constructed floating wetlands: A review of research, design, operation and management aspects, and data meta-analysis. *Chemical Engineering Journal*, *308*, 1120–1132.

Phyo, A. K., Jia, Y., Tan, Q., Sun, H., Liu, Y., Dong, B., & Ruan, R. (2020). Competitive growth of sulfate-reducing bacteria with bioleaching acidophiles for bioremediation of heap bioleaching residue. *International Journal of Environmental Research and Public Health*, *17*(8), 2715.

Rahman, M. A., & Hassler, C. (2014). Is arsenic biotransformation a detoxification mechanism for microorganisms? *Aquatic Toxicology*, *146*, 212–219.

Reis, A. C., Čvančarová, M., Liu, Y., Lenz, M., Hettich, T., Kolvenbach, B. A., . . . & Nunes, O. C. (2018). Biodegradation of sulfamethoxazole by a bacterial consortium of Achromobacter denitrificans PR1 and Leucobacter sp. GP. *Applied Microbiology and Biotechnology*, *102*(23), 10299–10314.

Saravanan, A., Kumar, P. S., Vo, D. V. N., Jeevanantham, S., Karishma, S., & Yaashikaa, P. R. (2021). A review on catalytic-enzyme degradation of toxic environmental pollutants: Microbial enzymes. *Journal of Hazardous Materials*, *419*, 126451.

Shahid, M. J., Arslan, M., Ali, S., Siddique, M., & Afzal, M. (2018). Floating wetlands: A sustainable tool for wastewater treatment. *Clean–Soil, Air, Water*, *46*(10), 1800120.

Sharma, A., Shukla, A., Attri, K., Kumar, M., Kumar, P., Suttee, A., . . . & Singla, N. (2020). Global trends in pesticides: A looming threat and viable alternatives. *Ecotoxicology and Environmental Safety*, *201*, 110812.

Sharma, B., Dangi, A. K., & Shukla, P. (2018). Contemporary enzyme based technologies for bioremediation: A review. *Journal of Environmental Management*, *210*, 10–22.

Sharma, P., Tripathi, S., & Chandra, R. (2021). Metagenomic analysis for profiling of microbial communities and tolerance in metal-polluted pulp and paper industry wastewater. *Bioresource Technology*, *324*, 124681.

Singh, J. S., Abhilash, P. C., Singh, H. B., Singh, R. P., & Singh, D. P. (2011). Genetically engineered bacteria: An emerging tool for environmental remediation and future research perspectives. *Gene*, *480*(1–2), 1–9.

Singh, P., Singh, V. K., Singh, R., Borthakur, A., Madhav, S., Ahamad, A., . . . & Mishra, P. K. (2020). Bioremediation: A Sustainable Approach for Management of Environmental Contaminants. In *Abatement of Environmental Pollutants* (pp. 1–23). Elsevier.

Tan, K. B., Vakili, M., Horri, B. A., Poh, P. E., Abdullah, A. Z., & Salamatinia, B. (2015). Adsorption of dyes by nanomaterials: Recent developments and adsorption mechanisms. *Separation and Purification Technology*, *150*, 229–242.

Tanner, C. C., & Headley, T. R. (2011). Components of floating emergent macrophyte treatment wetlands influencing removal of stormwater pollutants. *Ecological Engineering*, *37*(3), 474–486.

Tiwari, N., Santhiya, D., & Sharma, J. G. (2020). Microbial remediation of micro-nano plastics: Current knowledge and future trends. *Environmental Pollution*, *265*, 115044.

Vardhan, K. H., Kumar, P. S., & Panda, R. C. (2019). A review on heavy metal pollution, toxicity and remedial measures: Current trends and future perspectives. *Journal of Molecular Liquids*, *290*, 111197.

Wang, X., Li, D., Gao, P., Gu, W., He, X., Yang, W., & Tang, W. (2020). Analysis of biosorption and biotransformation mechanism of Pseudomonas chengduensis strain MBR under Cd (II) stress from genomic perspective. *Ecotoxicology and Environmental Safety*, *198*, 110655.

Yu, Z., Han, H., Feng, P., Zhao, S., Zhou, T., Kakade, A., . . . & Li, X. (2020). Recent advances in the recovery of metals from waste through biological processes. *Bioresource Technology*, *297*, 122416.

Zhang, F., Ge, Z., Grimaud, J., Hurst, J., & He, Z. (2013). Long-term performance of liter-scale microbial fuel cells treating primary effluent installed in a municipal wastewater treatment facility. *Environmental Science & Technology*, *47*(9), 4941–4948.

Zhang, L., Zhao, J., Cui, N., Dai, Y., Kong, L., Wu, J., & Cheng, S. (2016). Enhancing the water purification efficiency of a floating treatment wetland using a biofilm carrier. *Environmental Science and Pollution Research*, *23*(8), 7437–7443.

Zhang, X., Zhong, T., Liu, L., & Ouyang, X. (2015). Impact of soil heavy metal pollution on food safety in China. *PLoS One*, *10*(8), e0135182.

Zhao, J. L., Zhang, Q. Q., Chen, F., Wang, L., Ying, G. G., Liu, Y. S., . . . & Zhang, R. Q. (2013). Evaluation of triclosan and triclocarban at river basin scale using monitoring and modeling tools: Implications for controlling of urban domestic sewage discharge. *Water Research*, *47*(1), 395–405.

Zhuo, R., He, F., Zhang, X., & Yang, Y. (2015). Characterization of a yeast recombinant laccase rLAC-EN3-1 and its application in decolorizing synthetic dye with the coexistence of metal ions and organic solvents. *Biochemical Engineering Journal*, *93*, 63–72.

Sougata Ghosh*, Bishwarup Sarkar and Sirikanjana Thongmee

Chapter 7
Transgenic bacteria for bioremediation of mercury

Abstract: Bacteria-mediated bioremediation is a promising solution for addressing the hazardous effects of toxic metals on animals and plants. Among various heavy metals, mercury – in both its ionic and metallic forms – can accumulate in sediments that are slowly converted to highly toxic methyl mercury by bacteria. This poses a severe threat to human health due to the biomagnification of mercury through trophic levels via the consumption of seafood. Metagenomic-associated understanding of the metabolic activities of the bacteria has led to the development of genetically engineered bacteria with enhanced capability for mercury resistance and accumulation. Particularly, *mer* genes are integrated and expressed in bacteria for effective mercury bioremediation. Bacteria can synthesize and secrete mercuric ion reductase for conversion of the ionic mercury (Hg^{2+}) to elemental mercury (Hg^0), which is then volatilized from the cell into the environment resulting in the formation of highly toxic organomercurials. Hence, this chapter gives an elaborate account of the transgenic bacteria developed by recombinant DNA technology. Several bacterial strains such as *Escherichia coli, Deinococcus radiodurans, Bacillus cereus, Cupriavidus metallidurans, Rhodopseudomonas palustris,* and *Pseudomonas putida* are genetically engineered for the expression of *mer* genes. The rational expression of *mt* genes coded metallothioneins which are cysteine-rich low-molecular-weight metal-binding proteins can also help in effective sequestration of the Hg^{2+} ions in a biologically inactive form while other genes like the *ppk* that encodes an enzyme polyphosphate kinase are engineered and upregulated in bacteria for promoting biosynthesis of polyphosphates. Several molecular tools are employed to address the challenges of metallothionein instability in bacteria due to rapid degradation of the transcripts and small peptide, low protein expression, and interference with redox pathways. Thus, transgenic bacteria mediated mercury bioremediation seems to be a promising alternative to ensure clean environment.

Acknowledgment: Sougata Ghosh acknowledges Kasetsart University, Bangkok, Thailand, for Post Doctoral Fellowship and funding under Reinventing University Program (Ref. No. 6501.0207/10870 dated November 9, 2021).

***Corresponding author: Sougata Ghosh,** Department of Physics, Faculty of Science, Kasetsart University, Bangkok, Thailand; Department of Microbiology, School of Science, RK University, Rajkot, Gujarat, India, e-mail: ghoshsibb@gmail.com
Bishwarup Sarkar, College of Science, Northeastern University, Boston, MA, USA
Sirikanjana Thongmee, Department of Physics, Faculty of Science, Kasetsart University, Bangkok, Thailand

https://doi.org/10.1515/9783110793635-007

7.1 Introduction

The toxic, mobile, and long persistence in the atmosphere makes mercury (Hg) a worldwide concern to human health and the environment. Hg can be found in its elemental, inorganic, and organic forms in terrestrial as well as aquatic systems (Mahbub et al., 2017). Methylation of Hg results in the formation of mono- or di-methylmercury which can act as a neurotoxin. Moreover, organic Hg is extremely harmful because of its bioaccumulative nature. Thus, various industries and regulatory authorities focus on alleviating the Hg toxicity from contaminated water and soil using various Hg remediation strategies. Some of the physico-chemical conventional strategies for Hg remediation include adsorption, membrane filtration, precipitation, acid leaching, soil washing, and vitrification. However, unlike organic pollutants, Hg cannot be degraded or mineralized but can only be transformed into its less toxic form, these conventional remediation methods produce large volumes of Hg biomass (Kumari et al., 2020). Therefore, efficient strategies for effective and eco-friendly Hg remediation are required for the amelioration of Hg toxicity in the environment. Bioremediation is one such emerging strategy that applies low-cost biological processes for heavy metal removal. Several studies have shown promising results in removing Hg from contaminated environmental samples using bacteria (Dash & Das, 2012). The *mer* genes present in Hg-resistant bacteria are known to code for mercuric ion reductase and organomercurial lyase along with other enzymes which can effectively volatilize and facilitate bioaccumulation of toxic forms of Hg (Nascimento et al., 2003). This chapter gives an elaborate explanation of the application of genetically modified bacteria in the bioremediation of Hg. Several *mer* genes such as *mer*P, *mer*T, *mer*A, *mer*B, and *mer*R are expressed in different bacterial species using recombinant DNA technology of enhanced microbial Hg uptake and biosorption. Different bacterial strains such as *Escherichia coli* TOP10, *Bacillus cereus* BW-03, *Cupriavidus metallidurans* MSR33, *Rhodopseudomonas palustris* GIM1.167, *Pseudomonas putida* PpY101/pSR134, and *Deinococcus radiodurans* are genetically modified to express such *mer* genes, which in turn increased its Hg bioremediation potential. Thus, further exploration of the control of the detrimental environmental impact of applying such transgenic bacteria can provide useful insights into the feasibility and future strategies for the application of genetically modified bacteria in Hg removal processes.

7.2 Mercury toxicity

Almost all forms of Hg can interfere with normal cellular function through alteration in the protein structure along with excessive binding with sulfydryl and selenohydryl groups (Bernhoft, 2012). Exposure to food and water contaminated with methylmercury is the most common route of exposure to the organic form of mercury in humans whereas inorganic forms of mercury are mostly present in industrial emissions that are inhaled (Guzzi & La Porta, 2008). Acute exposures to Hg vapors can result in erosive bronchitis and bronchiolitis along with complications in the central nervous system (CNS) such as tremor and erethism. Moreover, these Hg vapors can easily pass through the blood-brain barrier and the placenta where chronic exposure can result in severe CNS defects (Andreoli & Sprovieri, 2017). Meanwhile, Hg salts such as $HgCl_2$ can create a negative impact on the gastrointestinal tract and kidneys. In addition, organic forms of Hg especially methylmercury are well-known neurotoxins (Abbott & Nigussie, 2021). A relatively long half-life of methylmercury in the human body facilitates easier retention of the mercurial species in vital organs such as the brain, liver, peripheral nerves, bone marrow, placenta, kidneys, and fetus (Clarkson et al., 2003). Further effects of Hg toxicity on the various human organ systems are summarized in Table 7.1.

Table 7.1: Hg associated toxicity affecting various organs.

Human organ system	Effects of Hg toxicity
Nervous system	Loss of memory, hypoesthesia, ataxia, impairment in sensory organs, fatigue, and dysarthria
Renal system	Increased levels of creatinine
Motor system	Reduced muscular strength, hindrance of fine motor function, and fatigue
Cardiovascular system	Alteration in homeostasis, hypertension, coronary heart disease, and atherosclerosis
Immune system	Reduced overall immunity, atopic eczema, induction of autoimmune disease, and hemolytic anemia
Reproductive system	Reduced fertility and abnormal offsprings
Embrional system	Cerebral hypoplasia, reduced brain weight, and abnormal neuron migration

7.3 Genetically engineered bacteria for Hg removal

Various genetically modified organisms are extensively under research and development for improvement in the current mercury bioremediation processes that are summarized in Table 7.2. Dash and Das (2015) in one such study reported the volatilization and biosorption of inorganic Hg using genetically engineered *Bacillus cereus* BW-03 strain. A plasmid construct harboring the *mer* operon of *Bacillus thuringiensis* PW-05 strain was transformed in the BW-03 strain which was resistant to Hg. The transformants obtained were then screened on a seawater nutrient medium supplemented with 50 ppm of $HgCl_2$ and ampicillin. Colony polymerase chain reaction (PCR) confirmed the presence of *merA* gene in the transformants which also efficiently volatilized mercury. The transformant strain had a minimum inhibitory concentration (MIC) of 50 ppm of Hg which was attributed to the Hg resistance conferred by the *mer* operon wherein the bacterium survived up to 7 days in the presence of 50 ppm of Hg. Moreover, the transformant exhibited tolerance against other salts of heavy metals such as 458, 1,614, 2,500, and 465 ppm of $CdCl_2$, $ZnSO_4$, $Pb(NO_3)_2$, and $Na_2HAsO_4{\cdot}7H_2O$, respectively. The transformant also showed resistance against 30, 30, 10, 5, and 30 μg of amoxycillin, cephradine, ampicillin, methicillin, and vancomycin, respectively. Thereafter, the biosorption of mercury by the transgenic bacteria was confirmed by hydrogen sulfide assay wherein the development of black-colored complex indicated successful Hg biosorption which was not observed in the presence of other toxic metals. The expression of *merA* gene was regulated under the influence of salt stress in which high mRNA levels of *merA* gene were obtained under low salinity thus, highlighting the potential application of this genetically modified in low saline environments. In addition, microcosms of the transgenic bacteria along with normal microbiota were established in contaminated sediment samples which showed 96.4% removal of Hg after 72 h as compared to 6.4% Hg removal by the normal microbiota under similar conditions. The transformants showed effective upregulation of Hg removal from the marine sediments. The encapsulated *B. cereus* BW-03 transformants expressing *mer* genes of *B. thuringiensis* PW-05 strain removed almost 100% of the 50 ppm of Hg after 48–72 h of incubation. Thus, the ability of the transformed strain to utilize both Hg resistance mechanisms was proposed to contribute to 99.3% Hg removal efficiency as compared to 71% and 95.6% by the parental *B. cereus* BW-03 and *B. thuringiensis* PW-05 strains, respectively. The decreased levels of pathogenicity of the transformed strain were evident from biofilm formation, motility, and cytotoxicity assessments.

In another study, Rojas et al. (2011) metabolically engineered *Cupriavidus metallidurans* MSR33 strain for Hg bioremediation. The IncP-1β plasmid pTP6 carrying *merB* and *merA* genes was successfully introduced in the heavy metal-resistant *C. metallidurans* CH34 strain using biparental conjugation with the *E. coli* JM109 strain. The incorporation of plasmid in the transconjugant was confirmed by plasmid isolation which was then visualized by agarose gel electrophoresis. The presence of the *merB* gene was confirmed by PCR while the stability of the plasmid under nonselective conditions was studied wherein the transgenic bacteria maintained Hg resistance even after 70 generations and possessed the pTP6 plasmid. The inorganic and organic Hg resistance of the MSR33 strain was evaluated wherein the engineered strain demonstrated resistance of 0.12 mM and 0.08 mM of Hg^{2+} and CH_3Hg^{2+} while the parental strain showed 0.05 mM resistance to Hg^{2+} and sensitivity to methylmercury. The transconjugant was cultivated in a low-phosphate Tris-buffered mineral salts (LPTMS) medium supplemented with Hg^{2+} and succinate as the primary source of carbon. Transmission electron microscopy (TEM) images in Figure 7.1

further revealed no change in cell morphology after Hg exposure whereas the parent CH34 strain displayed a fuzzy outer membrane after Hg^{2+} supplementation. The bioremediation potential of the MSR33 strain was further evaluated in a bioreactor containing Hg-polluted water. The transgenic *C. metallidurans* strain completely removed 0.1 and 0.15 mM of Hg from the polluted water samples after 2 h in the presence of thioglycolate whereas 88% of 0.1 mM was removed in the absence of thioglycolate after 3 h.

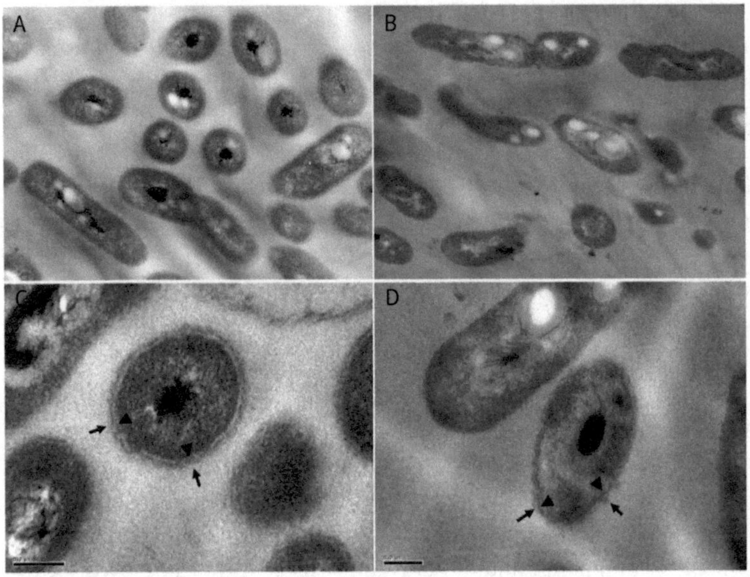

Figure 7.1: Ultrathin section of *Cupriavidus metallidurans* strain MSR33 exposed to Hg^{2+}. Cells grown in LPTMS medium with succinate until exponential phase were further incubated in presence of Hg^{2+} (0.04 mM) for 1 h. (A and C) MSR33 cells and (B and D) CH34 cells. Arrows and arrowheads indicate the outer and the inner membrane, respectively. The bars represent 0.2 mm. Reprinted from Rojas et al. (2011).

Deinococcus radiodurans was also genetically modified by Brim et al. (2000) for Hg bioremediation from radioactive mixed waste environments. The *merA* gene from an *E. coli* BL308 strain was successfully cloned into *D. radiodurans* which conferred Hg^{2+} resistance. A recombinant pMD727 plasmid was constructed by cloning the *mer* operon in the autonomously replicating plasmid where the operon was placed under the control of a constitutive P2 promoter. Moreover, the Southern blot analysis confirmed the presence of one copy of the *mer* operon per cell. The MerA protein activity was then spectrophotometrically evaluated using Hg^{2+}-stimulated oxidation of reduced nicotinamide adenine dinucleotide phosphate (NADPH) where the extracts of recombinant cells exhibited oxidation of Hg^{2+} in the presence of NADPH while no such oxidation was observed in the wild type *D. radiodurans* strain. It was assumed that the NADPH-mediated oxidation resulted in the formation of Hg(0) vapors which were confirmed using an X-ray film assay. The recombinant strains containing *mer* genes exhibited darkening of the X-ray film after 14 h of incubation in the presence of Hg^{2+} ions thus, confirming the formation of Hg(0) vapors which reacted with Ag^{1+} ions of the X-ray film.

In another similar study, Murtaza et al. (2002) showed biomolecular engineering of *E. coli* cells for the expression of *merB* genes. The *E. coli* strains used in this study were Hg resistant and isolated from various geographical locations of the Indian subcontinent. The *E. coli* strains were used for the isolation of plasmids which were then PCR amplified for confirming the presence of *merB* gene wherein all the five isolates demonstrated the presence of 0.64 kb bands that corresponding to the *merB* gene. The *E. coli* isolate G18 showed the highest tolerance towards phenyl mercuric acetate (PMA) whose *merB* gene was further cloned in pGEMT plasmid. The *E. coli* DH5α transformants having the recombinant pGEMT plasmid further displayed vigorous growth in the presence of 1–5 μg/mL whereas the growth of the control strains was inhibited even in the presence of 1 μg/mL of PMA. Hence, a fivefold increase in Hg resistance was attained after the pGEMT easy vector transformation which was attributed to the degradation of PMA into a less toxic Hg^{2+} form by the MerB protein which could be useful for bioremediation as well.

Deng et al. (2006) also reported the Hg bioaccumulation by a transgenic *E. coli* JM109 strain. The bacterial strain was genetically engineered to carry two compatible plasmids namely, pSTUP and pGPMT that had *merP*, *merT*, and glutathione-*S*-transferase -protein mannosyl-transferase (GST-PMT) fusion genes, respectively. The recombinant strains obtained were cultivated in a Luria-Bertani (LB) medium supplemented with 100 mg/L and 60 mg/L of ampicillin and kanamycin, respectively. The growth of the transgenic *E. coli* was affected in the presence of 2 mg/L of Hg^{2+} ions. However, the Hg removal efficiency remained more than 85% which further dropped down to 37% after supplementation of 4 mg/L of Hg^{2+} ions. On the contrary, the highest metal/biomass ratio of 25 mg of Hg^{2+}/g of the dry cell was observed at an Hg^{2+} concentration of 4 mg/L. Thereafter, simultaneous growth and bioaccumulation studies of the genetically engineered bacteria were carried out in the presence of 1 mg/L of Hg^{2+} ions. The presence of 0.3 mol/L of Mg^{2+}/Ca^{2+} ions further demonstrated slight inhibition of Hg^{2+} removal activity. Additionally, EDTA significantly affected the growth of transgenic *E. coli* JM109 cells. Moreover, the presence of other toxic metals showed little effect on the Hg removal capacity of the genetically modified bacteria with the retention of more than 90% Hg removal capacity.

Likewise, Bae et al. (2001) genetically engineered an *E. coli* JM109 strain for the heterologous co-expression of an Hg^{2+}-transport system along with $(Glu\text{-}Cys)_{20}Gly$ (EC20) phytochelatins and studied the Hg bioaccumulation capacity of the modified strain. In addition, MerP and MerT proteins were also co-expressed with maltose-binding protein-EC20 (MBP-EC20) for the intracellular transport of Hg^{2+} ions. Another strategy of directly anchoring EC20 on the cell surface was also examined using a lipoprotein outer membrane protein A (Lpp-OmpA) fusion system. Sodium dodecyl sulphate-polyacrylamide gel electrophoresis (SDS-PAGE) demonstrated the presence of 21 kDa and 47.5 kDa protein bands in the transgenic bacterial strain which corresponded with the Lpp-OmpA-EC20 and MBP-EC20, respectively. Thereafter, the binding stoichiometry of Hg^{2+} to EC20 was investigated wherein 5 nmol of the purified fusion protein were extracted and resuspended in 50 mM Tris-Cl buffer supplemented with 5 mM of dithiothreitol followed by incubation in the presence of 1–1,200 nmol of Hg^{2+} for 2 h. The saturating ratio of 20 Hg^{2+} per MBP-EC20 was achieved which was higher than the ratios obtained for MTs. Moreover, the bioaccumulation of Hg^{2+} was also evaluated which showed a sixfold increase in the Hg^{2+} uptake in the cells with heterologous expression of MBP-EC20 as compared to the control strain. Additionally, the Hg^{2+} uptake of cells with EC20 anchored on the cell surface was increased threefold as compared to the overexpression of EC20 in the cytoplasm. After 1 h of incubation, 100% of 5 µM of Hg^{2+} was successfully accumulated in the cells highlighting that the concentration of Hg^{2+} is the rate-limiting step in the bioaccumulation process. Analysis of the rate of bioaccumulation revealed an improved rate with co-expression of Hg^{2+} transporters and MBP-EC20 wherein more than 95% of Hg^{2+} was accumulated within 20 min of incubation which was further enhanced with surface EC20 expression. The co-expression of the Hg^{2+} transport system and MBP-EC20 also facilitated the specificity of the heavy metal bioaccumulation process.

Deng and Wilson (2001) also genetically engineered *E. coli* JM109 cells for the bioaccumulation of Hg from wastewater. The pSTUP and pGPMT compatible plasmids were incorporated in the *E. coli* JM109 cells which contained the *merP, merT,* and GST-PMT fusion genes, respectively. The maximum rate of Hg bioaccumulation was attained within 5 min of the operation using a cell concentration of 0.35 g/L. The accumulation of Hg was stimulated by EDTA and NaCl. The Hg bioaccumulation capacity of the recombinant strain was 30% lower in wastewater samples as compared to laboratory-scale Hg solutions which were attributed to the interference of other heavy metals present in the wastewater. Later, two continuous bioaccumulation experiments were carried out with a flow rate of 2 mL/min and 10 mL/min along with an initial induced cell concentration of 1.42 g and 1.56 g cell dry weight, respectively. Most of the Hg was bioaccumulated in both conditions with a final Hg bioaccumulation of 26.8 mg/g of dry cell weight.

Hg ion transporter from *Acidithiobacillus ferrooxidans* was also subjected to heterologous expression in *E. coli* cells (Sasaki et al., 2005). The Hg transporter *merC* was fused with *phoA* gene which encoded an alkaline phosphatase enzyme. *E. coli* LMG194 cells were incorporated with the pBAD plasmid containing *phoA* and *merC* fusion derivatives. The obtained transformants were hypersensitive to $CdCl_2$ which was alleviated after mutation of His72 of the heterologous MerC protein. The removal of a coding sequence of the carboxy-terminal cytoplasmic region of MerC protein resulted in Hg^{2+} hypersensitivity in the transformants along with 55% $HgCl_2$ bioaccumulation capacity as compared to transformants expressing the intact MerC protein. Therefore, a truncated MerC protein showed bioremediation potential in genetically engineered *E. coli* cells.

Xue et al. (2022) recently reported the generation of a transgenic bacterium with a self-controlled biosecurity circuit. *E. coli* cells were genetically engineered to express a heterologous protein that can efficiently adsorb Hg^{2+} ions. Moreover, a genetic circuit was constructed to alleviate the metabolic burden on the engineered cells because of the heterologous protein expression. Hence, a *MerR-*P_{merT} genetic circuit was developed which could sense the concentration of Hg^{2+} ions in the environment and trigger the expression of the protein only above a threshold concentration of Hg^{2+} ions. In addition, a negative regulatory protein TetR was also expressed by the P_{merT} promoter in the presence of Hg^{2+} ions which was a part of the P_{LtetO} promoter and facilitated the average growth of the bacteria. Further optimization modules were considered for maximum Hg recovery from water bodies. The engineered cells were magnetically immobilized using strontium ferrite ($SrFe_{12}O_{19}$) nanoparticles along with the magnetic surface display of artificial *E. coli* containing the complete genetic circuit along with the *INP-*CL_5 gene which expressed surface polypeptides using the N-terminal of ice-nucleation protein from *Pseudomonas syringae*. Almost 98% of 50 μM of initial Hg^{2+} ions were efficiently bound on the surface of the cells with an initial cell density of 10^{10} colony forming units (CFU) which reached adsorption equilibrium within 2 h.

In another study, Kiyono et al. (2003) reported the genetic modification of *E. coli* XL1-Blue cells through heterologous expression of pMKB18 plasmid which carried *mer-ppk* genes. The plasmid was created through the fusion of the *merR-p/p-T-P-B1* and *ppk* genes that were obtained from *Pseudomonas* K-62 and *Klebsiella aerogenes*, respectively. The transformants containing the fusion plasmid were then grown at 37 °C in LB medium supplemented with 100 μg/mL of ampicillin. The cells were then immobilized in calcium alginate beads and cultivated in the LB medium supplemented with 10 μM of Hg^{2+} and $C_6H_5Hg^+$ followed by incubation at room temperature. The recombinant cells exhibited almost 90% removal of both mercurial species from the LB medium which was significantly higher as compared to 10% removal by the immobilized parent strain. The Hg-removal activity was reduced with a concomitant increase in the concentration of mercurial species in the medium. The reusability of the transgenic bacteria for the removal of Hg was also examined wherein the Hg-removal capacity of the immobilized transgenic cells was analyzed using a wastewater sample. The immobilized recombinant cells showed negligible loss of Hg bioremediation activity after three successive cycles of Hg removal. The rate of Hg^{2+} and $C_6H_5Hg^+$ accumulations were 891 and 776 nmoles Hg/mg of cells, respectively. Furthermore, the presence of other heavy metals such as Cd^{2+}, Pb^{2+}, or Cr^{6+} showed no effect on the Hg removal capacity of the transgenic bacteria carrying the pMKB18 fusion plasmid.

E. coli cells were also genetically engineered for the production of polyphosphate (polyP) which conferred Hg bioaccumulation capability to the transgenic bacteria (Pan-Hou et al., 2002). *E. coli* XL1-Blue cells were incorporated with pMKB18 recombinant plasmid which carried the broad-spectrum Hg resistance determinants namely, *merR-o/p-T-P-A-G-B1* of *Pseudomonas* K-62 strain along with a *ppk* gene from *Klebsiella aerogenes*. The expression of *ppk* was directly investigated through observation of the biosynthesis of polyP wherein large amounts of polyP were obtained after Hg^{2+} and $C_6H_5Hg^+$ induction. Moreover, Northern blot analysis also showed the transcription of *merT, merP, merB1,* and *ppk* genes. Later, the Hg bioaccumulation capacity of the recombinant strain was also evaluated which showed a tenfold and a twofold increase in the Hg^{2+} and $C_6H_5Hg^+$ accumulation, respectively. Furthermore, the ratio of Hg (nmoles) accumulated to the polyP (μmol) produced in the cells was 16 and 24 for Hg^{2+} and $C_6H_5Hg^+$, respectively. Thus, these results highlighted the Hg^{2+} chelation activity of polyP.

Kiyono et al. (2009) also investigated the function of *merE* gene through the construction of transgenic *E. coli.* The *merE* gene of transposon Tn*21* from *Shigella flexneri* plasmid NR1 (R100) and *merR* gene of plasmid pMR26 from *Pseudomonas* strain K-62 were inserted into a pE4 plasmid which was then incorporated into *E. coli* XL1-Blue cells. The transformants showed the expression of MerE protein which was confirmed by Western blot analysis that showed the presence of a 15 kDa protein band in the membrane fraction of the transgenic bacteria which also specifically reacted with the anti-MerC antibody. Thereafter, the physiological importance of MerE was studied wherein transgenic bacteria containing pE4 plasmid having *merR*-o/p-*merE* genes exhibited CH_3Hg^+ and Hg^{2+} hypersensitivity as compared to the isogenic strains. The Hg bioaccumulation capacity of the recombinant cells was twofold higher as compared to the control which could be useful for bioremediation.

A genetically modified *Pseudomonas putida* PpY101/pSR134 strain was also applied by Okino et al. (2001) for the effective removal of Hg. A broad-host-range vector pSUP104 containing the Hg-resistant gene from *NR1* plasmid was constructed in this study which was then incorporated into the bacterial strain. The recombinant cells were then immobilized through entrapment in calcium alginate, strontium alginate, agar, as well as in a photo-cross-linkable resin prepolymer. Thereafter, the mercuric chloride removal capacity of the free and immobilized cells was compared which showed the same Hg removal rate in both types of cells in the presence of 10 mg/L of mercuric chloride. It is important to note that up to 100 mg/L of Hg was successfully removed by both free and immobilized cells. Calcium alginate immobilization exhibited the highest rate of Hg removal. The thermal stability for maintaining efficient Hg removal activity was then evaluated by assaying the Hg removal activity of both the free and immobilized cells after exposure to 40, 50, and 60 °C for 20 min. The immobilized cells retained higher activity of Hg removal as compared to the free cells at higher temperatures. The reusability of the immobilized cells was also significantly retained after four consecutive Hg removal cycles as compared to the free cells.

Deng and Jia (2011) also constructed a genetically engineered photosynthetic strain of *Rhodopseudomonas palustris* for enhanced Hg^{2+} uptake. *R. palustris* GIM1.167 strain was genetically modified by transforming two compatible plasmids namely, pSUTP and pGPMT for the expression of Hg^{2+} transport system and pea metallothionein (MT), respectively. The transformants were confirmed by PCR amplification of *merP-merT* and *mt* genes wherein 750 bp and 250 bp of the gene fragments were obtained in the recombinant pSUTP + pGMPT strain, respectively. The heavy metal sensitivity of the recombinant strain was then compared with the wild-type *R. palustris* strain in which the growth of the parent strain was completely inhibited in the presence of 5 mg/L of Hg^{2+} whereas the transgenic strain was capable of growth even in the presence of 50 mg/L of Hg^{2+} ions. This enhanced Hg resistance was attributed to the intracellular expression of MT which alleviated the toxic effects of the heavy metal. Thereafter, the bioaccumulation capacity of the recombinant strain was also evaluated which showed a threefold increase in the Hg uptake capacity when compared with the parent strain. Hence, the heterologous expression of the Hg transport system and MT facilitated increased binding of the metal ions. The effect of change in pH on the Hg bioaccumulation capacity was negligible in recombinant strain whereas the bioaccumulation of Hg significantly fluctuated in the wild-type strain. Likewise, the presence of cations such as Mg^{2+} and Ca^{2+} showed no significant change in the bioaccumulation capacity of the strain while the Hg^{2+} uptake capacity was reduced by 20% in the presence of 10 mg/L of Na^{+} ions. Furthermore, more than 85% of Hg^{2+} bioaccumulation capacity was retained in the presence of Zn^{2+} and Ni^{2+} ions. The intracellular bioaccumulation of Hg^{2+} by the recombinant strain was further confirmed in the presence of 0.1 M of ethylenediaminetetraacetic acid (EDTA) wherein only 13% of Hg^{2+} on the surface of recombinant *R. palustris* strain was removed by EDTA as compared to 84% of the parental strain. Therefore, 87% of Hg^{2+} was successfully transported into the recombinant *R. palustris* cells in this study, which expressed the *merP-merT* and *mt* genes.

Table 7.2: Genetically modified bacteria with Hg removal efficiency.

Name of bacteria	Heterologous gene expression	Efficiency of Hg removal (%)	References
Bacillus cereus BW-03	*mer* operon of *Bacillus thuringiensis* PW-05 strain	99.3	Dash and Das (2015)
Cupriavidus metallidurans MSR33	*merA* and *merB* genes	100	Rojas et al. (2011)
Deinococcus radiodurans	*merA* gene from *Escherichia coli* BL308	–	Brim et al. (2000)
Escherichia coli DH5α	*mer*B gene from *Escherichia coli*	–	Murtaza et al. (2002)
Escherichia coli JM109	*merP*, *merT*, and GST-PMT fusion genes	>90	Deng et al. (2006)
Escherichia coli JM109	*merP* and *merT* genes with EC20 and Lpp-OmpA	100	Bae et al. (2001)
Escherichia coli JM109	*merP*, *merT*, and GST-PMT fusion genes	–	Deng and Wilson (2001)

Table 7.2 (continued)

Name of bacteria	Heterologous gene expression	Efficiency of Hg removal (%)	References
Escherichia coli LMG194	*merC* gene from *Acidithiobacillus ferrooxidans* fused with *phoA* gene	55	Sasaki et al. (2005)
Escherichia coli TOP10	*MerR-P*$_{merT}$ genetic circuit along with *INP-CL*$_5$	98	Xue et al. (2022)
Escherichia coli XL1-Blue	*merR-p/p-T-P-B1* and *ppk* genes	90	Kiyono et al. (2003)
Escherichia coli XL1-Blue	*merR-o/p-T-P-A-G-B1* of *Pseudomonas* K-62 strainalong with a *ppk* gene from *Klebsiella aerogenes*	–	Pan-Hou et al. (2002)
Escherichia coli XL1-Blue	*merE* gene of transposon Tn*21* from *Shigella flexneri* plasmid NR1 (R100) and *merR* gene of plasmid pMR26 from *Pseudomonas* strain K-62	–	Kiyono et al. (2009)
Pseudomonas putida PpY101/pSR134	Hg-resistant gene from *NR1* plasmid	100	Okino et al. (2001)
Rhodopseudomonas palustris GIM1.167	*merP-merT* and *mt* genes	87	Deng and Jia (2011)

7.4 Conclusions and future perspectives

Bioaccumulation of hazardous Hg and bioconversion to nontoxic forms by microbes is considered as an attractive technology for treating Hg contaminated wastes. Development of genetically modified microbes employing recombinant DNA technology can further enhance the bioremediation ability of the microbes. Rational modification of the microbial mechanism for production of metallothionein can significantly enhance the uptake and intracellular accumulation of Hg. Engineered microbes with Hg^{2+} transport system will selectively overcome the limited uptake through bacterial cell membranes. Similarly, transformants expressing both metallothionein and Hg^{2+} transport systems can be of utmost significance due to their notable Hg removal efficiency. These genetically modified bacteria can be further encapsulated in the polymeric beads for ensuring their reusability and recyclability. Optimization of the process parameters like bacterial counts, contact time, initial Hg^{2+} concentration, temperature, pH, and precursors for more metallothionein production can be optimized to develop highly efficient, rapid and economical microbial processes for Hg removal in future.

References

Abbott, L. C., & Nigussie, F. (2021). Mercury toxicity and neurogenesis in the mammalian brain. *International Journal of Molecular Sciences*, *22*(14), 7520.

Andreoli, V., & Sprovieri, F. (2017). Genetic aspects of susceptibility to mercury toxicity: An overview. *International Journal of Environmental Research and Public Health*, *14*(1), 93.

Bae, W., Mehra, R. K., Mulchandani, A., & Chen, W. (2001). Genetic engineering of *Escherichia coli* for enhanced uptake and bioaccumulation of mercury. *Applied and Environmental Microbiology*, *67*(11), 5335–5338.

Bernhoft, R. A. (2012). Mercury toxicity and treatment: A review of the literature. *Journal of Environmental and Public Health*, *2012*, 460508.

Brim, H., McFarlan, S. C., Fredrickson, J. K., Minton, K. W., Zhai, M., Wackett, L. P., & Daly, M. J. (2000). Engineering *Deinococcus radiodurans* for metal remediation in radioactive mixed waste environments. *Nature Biotechnology*, *18*(1), 85–90.

Clarkson, T. W., Magos, L., & Myers, G. J. (2003). The toxicology of mercury – Current exposures and clinical manifestations. *New England Journal of Medicine*, *349*, 1731–1737.

Dash, H. R., & Das, S. (2012). Bioremediation of mercury and the importance of bacterial mer genes. *International Biodeterioration and Biodegradation*, *75*, 207–213.

Dash, H. R., & Das, S. (2015). Bioremediation of inorganic mercury through volatilization and biosorption by transgenic *Bacillus cereus* BW-03 (pPW-05). *International Biodeterioration and Biodegradation*, *103*, 179–185.

Deng, X., & Jia, P. (2011). Construction and characterization of a photosynthetic bacterium genetically engineered for Hg^{2+} uptake. *Bioresource Technology*, *102*(3), 3083–3088.

Deng, X., & Wilson, D. (2001). Bioaccumulation of mercury from wastewater by genetically engineered *Escherichia coli*. *Applied Microbiology and Biotechnology*, *56*(1), 276–279.

Deng, X., Zheng, Y., & Li, Q. (2006). Effect of ambient conditions on simultaneous growth and bioaccumulation of mercuric ion by genetically engineered *E. coli* JM109. *Journal of Hazardous Materials*, *136*(2), 233–238.

Guzzi, G., & La Porta, C. A. (2008). Molecular mechanisms triggered by mercury. *Toxicology*, *244*(1), 1–12.

Kiyono, M., Omura, H., Omura, T., Murata, S., & Pan-Hou, H. (2003). Removal of inorganic and organic mercurials by immobilized bacteria having mer-ppk fusion plasmids. *Applied Microbiology and Biotechnology*, *62*(2), 274–278.

Kiyono, M., Sone, Y., Nakamura, R., Pan-Hou, H., & Sakabe, K. (2009). The MerE protein encoded by transposon Tn21 is a broad mercury transporter in *Escherichia coli*. *FEBS Letters*, *583*(7), 1127–1131.

Kumari, S., Jamwal, R., Mishra, N., & Singh, D. K. (2020). Recent developments in environmental mercury bioremediation and its toxicity: A review. *Environmental Nanotechnology, Monitoring and Management*, *13*, 100283.

Mahbub, K. R., Bahar, M. M., Labbate, M., Krishnan, K., Andrews, S., Naidu, R., & Megharaj, M. (2017). Bioremediation of mercury: Not properly exploited in contaminated soils! *Applied Microbiology and Biotechnology*, *101*(3), 963–976.

Murtaza, I., Dutt, A., & Ali, A. (2002). Biomolecular engineering of *Escherichia coli* organo-mercurial lyase gene and its expression. *Indian Journal of Biotechnology*, *1*, 117–120.

Nascimento, A. M., & Chartone-Souza, E. (2003). Operon mer: Bacterial resistance to mercury and potential for bioremediation of contaminated environments. *Genetics and Molecular Research*, *2*(1), 92–101.

Okino, S., Iwasaki, K., Yagi, O., & Tanaka, H. (2001). Removal of mercuric chloride by immobilized cells of genetically modified *Pseudomonas putida* PpY101/pSR134. *Journal of Environmental Biotechnology*, *1*(1), 41–47.

Pan-Hou, H., Kiyono, M., Omura, H., Omura, T., & Endo, G. (2002). Polyphosphate produced in recombinant *Escherichia coli* confers mercury resistance. *FEMS Microbiology Letters*, *207*(2), 159–164.

Rojas, L. A., Yáñez, C., González, M., Lobos, S., Smalla, K., & Seeger, M. (2011). Characterization of the metabolically modified heavy metal-resistant *Cupriavidus metallidurans* strain MSR33 generated for mercury bioremediation. *PloS One*, *6*(3), e17555.

Sasaki, Y., Minakawa, T., Miyazaki, A., Silver, S., & Kusano, T. (2005). Functional dissection of a mercuric ion transporter, MerC, from *Acidithiobacillus ferrooxidans*. *Bioscience, Biotechnology, and Biochemistry*, *69*(7), 1394–1402.

Xue, Y., Du, P., Shendi, A. A. I., & Yu, B. (2022). Mercury bioremediation in aquatic environment by genetically modified bacteria with self-controlled biosecurity circuit. *Journal of Cleaner Production*, *337*, 130524.

Raturi Manisha Hemwhati Nandan, Vijaylakshmi and Santosh Kumar Karn*

Chapter 8
Bioinformatics software on taxonomic profiling

Abstract: The taxonomic profiling tool is designated to determine and classify the organisms, still we have least information about the microbial world as much the organism are present, but due to intervention of metagenomics we are able to classify the uncultured organism. The goal of metagenomics is to investigate the makeup and operation of microbial communities. There are currently many techniques for profiling environmental sample metagenomes and classifying sequences. Binning techniques and taxonomic profiling in particular are frequently utilized. These two groups of tools use a variety of methods, including read mapping, k-mer alignment, and composition analysis. For taxonomic profiling of microbial ecosystems, the 16S rRNA gene is frequently utilized. Recent developments in sequencing chemistry have made it possible to obtain exceptionally high numbers of sequences from little amounts of biological samples. Large-scale studies of complex microbiomes utilizing 16S sequencing depend on a number of significant variables, including analysis speed and data resolution to species-level taxa. There are numerous uses of various next-generation sequencing systems for identifying the microbiome, but the most well-liked technique is profiling selected regions utilizing short-read sequencing technology. Although there are numerous uses of various next-generation sequencing technologies for detecting microbiomes, short-read sequencing technology is the most widely used approach due to its low cost and typically dependable performance of determining overall microbiome compositions. Additionally, a large number of bioinformatics tools have been developed for the analysis of metagenomic and amplicon datasets, which makes choosing the right tools a difficult task. The creation of functional prediction tools employing taxonomic gene information and DNA metabarcoding data has been made possible in large part by recent breakthroughs in bioinformatics. In order to find software tools that are consistently accurate for mapping DNA sequences to taxonomic hierarchies, we used a network meta-analysis method and a *Z*-score ranking procedure. Shotgun sequencing is an alternative, and this method is more affordable over-

Acknowledgment: Authors are thankful to Dr. Gaurav Deep Singh, Chancellor, SBS University, for providing, space, facility, and resource to complete this work.

***Corresponding author: Santosh Kumar Karn,** Department of Biochemistry and Biotechnology, Sardar Bhagwan Singh University, Balawala, Dehradun 248001, Uttarakhand, India, e-mail: SantoshKarn@gmail.com

Raturi Manisha Hemwhati Nandan, Vijaylakshmi, Department of Biochemistry and Biotechnology, Sardar Bhagwan Singh University, Balawala, Dehradun 248001, Uttarakhand, India

https://doi.org/10.1515/9783110793635-008

all. Although ecologists utilizes these tools more frequently, their modularity, portability, and resilience have not been objectively assessed. The automatic taxonomic and functional analysis and interpretation of metagenome datasets without manual effort requires a bioinformatics software platform. Here, we offer phylogenetic tools for analysis of species-level taxa, MetaSAMS, and microbial community analysis (MICCA) (PhyloToAST), MetAnnotate, MetaMeta, taxonomic identification of phylogenetic profiling (TIPP), MetaPhyler, Metagenomic Phylogenetic Analysis (MetaPhlAn), PhymmBL, mOTU and nonexistant biological and chemical (NBC), VirtualBox, Quantative Insight into Microbial Ecology (QIIME2 core) VirtualBox, Miniconda, Greengenes (16S rRNA) data, CLARK, Kraken, and Treephyler. This chapter provides an insight into the Bioinformatic Software used to know the taxonomic composition and taxonomic profiling.

8.1 Introduction

Classification, nomenclature, and identification are the three independent but closely connected fields that make up the biological science field of taxonomy. The taxonomy of all living things offers a standardized method for naming and classifying creatures. Because of this uniformity, biologists from different disciplines can use the same name for every organism they study. Taxonomy can be used, among other things, to index and organize knowledge (stored as documents, articles, videos, etc.), such as in the form of search engine taxonomy or a library categorization system, so that users can more quickly discover the material they're looking for. The common language that taxonomy offers reduces name ambiguity and allows focus to be placed on more significant scientific topics and phenomena. Organisms that have similar morphological, physiological, and genetic features are organized into distinct groups or taxa through classification. The acknowledged labels by which organisms are globally recognized are provided by nomenclature, the naming of microorganisms in accordance with established norms and guidelines. Microbes are categorized based on their appearance and functional capabilities. For scientists working in many different fields of applied research and industry, including microbial systematists, the accurate identification of microorganisms is crucial. These technologies are becoming readily available as automation and user-friendly software are used more frequently. Overall, clinical chemistry and clinical microbiology methods are truly combined in the detection of infectious pathogens at the nucleic acid level. A reliable categorization or system of grouping species together is necessary for accurate identification, as is an unambiguous nomenclature for identifying them.

8.2 Latest microbial taxonomy

In order to characterize microbial genotypes, molecular approaches offer a potential basis for defining a microbial species. Growth is no longer necessary for microbial identification because deoxyribonucleic acid (DNA) amplification technology has created new pathways for microbial detection and characterization. The two main categories of microbial identification techniques are genotypic approaches, which profile an organism's genetic material, and phenotypic techniques, which profile either an organism's metabolic characteristics or a component of its chemical makeup. Microbes can be categorized based on both their genotypic and phenotypic traits (Bisen, 2014). The vast diversity of microorganisms on Earth represents the unexplored biological diversity that is still present. Microbial systematics is undergoing a revolutionary transition in its methodology to define novel taxa based on DNA/enviroDNA information in order to comprehend this immense, but generally unappreciated variety with its untapped genetic, enzymatic, and industrial potential (Móra & Whitman, 2019).

Methodological hurdles are no longer a barrier to the characterization of an organism, and it is now possible to fully sequence the entire genome of a strain to investigate specific genes or to analyze the genetic information using various approaches (Mahato et al., 2017). In reality, the use of genomics is assisting in improving our grasp of the definition of the term "species," as well as the limits of genera and higher levels of taxonomy. Additionally, a better grasp of phylogeny makes it possible to forecast the genetic potential of microbes for biotechnological uses and environmental adaptation.

8.3 Metagenomic and their importance in taxonomy

Ultrafast DNA sequencing has developed into a routine procedure for many scientific applications thanks to the development of new high-throughput technology. This method generates significant amounts of data quickly and at steadily falling costs. Directly extracted DNA from the environment is sequenced in metagenomic research. Metagenome sequencing gives access to the non-culturable portion of microbial populations, in contrast to traditional whole genome sequencing of cultivated species.

Finding the precise species of origin for metagenomic sequences is the goal of taxonomic classification, which is also done largely with the aim of cataloging/classifying different microbial groups that exist in a certain habitat. As a result, taxonomic categorization is an essential step in many metagenomic applications, including disease diagnostics, microbiome analysis, and outbreak tracing, among others. According to Rosen et al. (2011), current taxonomy classification algorithms mostly make use of manually created sequence composition parameters such oligonucleotide frequency (Vervier et al., 2016). Given the enormous amount of bio-sequencing data, it is crucial to quickly and effectively assess the many biological sequence categories in order to identify the aspects of the microbial community that researchers are most interested in. After that, by classifying sequencing data from environmental samples using metagenomics sample classification methods, researchers can further explore the species diversity of microbial communities. By identifying or forecasting categories of metagenomics samples, the sample classification methods investigate the relationship between changes in the environment or the host status and the microbial population. According to various sample data processing techniques, the methods for classifying metagenomics sequencing data may be classified into two groups: one group is based on the sequence of marker genes like 16S rRNA, and the other group is based on fragments from whole-genome sequencing.

Millions to hundreds of millions of sequencing reads are frequently produced during metagenomic analyses of microbial populations.

The majority of the organisms found in ambient or host-associated communities cannot be easily grown in a lab, and of the organisms that can be cultured, an even smaller number have been sequenced, even partially, making it difficult to assign appropriate taxonomic names to these sequences. As a result, these frequently encountered organisms are mainly missing from known genome and gene databases. To assign taxonomic labels to metagenomic sequences, estimation of the information in sequence databases to previously unknown DNA strings is necessary (Koski & Golding, 2001). DNA extraction from samples is not necessary for microbial isolation or culture in metagenomics studies. The extracted metagenome, which contains the entire DNA contained in the sample, is analysed downstream using next-generation sequencing techniques to determine the nucleotide sequences. For metagenome-based microbiological research, there are now two main methods available: shotgun metagenome sequencing and targeted amplicon sequencing.

Metagenomic next-generation sequencing (mNGS) is utilized to determine the DNA/RNA sequences in a variety of clinical sample types from patients with infectious disorders (Carbo et al., 2020; Chiu et al., 2019; Kufner et al., 2019). Any pathogen, including variations that have diverged at traditional PCR amplification sites, pathogens not previously known to be linked to a particular clinical syndrome, and novel pathogens that may go unnoticed by target-based approaches, can be identified using mNGS (Carbo et al., 2020). The introduction of NGS technology made it possible for researchers studying microbes to get massive amounts of sequencing data, which increased the importance of the bioinformatics analysis methods that analyze the sequencing data and interpret it through downstream analysis.

8.4 Scope of bioinformatics

Microbial communities are essential to industrial processes, and recent developments in computational biology and sequencing have greatly improved our ability to examine the taxonomic and functional makeup. Similar to this, microbial communities produce significant contributions for fields that have attracted economic attention, such as food production, probiotics, cosmetics, and enzyme development. Software that uses the user's gut microbiome data as one of its inputs and produces evidence-based, automated, and tailored food recommendations for stable blood sugar levels is another example of a commercial application. These applications represent a variety of bioinformatic and data science debates, ranging from the need for strain-level resolution in community profiles to the incorporation of huge datasets for predictive machine learning. The administration of large amounts of genomic and proteomic data, drug design, homology modeling, motif/domain prediction, docking , annotation, and dynamic simulation are all common uses for bioinformatic tools. Numerous areas, including genomics, proteomics, comparative genomics, nutrigenomics, microbial genome, biodefense, forensics, etc., can benefit greatly from bioinformatics. Applications of bioinformatics are mentioned **below**.

8.4.1 Phylogenetic studies

It is simple to do phylogenetic analysis of the molecular data using a variety of bioinformatics tools in a short amount of time by building phylogenetic trees to investigate its evolutionary relationship based on sequence alignment (Baxevanis & Ouellette, 2004).

8.4.2 Microbial genome applications

With the use of bioinformatics tools needed to analyze their special gene set that allows them to live in unfavorable settings, microbes have been examined at a very fundamental level. For bioinformatics, advances in science have created new problems and opportunities. The mapping of the entire human genome and the genomes of many other creatures took slightly over 10 years, thanks to advancements in bioinformatics. These discoveries have enhanced our capacity to comprehend the underlying causes of human, animal, and plant diseases and uncover novel treatments, coupled with ongoing efforts to ascertain gene and protein activities. The need for data and analysis in the biological sciences will also likely be a major driver of much future advancement in bioinformatics.

8.4.3 Predicting protein structure and function

Thanks to bioinformatics, it is now much simpler to predict protein topology, which provides insight into a protein's function as well as its 3D structure (Kelley & Sternberg, 2009).

8.4.4 Forensic science

Investigations into the identity and relationships between people are part of forensic science. Given that both bioinformatics and computer science depend on statistics and computer science, it is essentially interdisciplinary. Since the foundation of this sector is molecular data, numerous databases are being created to record the DNA profiles of known criminals. Technological and statistical advancements in microarray, Bayesian networks, machine learning algorithms, thin-film transistors biosensors, and other areas are driving this discipline forward. This offers an efficient method of organizing the evidence and drawing conclusions (Bianchi & Lio et al., 2007).

8.4.5 Biodefense

Though there is a need for more sophisticated algorithms and computational applications so that the current databases can display interoperability with one another, bioinformatics has a limited impact on forensics (Valdivia Granda, 2010).

8.4.6 Gene expression

The foundation of functional genomics is the regulation of gene expression, which enables researchers to use molecular technologies to measure the number of actively transcribing genes in each cell at any moment, such as gene expression arrays (Hoon et al., 2004).

8.4.7 Nutrigenomics

The development of molecular technologies such as genome sequencing and DNA microarrays, as well as structural and functional genomics, has produced important knowledge that explains nutrition in relation to a person's genetic makeup, which directly affects their metabolism (Corthésy-Theulaz I et al., 2005). Research on nutrition has greatly raised as a result of the introduction of bioinformatics technologies (Müller & Kersten, 2003; Werf et al., 2001).

8.4.8 Cheminformatics

Cheminformatics is primarily concerned with storing, indexing, searching for, obtaining, and using data about chemical substances. It entails the logical organizing of chemical data to make it easier to retrieve chemical properties, structures, and interactions. Theoretically, using bioinformatics, one could detect and structurally alter a natural product, design a molecule with the necessary qualities, and evaluate its medicinal effects. Analyses such as similarity searching, clustering, quantitative structure-activity relationship (QSAR) modeling, and virtual screening are all included in cheminformatics analysis (Fourches et al., 2010).

8.4.9 Transcriptomics

Transcriptomics, the study of collections of all messenger RNA molecules in the cell, is also known as expression profiling, and it makes use of DNA microarrays to control the level of mRNA expression in a particular cell population. Transcriptome analysis using bioinformatics allows the measurement of mRNA expression levels (Aplin & Singh, 2008). The presence and amount of RNA in a particular sample can be determined using RNA sequencing (RNAseq), which uses next-generation sequencing. It has also been categorized as part of transcriptomics and is used to examine how the cellular transcriptome is constantly evolving.

8.4.10 Genomics

The study of genes and gene expression is known as genomics. It generates a large amount of data about gene sequences, relationships between them, and functions. With the complete genome sequences for an increasing number of organisms, bioinformatics is beginning to provide both conceptual bases and practical methods for detecting systemic functional behaviors of the cell and the organism. Bioinformatics plays a very important role in controlling the enormous data (Kanehisa & Bork, 2003).

8.4.11 Proteomics

Proteomics, which uses methods from genetics, biochemistry, and molecular biology, is the study of protein structure, function, and relationships created by a particular cell, tissue, or organism. Modern biological techniques allow the collection of enormous amounts of information about protein-protein interactions, protein profiles, protein activity patterns, and organelle compositions. Huge amounts of data may be easily controlled and accessed using bioinformatics tools, software, and databases (Tyers & Mann, 2003).

8.5 Tools for taxonomy

Metagenomic and metatranscriptomic sequencing methods are more frequently utilized to connect the microbiome to severe diseases and ecological changes. The taxonomic and functional characteristics of the microbiota across ecosystems or individuals have been the subject of numerous analyses. Although a large portion of metagenomic analyses concentrate on species-level profiling, other research use strain-level metagenomic investigations to look into the connection between particular strains and particular conditions. Metatranscriptomic analysis, which looks at the levels of gene expression in the microbiome, offers yet another important insight into the functions of genes.

8.5.1 Live evaluation of computational methods for metagenome investigation

In order to guarantee that published methods have a trustworthy benchmark on uniform hardware, a Live Evaluation of Computational Methods for Metagenome Investigation (LEMMI) begins continuous and generalizable comparisons of individual tools or heterogeneous multistep pipelines in a controlled environment. It offers users more flexibility to weigh the benefits and drawbacks of new entries than a static report would allow by hosting in its infrastructure a semiautomated evaluation and presenting detailed results and rankings for multiple analytic objectives and their computational costs (such as accuracy vs. memory usage) in a web-based interface. The main objective of this chapter is to evaluate the first LEMMI component, which rates taxonomic binners and profilers. LEMMI complies with a number of recommendations for effective "omics" tool benchmarking (Gutierrez et al., 2017; Hein et al., 2018; Mangul et al., 2019a) including: (1) using and disseminating containers (isolated software packages) to allow reproducibility and long-term availability of tools (Mangul et al., 2019b); (2) reporting the consumption of computational resources, which is crucial as available hardware restricts the choice to otherwise. In the specific situation of taxonomic classification, a benchmark that employs the same reference for all approaches is required to carry out a reliable evaluation of their individual algorithms as well as to evaluate the variety of databases accessible. LEMMI therefore employs both methodologies.

8.5.2 Open-community Profiling Assessment tooL (OPAL)

The Open-community Profiling Assessment tooL (OPAL) implements widely used performance indicators, such as those of the first challenge of the Critical Assessment of Metagenome Interpretation (CAMI) effort, along with practical visualizations. There is a need for systematic comparisons utilizing pertinent performance criteria because of the rapid development of taxonomic metagenome profiling techniques in recent years (Fernando Meyer et al., 2019).

8.5.3 Statistical analysis of taxonomic and functional profiles (STAMP)

It is a software program that encourages "best practices" in selecting appropriate statistical methods and achieving outcomes. It is employed to examine metabolic or taxonomic profiles. Along with a variety of exploratory plots, statistical hypothesis testing for pairs of samples or groups of data are supported. For the purpose of determining biological significance, statistical analysis of taxonomic and functional profile (STAMP) recommends the use of effect sizes and confidence intervals. In order to determine the biological importance of elements in a metagenomic profile, a user-friendly, graphical interface makes it simple to investigate statistical data and create plots of publishing quality. STAMP is available for all main platforms, open source, and extensible through a plugin system (Parks et al. 2014).

8.5.4 DMET-Analyzer

The Java-based DMET-Analyzer program provides the automatic statistics examination of the connection between SNP and sample circumstances. It is platform-independent. Users can submit DMET files created by DMET Console using its straightforward graphical user interface, and it outputs a list of potential SNPs. To provide users with quick visual feedback, DMET-Analyzer enables heatmap visualization of the SNPs found across the whole dataset. The genetic model can be tested using the Hardy-Weinberg equilibrium calculator that is implemented. In addition, it links to the dbSNP database (for basic information about SNP) and the PharmaGKB pharmacogenomics knowledge base, which provides diverse information (such as pathways) connected to important SNP. Finally, it annotates significant SNP with information provided by Affymetrix libraries.

8.5.5 Apt-dmet

Affymetrix makes available the command-line program apt-dmet genotyping, which facilitates the summarization of probe sets in binary files (.CEL) and the handling of the resulting preprocessed files (.CHP). The final tabular dataset comprising the genotyping call for each probeset and each sample is not constructed. As a result, the user must conduct this phase via the DMET Console.

8.5.6 DMET

The handling of the resulting preprocessed files (.CHP) and the construction of a tabular dataset comprising the genotype call for all the probe sets and all the samples are supported by the GUI software DMET Console (Burmester et al., 2010; Sissung et al., 2010).

8.5.7 MetaMeta

MetaMeta is a brand-new pipeline for the cooperative execution and integration of technologies for classifying metagenomic sequences. Simple installation and setup, support for numerous tools, samples, and databases, an improved final profile combining multiple results, out-of-the-box parallelization and high-performance computing (HPC) integration, automated database download and setup, custom database creation, an integrated preprocessing step (read trimming, error correction, and subsampling), as well as standardized rules for integrating new tools are just a few of MetaMeta's advantages. By combining their accurate identifications and appropriately weeding out incorrect identifications, MetaMeta produces more sensitive profiling results than a single tool alone. Snake Make (Koster & Rahmann, 2012) was used to create the open-source MetaMeta application.

8.5.8 Microbial community analysis (MICCA)

Microbial community analysis (MICCA) is a software pipeline that effectively combines quality filtering, clustering of operational taxonomic units (OTUs), taxonomy assignment, and phylogenetic tree inference for the processing of amplicon and metagenomic datasets. With its precise results, MICCA successfully balances the modularity and usability of metagenomic data. Additionally, we provide a brand-new clustering approach created exclusively for the inference of OTUs. Tests on real and synthetic datasets demonstrate that MICCA offers estimates of the number of OTUs and of other common ecological indices that are more accurate and robust than currently available pipelines. This is made possible by the optimized reads filtering process and the new clustering algorithm. The structure of microbial communities that are related with the environment and humans is better understood thanks to the improved consistency of results, according to analysis of public metagenomic datasets. A project that is open source is MICCA.

8.5.9 Quantitative insights into microbial ecology (QIIME)

In order to organize OTU data into per-sample abundance, in addition to creating phylogenetic trees and computing useful statistics, Quantitative Insights into Microbial Ecology (QIIME) software was created. However, it has been discovered that there is room for development in terms of processing speed and species-level analysis, which is crucial when looking at extremely complicated human microbiomes.

8.6 Conclusion

By using the taxonomic classification pipeline, which incorporates many methodologies and is based on multiple BLAST results, it is possible to explore the taxonomic composition of the underlying community in great detail. It is possible to compare the taxonomic outcomes produced by various classifiers using bioinformatics software. Analytical pipelines is provided by bioinformatics software for both the processing and annotation of metagenome sequence data. It is envisaged that future advances will support the gradual and widespread implementation of metagenomic sequencing and Public Health laboratories. Bioinformatic software tools and platforms will advance extremely quickly.

References

Aplin, J. D., & Singh, H. (2008). Bioinformatics and transcriptomics studies of early implantation. *Annals of the New York Academy of Sciences*, *1127*, 116–120. https://doi.org/10.1196/annals.1434.005

Baxevanis, A. D., & Ouellette, B. F. (2004). *Bioinformatics: A Practical Guide to the Analysis of Genes and Proteins*. JohnWiley&Sons (vol. 43).

Bianchi, L., & Liò, P. (2007). Forensic DNA and bioinformatics. *Briefings in Bioinformatics*, *8*(2), 117–128. https://doi.org/10.1093/bib/bbm006

Bisen, P. S. (2014). Microbial Taxonomy Microbes in Practice IK International New Delhi pp 196–259 Chapter 8 pgno. 1.

Burmester, J. K., Sedova, M., Shapero, M. H., & Mansfield, E. (2010). DMET microarray technology for pharmacogenomics-based personalized medicine. *Methods in Molecular Biology (Clifton, N.J.)*, *632*, 99–124. https://doi.org/10.1007/978-1-60761-663-4_7

Capella-Gutierrez, S., de la Iglesia, D., Haas, J., Lourenco, A., Fernández, J., Repchevsky, D., Dessimoz, C., Schwede, T., Notredame, C., Gelpí, J., & Valencia, A. (2017). Lessons learned: Recommendations for establishing critical periodic scientific benchmarking. *bioRxiv*. https://doi.org/10.1101/181677

Carbo, E. C., Buddingh, E. P., Karelioti, E., Sidorov, I. A., Feltkamp, M., Borne, P., Verschuuren, J., Kroes, A., Claas, E., & de Vries, J. (2020). Improved diagnosis of viral encephalitis in adult and pediatric hematological patients using viral metagenomics. *Journal of Clinical Virology: The Official Publication of the Pan American Society for Clinical Virology*, *130*, 104566. https://doi.org/10.1016/j.jcv.2020.104566

Carbo, E. C., Sidorov, I. A., Zevenhoven-Dobbe, J. C., Snijder, E. J., Claas, E. C., Laros, J., Kroes, A., & de Vries, J. (2020). Coronavirus discovery by metagenomic sequencing: A tool for pandemic preparedness. *Journal of Clinical Virology: The Official Publication of the Pan American Society for Clinical Virology*, *131*, 104594. https://doi.org/10.1016/j.jcv.2020.104594

Chiu, C. Y., & Miller, S. A. (2019). Clinical metagenomics. *Nature Reviews Genetics*, *20*(6), 341–355. https://doi.org/10.1038/s41576-019-0113-7

Corthésy-Theulaz, I., den Dunnen, J. T., Ferré, P., Geurts, J. M., Müller, M., van Belzen, N., & van Ommen, B. (2005). Nutrigenomics: The impact of biomics technology on nutrition research. *Annals of Nutrition & Metabolism*, *49*(6), 355–365. https://doi.org/10.1159/000088315

de Hoon, M. J., Imoto, S., Nolan, J., & Miyano, S. (2004). Open source clustering software. *Bioinformatics (Oxford, England)*, *20*(9), 1453–1454. https://doi.org/10.1093/bioinformatics/bth078

Fourches, D., Muratov, E., & Tropsha, A. (2010). Trust, but verify: On the importance of chemical structure curation in cheminformatics and QSAR modeling research. *Journal of Chemical Information and Modeling*, *50*(7), 1189–1204. https://doi.org/10.1021/ci100176x

Kanehisa, M., & Bork, P. (2003). Bioinformatics in the post-sequence era. *Nature Genetics*, *33(*Suppl*)*, 305–310. https://doi.org/10.1038/ng1109

Kelley, L. A., & Sternberg, M. J. (2009). Protein structure prediction on the Web: A case study using the Phyre server. *Nature Protocols*, *4*(3), 363–371. https://doi.org/10.1038/nprot.2009.2

Koski, L. B., & Golding, G. B. (2001). The closest BLAST hit is often not the nearest neighbor. *Journal of Molecular Evolution*, *52*(6), 540–542. https://doi.org/10.1007/s002390010184

Köster, J., & Rahmann, S. (2012). Snakemake–a scalable bioinformatics workflow engine. *Bioinformatics (Oxford, England)*, *28*(19), 2520–2522. https://doi.org/10.1093/bioinformatics/bts480

Kufner, V., Plate, A., Schmutz, S., Braun, D. L., Günthard, H. F., Capaul, R., Zbinden, A., Mueller, N. J., Trkola, A., & Huber, M. (2019). Two years of viral metagenomics in a tertiary diagnostics unit: Evaluation of the first 105 cases. *Genes*, *10*(9), 661. https://doi.org/10.3390/genes10090661

Mahato, N. K., Gupta, V., Singh, P., Kumari, R., Verma, H., Tripathi, C., Rani, P., Sharma, A., Singhvi, N., Sood, U., Hira, P., Kohli, P., Nayyar, N., Puri, A., Bajaj, A., Kumar, R., Negi, V., Talwar, C., Khurana, H., Nagar, S., & Lal, R. (2017). Microbial taxonomy in the era of OMICS: Application of DNA sequences, computational tools and techniques. *Antonie van Leeuwenhoek*, *110*(10), 1357–1371. https://doi.org/10.1007/s10482-017-0928-1

Maier-Hein, L., Eisenmann, M., Reinke, A., Onogur, S., Stankovic, M., Scholz, P., Arbel, T., Bogunovic, H., Bradley, A. P., Carass, A., Feldmann, C., Frangi, A. F., Full, P. M., van Ginneken, B., Hanbury, A., Honauer, K., Kozubek, M., Landman, B. A., März, K., Maier, O., & Kopp-Schneider, A. (2018). Why rankings of biomedical image analysis competitions should be interpreted with care. *Nature Communications*, *9*(1), 5217. https://doi.org/10.1038/s41467-018-07619-7

Mangul, S., Martin, L. S., Eskin, E., & Blekhman, R. (2019). Improving the usability and archival stability of bioinformatics software. *Genome Biology*, *20*(1), 47. https://doi.org/10.1186/s13059-019-1649-8

Meyer, F., Bremges, A., Belmann, P., Janssen, S., McHardy, A. C., & Koslicki, D. (2019). Assessing taxonomic metagenome profilers with OPAL. *Genome Biology*, *20*(1), 51. https://doi.org/10.1186/s13059-019-1646-y

Müller, M., & Kersten, S. (2003). Nutrigenomics: Goals and strategies. *Nature Reviews Genetics*, *4*(4), 315–322. https://doi.org/10.1038/nrg1047

Parks, D. H., Tyson, G. W., Hugenholtz, P., & Beiko, R. G. (2014). STAMP: Statistical analysis of taxonomic and functional profiles. *Bioinformatics (Oxford, England)*, *30*(21), 3123–3124. https://doi.org/10.1093/bioinformatics/btu494

Rosen, G. L., Reichenberger, E. R., & Rosenfeld, A. M. (2011). NBC: The Naive Bayes Classification tool webserver for taxonomic classification of metagenomic reads. *Bioinformatics (Oxford, England)*, *27*(1), 127–129. https://doi.org/10.1093/bioinformatics/btq619

Rosselló-Móra, R., & Whitman, W. B. (2019). Dialogue on the nomenclature and classification of prokaryotes. *Systematic and Applied Microbiology*, *42*(1), 5–14. https://doi.org/10.1016/j.syapm.2018.07.002

Sissung, T. M., English, B. C., Venzon, D., Figg, W. D., & Deeken, J. F. (2010). Clinical pharmacology and pharmacogenetics in a genomics era: The DMET platform. *Pharmacogenomics*, *11*(1), 89–103. https://doi.org/10.2217/pgs.09.154

Tyers, M., & Mann, M. (2003). From genomics to proteomics. *Nature*, *422*(6928), 193–197. https://doi.org/10.1038/nature01510

Valdivia-Granda, W. A. (2010). Bioinformatics for biodefense: Challenges and opportunities. *Biosecurity and Bioterrorism: Biodefense Strategy, Practice, and Science*, *8*(1), 69–77. https://doi.org/10.1089/bsp.2009.0024

Vervier, K., Mahé, P., Tournoud, M., Veyrieras, J. B., & Vert, J. P. (2016). Large-scale machine learning for metagenomics sequence classification. *Bioinformatics (Oxford, England)*, *32*(7), 1023–1032. https://doi.org/10.1093/bioinformatics/btv683

Werf, M. V. D., Schuren, F. H. J., Bijlsma, S., Tas, A. C., & Ommen, B. V. (2001). Nutrigenomics: Application of genomics technologies in nutritional sciences and food technology. *Journal of Food Science*, *66*(6), 772–780.

Debapriya Maitra, Bedaprana Roy, Debdatta Das, Ananya Datta, Ruchira Pal, Sudeshna Shyam Choudhury and Arup Kumar Mitra

Chapter 9
Metagenomic analysis of microbial diversity in the Rarh region of West Bengal

Abstract: Rarh or the red lateritic region capping from the Chota Nagpur Plateau in the west to the Ganges in the east scales around 10–13.4% of cropping land in the state of West Bengal. Standing at the edge of Malthusian precipice about diminishing man-land ratio, the majorly barren fallows and agronomically poor lands of the four significant districts (approximate area: 17,131 km^2) shepherds to the poor GDP of these regions. Owing to the undulating, lateritic soil conditions with low pH (5.5–6.2), ferruginous concentration at the depth of 15–30 cm, deficient NPK concentrations, sporadically low micro-nutrients, low moisture holding capacity, erosion-prone nature and coarse soil texture the fallow land pockets can barely grow rice, millets and kharif vegetables. Interestingly, in recent years, with the advancement in metagenomics and soil-microbial-spatial-ecology researchers have tried to map the microbial flora as per their diversity in the Rarh regions. These metagenomic mapping have helped in our understanding of the microbial diversity, thereby, benefiting in selection of viable bioremediating and plant growth promoting agents. Further utilization of high throughput comparative analysis of the microbial alpha and beta diversity can help in the identification of dominant flora, and also can predict futuristic models for sustainable agricultural practices. Therefore, this study focuses on the microbial diversity of the Rarh regions to understand the beneficiary alpha and beta microbial range and explores the futuristic use of metagenomic tools for predicting sustainable agronomic models.

9.1 Introduction

Red soil, also known as the omnibus group, is the second largest soil cover of the Indian subcontinent covering over an area of approximately 18.5%. The soil developed from prolonged erosion of the Archean granite has the texture of being sandy to clayey to loamy. This red lateritic region capping from the Chota Nagpur Plateau in the west to the Ganges in the east scales around 10–13.4% of cropping land in the state of West Bengal. Standing at the edge of Malthusian precipice about diminishing man-land ratio, the majorly barren fallows and agronomically poor lands of the significant

Debapriya Maitra, Bedaprana Roy, Debdatta Das, Ananya Datta, Sudeshna Shyam Choudhury, Arup Kumar Mitra, Department of Microbiology, St. Xavier's College (Autonomous), Kolkata
Ruchira Pal, Department of Biotechnology, St. Xavier's College (Autonomous), Kolkata

https://doi.org/10.1515/9783110793635-009

Rarh districts (approximate area: 17,131 km^2) shepherds to the poor GDP of these regions. Owing to the undulating, lateritic soil conditions with low pH (5.5–6.2), phosphate, lime, magnesium and nitrogen deficiency, and ferruginous concentration at the depth of 15–30 cm, the gross agricultural productivity of the region stays quite low. In these regions, low moisture-holding capacity generally prevails with erosion-prone nature and coarse soil texture, whereby the fallow land pockets can barely grow rice, millets, and kharif vegetables. Conceptually, the soils having hue of the grade 5YR or redder with respect to the series control section having a soil depth of around 25–150 cm can be considered as red soils. Within the subset of different kinds of red lateritic soil conditions a number of variations are observed including red loams, latosolic soils, red earth soils, red gravel filled soils etc. These soils are the result of high-end weathering, making the soil enriched with various secondary grades of iron, aluminum, or sometimes even both (Ghosh, 2019). The soil microbiomes in this context play the role of being a pivotal and complex component for functionality of the red lateritic soil ecosystem. However, the limitations of culture-dependent techniques to understand, unravel, and explore the taxonomic, functional, and communal structures of these microbes in the soil and the role they play in maintaining the niche dynamics incurs the necessity of high-throughput metagenomic studies (Azeem, 2021). Different grades of metagenomic analysis like the function-based sequencing, sequence-based screenings, assembly, functional annotations, and ultimately prediction are few of the many steps in metagenomic profiling of a soil sample. Recently, a newer dawn in the world of omics, came with the inclusion of rapidly emerging area of phenomics. Here, the characteristics of various phenotypic traits are linked with their associated alleles or genes. The transition from metagenomics to metaphenomics in this context becomes very important. Metagenomic analysis enables the elucidation of the makeup and variety of microbial communities within soil flora; however, it does not facilitate the identification of functional genes in situ (Azeem, 2021). Therefore, in this study an attempt has been taken to understand the dynamics of red lateritic soil flora and the impact of metagenomic analysis on unraveling them.

9.2 Soils of Rarh region

The soils of the Rarh region refers to the red soil or laterite soil which is distributed across the region of the Indian subcontinent from the Chhota Nagpur Plateau up to the Ganges delta situated in Eastern India. The Rarh region is mostly located in the Indian state of West Bengal. Some parts are also present in the states of Bihar and Jharkhand. In Eastern India, Rarh region is mostly comprised of the lower Gangetic Plains. It extends to the Ganges in the south and the Bhagirathi-Hooghly distributary on the west. The soils of the Rarh region mainly comprise of rich alluvial deposits with an elevation of 75–150 m. Rarh region has low-level laterite badlands which are unsuitable for agriculture. The local name for these is Khoai (Herbilion, 1988).

9.3 Origin of special soil conditions

Laterite soil derives its name from the Latin word "later," which means a brick. Being rich in iron oxide, these soils are rusty-red in color (Bawa, 1957). Lateritic soil is both a soil and a rock type which is common in tropical humid conditions (Hill, 2000). Several of these regions were formed due to erosion and other tectonic movements. Alternate weather conditions of heavy rainfall and high temperatures along with alternate dry periods promote the prolonged and extensive weathering of the underlying rock, leading to the formation of the laterite soil (Winterkorn, 1951). This process is known as laterization. Laterite soil has a variable composition with varying degrees of thickness of soil layer, grade, chemical composition and deposition of the mineral ores (Dalvi, 2004) Based on the extent of weathering, the laterite soil can be firm and resistant or it can have a soft and fragile composition which can be easily broken down into pieces.

Aluminum ore comprises a major portion of the mineral ores present in laterite soil, which exists mostly as clay minerals and the hydroxides, gibbsite, diaspora, and boehmite. This composition is very similar to that of bauxite (Sarkar, 2020).

Laterites are mainly formed due to the leaching of the parent sedimentary rock, igneous rock, metamorphic rock, and some mineralized proto-ores (Tardy, 1997). This leaves behind the more insoluble ions, iron and aluminum (Whittington, 2000). Alternate wet and dry periods are essential to the formation of this soil. The leaching occurs due to rainfall during the monsoon season and the leachate containing the leached ions reaches the surface when the dry season sets in due to capillary action, leading to a high salt content on the surface layer (Yamaguchi, 2010). The chemical and mineral composition of laterite soil is mostly dependent on the parent rock from which the soil has been derived due to weathering. Laterites may be composed of zircon, quartz, titanium, iron, tin, manganese, and aluminum. Laterites also vary widely based on their location, climate, and depth.

9.4 Economically less viability of Rarh region

Geotourism is one of the most promising aspects of tourism and is crucial to the economic development of a country. It is thus being practiced by many countries, including India. Of all the Indian states, West Bengal being blessed with great natural resources holds great promise. West Bengal attracts a huge number of national and international tourists every year, with the numbers increasing every day. One such region includes the Rarh region of West Bengal. This region has the potential to become a geotourism hotspot if planning is done properly (Mandal, 2021).

The Rarh Plains are prone to riverine floods, which make the area economically vulnerable. The floods are variable in their magnitude. The rural economy of the Rarh region is mainly agriculture-based, which faces a huge loss due to the prevalence of floods in that region. Due to flooding the agriculture and thus the livelihood of the local people of the Rarh region are heavily affected. Farmers who grow traditional crops are the ones most affected and thus diversification of crops can be thought of as an important step in stabilizing the losses incurred by the Rarh localites (Islam, 2021).

Due to this flood hazard the lives of the local community is very vulnerable. Data surveys which are done at the primary levels followed by statistical analysis reveal a strongly negative correlation with economic vulnerability and vice-versa. These studies have proven that the main driving force leading to differential vulnerability is the agricultural distress. This results in the labor migration that ultimately brings higher per capita through the foreign remittances (Islam, 2022).

As discussed earlier, the primary source of income for the local people of the Rarh region include agriculture and forestry. These driving forces of livelihood help improve the sustainability of livelihood of the locals of the Rarh region, in the face of economic and environmental hazards (Satpati, 2019).

Thus, there exists poor economic sustainability among the locals of the Rarh region mainly due to the high prevalence of flash floods. Alternative agriculture strategies such the use of flood resistant crop varieties and crop diversification has shown promising results in order to make the Rarh region economically sustainable and viable.

9.5 Soil-microbial ecology of Rarh region

Soil-microbial ecology is defined as the study of microorganisms within the soil environment (Yarwood, 2021). The significance of soil ecology increased exponentially with the appreciation of soil as a major constituent of the global C cycle and the fact that the current scenario of climate change may alter the C balance and make the soil either sinks or, sources of CO_2 or other significant greenhouse gases (Baldrian, 2019). Mukherjee (1983) pioneered in geographically recognizing the north-southern belt of lateritic soil in West Bengal as the "Rarh Bengal." The reddish-brown soil or laterite in this region is created as a result of intense weathering of rocks and is made up of a combination of minerals that comprise aluminum or iron oxides, hydroxides or oxyhydroxides, quartz and kaolinite. The soil is typically infertile, gravelly and coarse textured, with low humus content and well-drained with low water holding capacity. These typical geomorphic features of the Rarh belt make it an interesting area of study especially in the field of biological features (Ghosh et al., 2015). In this context, the following sections explore the impact of few important factors upon the soil-

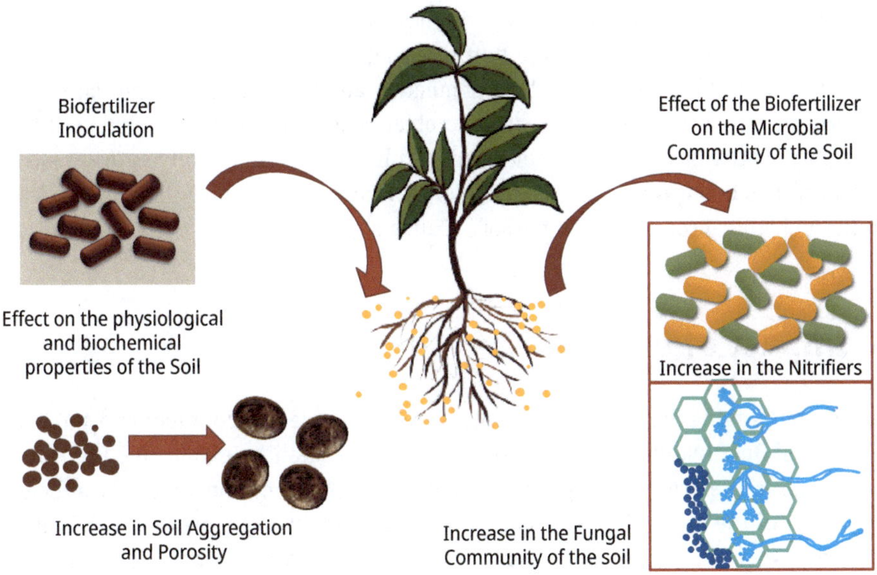

Figure 9.1: Schematic representation of effect of biofertilizer inoculation on the soil-microbial ecology in the Rarh plain (illustration courtesy: Author Bedaprana Roy).

microbial ecology indigenous to Rarh Bengal A brief schematic representation of effect of biofertilizer inoculation on soil-ecology in rarh plain has been given in Fig 9.1.

9.6 Impact of depletion

The balance and activity of a soil ecosystem depends on the cycling of nutrients by the microorganisms thriving in the soil (Zheng et al., 2005). Microorganisms of the soil comprise a huge dynamic sink and source of valuable nutrients that play a significant role in decomposition of plant waste products and cycling of nutrients, structure of soil particles, fixation of nitrogen, mycorrhizal consortia, diminution in plant pathogens and other variation in soil properties that influence plant survival. In addition, they are extremely sensitive to changes in the environment. Not only do these changes influence soil fertility but they also influence the microbial flora viability, biomass turnover, and utilization efficiency of organic carbon by the microbes, which are all important indicators of soil quality (Liao & Xie, 2007). Red soil is typically not only thoroughly weathered but are also often eroded resulting in the topsoil being underdeveloped such that their innate fertility is diminished and plant nutrients, derived essentially from minerals are limited (Yao et al., 2000). Studies have found that microbial growth rates decreased in case of lower nutrient soil samples (Merckx et al., 1987). Laterite soils such as those found in the Rarh are usually rich in kaolinite, which are deficit in available K, and crops grown in these soils more often than not suffer from K deficiency (Das et al., 2019). Both organic and inorganic fertilizers are used to increase nutrient availability to plants, but they have also been observed to affect the composition, population and functionality of the soil microflora. Organic fertilizers tend to improve microbial biomass of the soil, evolution of CO_2 and enzymatic activities. In contrast, inorganic fertilizers have comparatively lesser effect on soil microbial biomass and activities than organic fertilizers. Fertilization can also lead to shifts in microbial community of the soils (Chu et al., 2007).

9.7 Impact of erosion

Soil erosion is among the biggest challenges for sustainable development and soil constitution. Soil degradation by erosion has been exacerbated by exhaustive anthropogenic factors and extreme changes in the climate. Soil erosion has not only affected around 84% of the global land surfaces but has also resulted in the deterioration of more than 33% of Earth's soils. Erosion leads to the deprivation of the fertile topsoil which contains the nutrients, and which also support microbial habitats, leading to reduction in output. Erosion has led to dramatic loss of microbial diversity of the soil. Previous studies have observed that decrease in microbial diversity in the degraded soils were analogous with exacerbation in functions of the soil ecosystem (Qiu et al., 2021). For the global semi-arid and tropical regions, soil erosion poses a serious issue. A country of the monsoonal tropics, India is also plagued with this environmental menace. Erosion by water is so far the gravest reason behind degradation of land in India and occurs at a threatening rate of 16.4 tons per hectare per year resulting in a total of around 5 billion tons of soil loss a year. Gully erosion is a significant reason behind local loss of soil, since gullies promote close to 94% of soil lost by water erosion in spite of covering extremely small areas (ca. 1–5%) of a terrain. Gullies can be defined as erosional features of topography created by intermittent yet condensed and erosive outflow, and are classified as temporary or everlasting while those in the Rarh are permanent (Majhi et al., 2021). The formation of gullies leading to erosion in rarh plains have been schematically represented in fig 9.2.

Out of a total 8.87 million hectares, West Bengal alone has contributed to about 2.2 million hectares of soil erosion in India (Bhunia et al., 2018). The plain of Rarh located in the Lower Gangetic Basin, West Bengal, is famous for its characteristic lateritic terrain. This basin faces serious issues contributed by numerous patterns of soil erosion, and among these gully erosion is highly prominent (Majhi et al., 2021).

The total annual rainfall in the Rarh plain is about 1,500 mm, of which 75–78% is usually experienced in the monsoon months of June to September, as well as from frequent cyclonic activity and weather depressions (Majhi et al., 2021). This is a prominent reason behind the formation of gullies in this area.

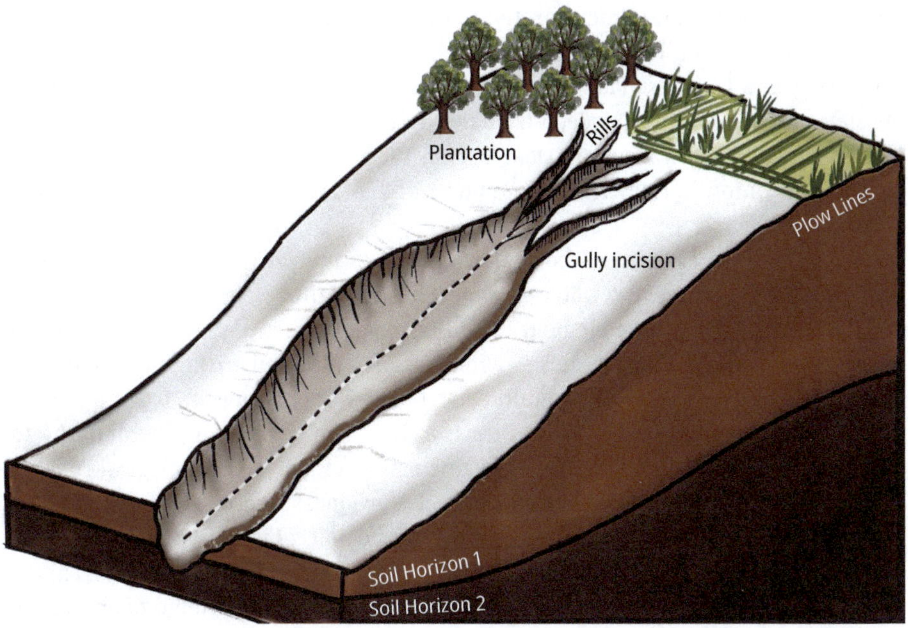

Figure 9.2: The formation of gullies in the Rarh plain (illustration courtesy: Author Bedaprana Roy).

Experiment results have reported that erosion of soil leads to depletion of microflora. Severe erosion by water can dramatically lessen microbial biomass. Water erosion leads to breakdown of soil aggregates, acceleration of transport of sediment and diminish the soil moisture-holding ability as seen in the gullies of Rarh. The degraded environmental condition at the eroded locations often alters the natural progression of soil microbes. Moreover, the surface soil is imperative, in which most microorganisms thrive. Severe erosion by water causes microbial exodus in the surface soil by means of outflow, and this ultimately affects the geographical dissemination of microbes and reduces the microbial abundance in the degraded sites (Li et al., 2015).

9.8 Impact of good agricultural practices in adjoining districts and introduction of alien microbial flora

In India, the red soils cover an area of around 110 million hectares. These soils have developed under varying physiographic and climatic conditions in the tropical and subtropical conditions with precipitation that ranges between 600 mm and 4,000 mm. The average yearly temperature ranges from 22 to 30 °C with little difference between the mean summer and winter soil temperatures. Red and lateritic soils have been found in eight agroecological regions of India as observed by the National Bureau of Soil Survey and Land Use Planning. These agroecological regions are distributed over the states of Tamil Nadu, Kerala, Karnataka, Telengana, Andhra Pradesh, Chattisgarh, Maharashtra, Madhya Pradesh, Goa, Orissa, West Bengal, North Eastern States, as well as the Union Territories of Andaman and Nicobar Islands and Pondicherry (Ghosh, 2021a).

These soils, as established before, are prone to nutrient losses due to excessive rainfall, leaching and severe water erosion (Ghosh, 2021b).

Conservation and improvement of soil productivity hence play a pivotal part in the maintenance of the soil resources. Biotic factors of the soil are an important factor of soil condition by virtue of its multifaceted role. Not only does it act as a basin of nutrients for the plants but it also serves as a supporting media for growth of the soil microorganisms. The function of the organic matter of soil in the biological roles, cycling of nutrients and structure of soil particles are measured by some criteria, such as total organic carbon and nitrogen and their ratio, microbial activities and biomass. The fluctuating basin of plant nutrients, gaseous transaction between the soils and its physical properties is called microbial biomass. A study was attempted to analyze the consequence of vermicompost and decayed cow manure, either by itself or in association with urea in the Rarh of West Bengal (Dutta et al., 2003). Now, it has already been established that microflora of cow dung comprises around 60 different species of bacteria dominated by *Corynebacterium* sp., *Lactobacillus* sp., and *Bacillus* sp., and a few fungal species such as *Aspergillus* sp. and *Trichoderma* sp., about 100 species of protozoa, and 2 different yeast species (Bhatt & Maheshwari, 2019), while Actinobacteria and Gammaproteobacteria are found in abundance in vermicompost (Pathma & Sakthivel, 2012). The research team observed the most microbial biomass carbon in the soil treated with decayed cow manure along with urea. The next most effective treatment was vermicompost along with urea followed by decomposed cow manure, vermicompost, and chemical fertilizers (Dutta et al., 2003).

9.9 Comparative analysis of the microbial diversity of the four districts

The Rarh region in West Bengal spans over the districts of Bankura, Birbhum, Jhargram and Purulia. A study obtained the metagenomic sequence of soil obtained from the Rarh plain, more specifically from the district of Birbhum. Results so obtained indicated towards the affluence of Proteobacteria at 61% and Bacterioidetes at 35%. Among the members of Proteobacteria, Gammaproteobacteria was the most numerous at 59% followed closely by Alphaproteobacteria and Betaproteobacteria, with abundance of Moraxallecaeae, Sphingobacteriaceae, Pseudomonadiales, and Enterobacteriacea. The dominating genera so identified were *Sphingobacterium* at 35% and *Acinetobacter* at 31%, and followed by *Pseudomonas* at 7%. Only about 2% of the observed results represented members of the Archaea and the remaining 98% were from the Bacterial phylum. As part of the *Bacillus* clade, *Bacillus cereus* at 17% and *Bacillus thuringiensis* at 6% were found in abundance (Mukhopadhyay et al., 2021).

Another study conducted in the Beliatore and Joypur forests belonging to Bankura district found microorganisms belonging to four dominant groups, viz. fungi, bacteria, algae, and Actinomycetes, which are responsible for microbial decomposition and contributing to the microbial biomass present in the floor of the forests. This microbial biomass has contributed around 1.8–3.5% to total organic C, 1.9–4.6% to the total N, and 4.8–12.3% to total P. The research team concluded that the microbial biomass and nutrients available are precisely related in the fluctuation of the health of soil in the forest soils (Das, 2021).

An extensive research was conducted in selected agroecological subregions to calculate the abundance of certain agriculturally significant microorganisms from the different terrains of West Bengal. The team collected soil samples from various crop fields to analyze their microbial diversity. In the Rarh region, the study covered the districts of Jhargram, Bankura and Birbhum (Dasgupta & Brahmaprakash, 2016). The results so obtained are indicated through the tables 9.1–9.3.

Table 9.1: Microbial diversity of Jhargram's arecanut, gamhar, and guava fields.

Microorganism	Arecanut (cfu/g)	Gamhar (cfu/g)	Guava (cfu/g)
Bacteria	2.7×10^5	1.76×10^5	2.27×10^5
Fungi	1.9×10^4	1.06×10^4	1.97×10^4
Actinomycetes	1.16×10^5	2.9×10^4	1.8×10^4
Rhizobium	5.67×10^2	3.33×10^2	2.97×10^2
Azotobacter	5×10^2	1.95×10^3	1×10^3
Azospirillum	7×10^3	4×10^2	4×10^2
Phosphate solubilizers	6.33×10^2	5.67×10^2	2.8×10^3
Nitrifiers			
NH_4^+ oxidizers	4×10^3	4×10^2	10^3
NO_2^- oxidizers	10^2	10^2	4×10^2
Index			
Proportion of bacteria	0.660	0.816	0.858
Proportion of fungi	0.047	0.049	0.074
Proportion of Actinomycetes	0.286	0.134	0.068
Shannon-Wiener index	0.333	0.253	0.219
Simpson's index	0.471	0.314	0.254

The Indian state of West Bengal shares its borders with Bihar and Jharkhand in the west and is characterized by diverse geomorphology and climate in addition to exclusive natural and anthropogenic resources. The "Rarh" region refers to the laterite badlands existing in the western part of West Bengal where the Chhota Nagpur Plateau

Table 9.2: Microbial diversity of Bankura's rice field.

Microorganism	Rice (cfu/g)
Bacteria	1.96×10^5
Fungi	4.33×10^4
Actinomycetes	2.5×10^4
Rhizobium	3.33×10^2
Azotobacter	11×10^3
Azospirillum	7×10^3
Phosphate solubilizers	1.43×10^3
Nitrifiers	
NH_4^+ oxidizers	4×10^2
NO_2^- oxidizers	10^2
Index	
Proportion of bacteria	0.741
Proportion of fungi	0.163
Proportion of Actinomycetes	0.095
Shannon-Wiener index	0.321
Simpson's index	0.415

Table 9.3: Microbial diversity of Birbhum; rice field.

Microorganism	Rice (cfu/g)
Bacteria	3.17×10^5
Fungi	3.73×10^4
Actinomycetes	8.3×10^3
Rhizobium	7×10^2
Azotobacter	1.87×10^3
Azospirillum	3.9×10^3
Phosphate solubilizers	1.47×10^3
Nitrifiers	
NH_4^+ oxidizers	10^2
NO_2^- oxidizers	7×10^1
Index	
Proportion of bacteria	0.874
Proportion of fungi	0.103
Proportion of Actinomycetes	0.023
Shannon-Wiener index	0.190
Simpson's index	0.225

continues. A massive erosional phase that occurred underneath the plateau-fringe area here had transformed its geomorphology into a rolling peneplain of small monadnocks known as "tila" – the features of which could be traced to the western parts of Birbhum, Purulia, Bankura, Bardhaman, and Paschim Medinipur districts (Mukhopadhyay, 2021; Jana, 2021).

The laterite soil in this region has been observed to be deficient in organic carbon and mineral nutrient content (NPK), besides having a low water retention capacity, thereby posing difficulty to carry out agricultural practices (Mukhopadhyay, 2021).

With an increase in latitude in Birbhum, it is seen that the pH of the soil, accessible potassium (K), boron (B), iron (Fe), and organic carbon (HC) content composing the soil organic matter decreases sharply (Mukhopadhyay, 2021).

A soil sample (pH = 5.99) from Birbhum's Sadaipur area had been collected and analyzed for determining the constituting organic carbon and NPK content present in that soil. (Table 9.4) (Mukhopadhyay, 2021)

Table 9.4: Chemical composition of soil in Birbhum, West Bengal (Mukhopadhyay, 2021).

Soil chemical composition	Content result (with units)
Total organic carbon (HC)	**0.57 ± 0.011 (%)**
Available nitrogen (N)	**81.91 ± 0.015 (mg/kg)**
Available phosphorus (P)	**6.90 ± 0.01 (mg/kg)**
Available potassium (K)	**126.57 ± 0.021 (mg/kg)**

While mineral nutrients like NPK are found to be deficient mainly in the Birbhum region, excessive fluoride content seems to exist in regions like Bankura and Purulia.

In the West Bengal districts of Purulia and Bankura, the average fluoride (F^-) quantity in agronomic soil had been measured to be 126 ± 65 mg/kg and 114 ± 59 mg/kg, respectively (De, 2021). Even metalloid toxicity like that of arsenic seems to be rampant in the "Rarh" region (De, 2021).

The Sadaipur soil sample, after being analyzed using the Illumina Miseq platform showed that only 2% of the reads obtained represented members from Archaea, while the rest 98% represented the Bacterial phylum members. Data profiling indicated the Proteobacteria abundance to be high. In the sample's total distribution of genera, *Sphingobacterium* (35%), followed by *Acinetobacter* (31%), and then *Pseudomonas* (7%) were discovered have the highest percentages (Mukhopadhyay, 2021). Among the other soil microorganisms found, nitrogen-fixing cyanobacteria like *Anacystis* (*Chroococcus*) sp., could be used for bioremediation to improve nitrogen content (Odokuma, 2002).

Additionally, cadmium-resistant gram-negative bacteria *Ochrobactrum* sp., and lead and arsenic-resistant gram-positive ones such as *Bacillus* spp. have demonstrated crop growth-promoting abilities while mitigating metal toxicity with the help of energy-dependent and plasmid-encoded metal efflux systems (like ATPases) (Verma, 2019; Pandey, 2013; Fabian, 2020).

The purpose of exploring the different microbial consortia present in the soil is to enable an understanding of the microflora that could be utilized to increase soil productivity. This is vital for developing desired agricultural practices using bioremediation, which involves accumulation, immobilization, oxidation, transformation, and volatilizing of heavy metals and bioaugmentation of mineral nutrients (Verma, 2019).

The average performance of a crop with respect to its output per hectare is given by agricultural efficiency (Roy, 2019). Common agricultural crops that are grown in Purulia include rice, maize, wheat, bajra, gondly, arhar, gram, kulthi, surguja, and kudrum (Roy, 2019). Their cultivation success or agricultural efficiency is often jeopardized by the anomaly in precipitation (Goswami, 2018). Soil erosion is an equally contributing factor in affecting crop yield.

In the agriculturally developed areas of Bardhaman district, rice, paddy, jute, and potato are cultivated. However, seasonal drought in these regions often affected the agroeconomy and forced outmigration (Deshingkar, 2003; Debnath, 2021).

The soil in Bankura is deficient in micronutrients such as calcium and sulfur, which makes it suitable for the production of groundnut (*Arachis hypogaea*) during the Kharif season (Das, 2015). Bankura district, however, also receives variable amounts of precipitation during the cropping season characterized by frequent and longer dry spells. This results in agricultural drought, increment in insect infestation, plant disease, and wind erosion which ultimately leads to crop harvest failure (Das, 2013).

Fluoride accumulation in both soil and water of the "Rarh" region also poses a threat to biodiversity and affects crop yield. Owing to geogenic and anthropogenic events, the mean fluoride concentration from a study found Purulia's soil sample to be 0.04× than that of the Bankura district's soil (De, 2021). On this account, F^- accumulation was found to be significantly higher in crops such as vegetables compared to pulses and cereals. The highest accumulation was observed in onions, while the lowest was found in rice (De, 2021).

In Birbhum, the soil is characterized by high salinity and is commonly drought affected. Thus, trials have been made to grow sweet sorghum (i.e., *Sorghum bicolor* (L.) Moench) as it has a higher tolerance to salt and drought (Barik, 2015).

To combat and prevent plant disease and enhance nutrient availability, plant growth-promoting rhizobacteria (PGPR), which colonize the plant root, are to be a viable alternative to harmful chemical pesticides.

Biostimulants like *Fluorescent pseudomonads,* multi-beneficiary gram-positive bacteria like *Bacillus* spp., P-solubilizing bacteria (PSB) like *Burkholderia* spp., heavy metals remediating endophytic bacteria like *Paenibacillus* sp., sulfur-oxidizing bacterium like *Brevundimonas* sp. (obtained from Barjora coal mine in Bankura) are examples of few of the fundamental PGPR that could be utilized to promote and improve plant growth and development (Keerthana, 2022; Prabhukarthikeyan, 2022; Ghosh, 2020; Roy, 2015; Govarthanan, 2016; Roy, 2017). Hence, PGPRs are crucial for conducting sustainable agricultural practices.

9.10 Conclusion

The importance of soil fertility in agriculture is endless. The biological fertility of soil mainly depends on its prevalent flora, however, till the last decade very little was known about these microorganisms due to the limitations of culture based techniques. However, with the inclusion of high-throughput metagenomic tools study of uncultured organisms, understanding their true diversity, their evolution, their phylogeny, and the interdynamics between each other became easy. Going by soil geography, the red lateritic soil comprises of soil largest area in the subcontinent and in entire state of West Bengal it encapsulates approximately around 17,131 km^2. These majorly barren fallows and agronomically poor lands of the four significant districts contribute to the poor GDP of these regions. However, over the time no holistic attempt has been taken to mitigate the issue from the soil health level. Interestingly, in recent years, with the advancement in metagenomics and soil-microbial-spatial-ecology researchers have tried to map the microbial flora as per their diversity in the Rarh regions. These metagenomic mapping have helped in our understanding of the microbial diversity, thereby, benefiting in selection of viable bioremediating and plant growth promoting agents. Further utilization of high throughput comparative analysis of the microbial alpha and beta diversity can help in the identification of dominant flora, and also can predict futuristic models for sustainable agricultural practices. Although much work is still required in this context, this study makes an initial attempt toward addressing it in the future.

References

Azeem, M., Soundari, P. G., Ali, A., Tahir, M. I., Imran, M., Bashir, S., Irfan, M., Li, G., Zhu, Y.-G., & Zhang, Z. (2021). Soil metaphenomics: A step forward in metagenomics. *Archives of Agronomy and Soil Science*, 10.1080/03650340.2021.1921743(PDF), https://www.researchgate.net/publication/351639191_Soil_metaphenomics_a_step_forward_in_metagenomics (accessed Sep 13 2022).

Baldrian, P. (2019). The known and the unknown in soil microbial ecology. *FEMS Microbiology Ecology*, *95*(2), fiz005. https://doi.org/10.1093/femsec/fiz005

Barik, S., & Roy, P. (2015). Agronomic practices for the development of sweet sorghum [Sorghum bicolor (L.) Moench] crop in Birbhum district of West Bengal. *Research on Crops*, *16*, 230. 10.5958/2348-7542.2015.00034.0

Bawa, K. S. (1957). Laterite soils and their engineering characteristics. *Journal of the Soil Mechanics and Foundations Division*, *83*(4), 1428–1.

Bhatt, K., & Maheshwari, D. K. (2019). Decoding multifarious role of cow dung bacteria in mobilization of zinc fractions along with growth promotion of C. annuum L. *Scientific Reports*, *9*(1), 14232. https://doi.org/10.1038/s41598-019-50788-8

Bhunia, G. S., Shit, P. K., & Chattopadhyay, R. (2018). Assessment of spatial variability of soil properties using geostatistical approach of lateritic soil (West Bengal, India). *Annals of Agrarian Science*, *16*(4), 436–443. https://doi.org/10.1016/j.aasci.2018.06.003

Chu, H., et al. (2007). Soil microbial biomass, dehydrogenase activity, bacterial community structure in response to long-term fertilizer management. *Soil Biology and Biochemistry, 39*(11), 2971–2976. https://doi.org/10.1016/j.soilbio.2007.05.031

Dalvi, A. D., Bacon, W. G., & Osborne, R. C. (2004, March). The Past and the Future of Nickel Laterites. In *PDAC 2004 International Convention, Trade Show & Investors Exchange* (pp. 1–27). Toronto: The prospectors and Developers Association of Canada.

Das, D., et al. (2019). Potassium supplying capacity of a red soil from eastern India after forty-two years of continuous cropping and fertilization. *Geoderma, 341*, 76–92. https://doi.org/10.1016/j.geoderma.2019.01.041

Das, G. (2021). Forest integrity monitoring of Joypur and Beliatore forests in Bankura district, West Bengal. *Indian Science Cruiser, 35*, 53–63. https://doi.org/10.24906/isc/2021/v35/i3/209193

Das, S., Chowdhuri, T., & Singh, A. (2015). Effect of calcium and sulphur based fertilizers on yield and quality of ground nut in red and lateritic belt of Bankura district of West Bengal. *Journal of Krishi Vigyan, 3*, 35. 10.5958/2349-4433.2015.00007.0

Das, S., Roy Choudhury, M., & Nanda, S. (2013). Geospatial assessment of agricultural drought (A case study of Bankura district, West Bengal). *International Journal of Agricultural Science and Research, 3*, 1–28.

Dasgupta, D., & Brahmaprakash, G. P. (2016). Microbial diversity analysis of selected agro-ecological sub-regions of West Bengal. 1.

De, A., Mridha, D., Ray, I., Joardar, M., Das, A., Chowdhury, N. R., & Roychowdhury, T. (2021). Fluoride exposure and probabilistic health risk assessment through different agricultural food crops from fluoride endemic Bankura and Purulia districts of West Bengal, India. *Frontiers in Environmental Science, 9*. https://doi.org/10.3389/fenvs.2021.713148

Debnath, M. (2021). Exploring the behavior of seasonal agricultural migrants in the Rarh region of West Bengal, India. *Spatial Information Research, 29*, 83–96. https://doi.org/10.1007/s41324-020-00338-6

Deshingkar, P., & Start, D. (2003). Seasonal Migration for Livelihoods in India: Coping, Accumulation, and Exclusion. ISBN: 0 85003 670 4.

District-wise crop production in West Bengal: Major crops in West Bengal. https://www.agrifarming.in/district-wise-crop-production-in-west-bengal-major-crops-in-west-bengal#

Dutta, S., et al. (2003). Influence of integrated plant nutrient supply system on soil quality restoration in a red and laterite soil: Einfluss integrierter pflanzennährstoff versorgung auf die wiederherstellun der bodenqualität von rotem und laterit boden. *Archives of Agronomy and Soil Science, 49*(6), 631–637. https://doi.org/10.1080/03650340310001599722

Fabian, B. K., Tetu, S. G., & Paulsen, I. T. (2020). Application of transposon insertion sequencing to agricultural science. *Frontiers in Plant Science Crop and Product Physiology*. https://doi.org/10.3389/fpls.2020.00291

Ghosh, G. (2019). Red and Lateritic Soils and Agri-Productivity : Issues and Strategies.

Ghosh, G. (2021a). Red and Lateritic Soils and Agri-Productivity : Issues and Strategies.

Ghosh, G. (2021b). Red and Lateritic Soils and Agri-Productivity : Issues and Strategies.

Ghosh, R., & Mandal, N. C. (2020). Use of Plant Growth Promoting Burkholderia Species with Rock Phosphate Solubilizing Potential toward Crop Improvement. In *Microbial Services in Restoration Ecology*, Ch 10 (pp. 139–156). Elsevier, ISBN: 9780128199787, https://doi.org/10.1016/B978-0-12-819978-7.00010-5

Ghosh, S., Guchhait, S. K., & Hu, X.-F. (2015). Characterization and evolution of primary and secondary laterites in northwestern Bengal Basin, West Bengal, India. *Journal of Palaeogeography, 4*(2), 203–230. https://doi.org/10.3724/SP.J.1261.2015.00074

Goswami, A. (2018). Identifying the trend of meteorological drought in Purulia district of West Bengal, India. *Environment and Ecology, 37*(1B), 387–392, ISSN 0970-0420.

Govarthanan, M., Mythili, R., Selvankumar, T., et al. (2016). Bioremediation of heavy metals using an endophytic bacterium Paenibacillus sp. RM isolated from the roots of Tridax procumbens. *3 Biotech, 6*, 242. https://doi.org/10.1007/s13205-016-0560-1

Herbilion, A. J., & Nahon, D. (1988). Laterites and Laterization Processes. In *Iron in Soils and Clay Minerals* (pp. 779–796). Dordrecht: Springer.

Hill, I. G., Worden, R. H., & Meighan, I. G. (2000). Geochemical evolution of a palaeolaterite: The interbasaltic formation, Northern Ireland. *Chemical Geology, 166*(1–2), 65–84.

Islam, A., & Ghosh, S. (2021). Economic transformation in the wake of flood: A case of the lower stretch of the Mayurakshi River Basin, India. *Environment, Development and Sustainability, 23*(10), 15550–15590.

Islam, A., & Ghosh, S. (2022). Community-based riverine flood risk assessment and evaluating its drivers: Evidence from Rarh Plains of India. *Applied Spatial Analysis and Policy, 15*(1), 1–47.

Jana, D., Guchhait, K. C., Ballav, S., Panda, A. K., & Ghosh, C. (2021). Traditional Herbal Medicine Practiced in Plateau-Fringe and Rarh Districts of West Bengal, India. In S. C. Mandal, R. Chakraborty, & S. Sen (Eds.), *Evidence-Based Validation of Traditional Medicines*. Singapore: Springer. https://doi.org/10.1007/978-981-15-8127-4_48

Keerthana, U., Prabhukarthikeyan, S. R., et. al. (2022). Fluorescent Pseudomonads: A Multifaceted Biocontrol Agent for Sustainable Agriculture. In *New and Future Developments in Microbial Biotechnology and Bioengineering*, Ch 6 (pp. 83–92). Elsevier, ISBN: 9780323851633, https://doi.org/10.1016/B978-0-323-85163-3.00015-6

Li, Z., et al. (2015). Microbial responses to erosion-induced soil physico-chemical property changes in the hilly red soil region of southern China. *European Journal of Soil Biology, 71*, 37–44. https://doi.org/10.1016/j.ejsobi.2015.10.003

Liao, M., & Xie, X. M. (2007). Effect of heavy metals on substrate utilization pattern, biomass, and activity of microbial communities in a reclaimed mining wasteland of red soil area. *Ecotoxicology and Environmental Safety, 66*(2), 217–223. https://doi.org/10.1016/j.ecoenv.2005.12.013

Majhi, A., Nyssen, J., & Verdoodt, A. (2021). What is the best technique to estimate topographic thresholds of gully erosion? Insights from a case study on the permanent gullies of Rarh plain, India. *Geomorphology, 375*, 107547. https://doi.org/10.1016/j.geomorph.2020.107547

Mandal, T. (2021). Geotourism potentiality of the rarh region of West Bengal: A geographical outlook. *Research Review International Journal of Multidisciplinary, 6*(12), 90–97.

Merckx, R., et al. (1987). Production of root-derived material and associated microbial growth in soil at different nutrient levels. *Biology and Fertility of Soils, 5*(2), 126–132. https://doi.org/10.1007/BF00257647

Mukherjee, K. N. (1983). Diagnostic Survey Of West Bengal. Available at: http://archive.org/details/in.ernet.dli.2015.32866 (Accessed: 8 August 2022).

Mukhopadhyay, M., Kumar Mitra, A., Choudhury, S. S., & Ganguli, S. (2021). Metagenome dataset of lateritic soil microbiota from Sadaipur, Birbhum, West Bengal, India. *Data in Brief, 36*, ISSN 2352-3409, https://doi.org/10.1016/j.dib.2021.107041

Odokuma, L. O., & Inor, M. N. (2002). Nitrogen fixing bacteria enhanced bioremediation of a crude oil polluted soil. *Global Journal of Pure and Applied Sciences, 8*(4). 10.4314/gjpas.v8i4.15993

Pandey, S., Ghosh, P., Ghosh, S., De, T., & Maiti, T. (2013). Role of heavy metal resistant Ochrobactrum sp. and Bacillus spp. strains in bioremediation of a rice cultivar and their PGPR-like activities. *Journal of Microbiology (Seoul, Korea), 51*, 11–17. 10.1007/s12275-013-2330-7

Pathma, J., & Sakthivel, N. (2012). Microbial diversity of vermicompost bacteria that exhibit useful agricultural traits and waste management potential. *SpringerPlus, 1*, 26. https://doi.org/10.1186/2193-1801-1-26

Prabhukarthikeyan, S. R., Keerthana, U., Baite, P. M., et. al. (2022). Bacillus Rhizobacteria: A Versatile Biostimulant for Sustainable Agriculture. In *New and Future Developments in Microbial Biotechnology and Bioengineering*, Ch 3 (pp. 33–44). Elsevier, ISBN: 9780323851633, https://doi.org/10.1016/B978-0-323-85163-3.00009-0

Qiu, L., et al. (2021). Erosion reduces soil microbial diversity, network complexity and multifunctionality. *The ISME Journal*, *15*(8), 2474–2489. https://doi.org/10.1038/s41396-021-00913-1

Roy, S., Nayak, B., & Mitra, A. (2017). Isolation and characterization of plant growth promoting obligate chemolithoautotrophic sulphur-oxidizing bacterium from Barjora coal mine soil sample. *Journal of Science, Engineering, Health, and Management*, *1*(4).

Roy, S. S., & Jana, N. C. (2019). Agricultural productivity and efficiency in Purulia district, West Bengal. *International Journal of Reviews and Research in Social Sciences*, *7*(2), 283–292. 10.5958/2454-2687.2019.00016.9

Roy, T., & Biswas, D. R. (2015). Solubilization of Purulia rock phosphate through organic acid loaded nanoclay polymer composite and phosphate solubilizing bacteria and its effectiveness as P-fertilizer to wheat. *Journal of the Indian Society of Soil Science*, *63*(3), 327–338, Print ISSN: 0019-638X. Online ISSN: 0974-0228, 10.5958/0974-0228.2015.00043.2

Sarkar, T., Mishra, M., & Chatterjee, S. (2020). On detailed field-based observations of laterite and laterization: A study in the Paschim Medinipur lateritic upland of India. *Journal of Sedimentary Environments*, *5*(2), 219–245.

Satpati, S. (2019). Forest resource and agriculture option for livelihoods among schedule tribe in RARH region of West Bengal in India. *Research Journal of Humanities and Social Sciences*, *10*(3), 899–806.

Tardy, Y. (1997). *Petrology of Laterites and Tropical Soils*. AA Balkema.

Verma, S., & Kuila, A. (2019). Bioremediation of heavy metals by microbial process. *Environmental Technology & Innovation*, *14*, 100369, ISSN 2352-1864, https://doi.org/10.1016/j.eti.2019.100369

Whittington, B. I., & Muir*, D. (2000). Pressure acid leaching of nickel laterites: A review. *Mineral Processing and Extractive Metullargy Review*, *21*(6), 527–599.

Winterkorn, H. F., & Chandrasekharan, E. C. (1951). *Laterite Soils and Their Stabilization*. Winterkorn Road Research Institute.

Yamaguchi, K. E., & Kosei, E. (2010). Iron isotope compositions of Fe-oxide as a measure of water-rock interaction: An example from Precambrian tropical laterite in Botswana. *Frontier Research on Earth Evolution*, (IFREE Report for 2003–2004).

Yao, H., et al. (2000). Microbial biomass and community structure in a sequence of soils with increasing fertility and changing land use. *Microbial Ecology*, *40*(3), 223–237. https://doi.org/10.1007/s002480000053

Yarwood, S. A. (2021). 10 – Microbial Ecology. In T. J. Gentry, J. J. Fuhrmann, & D. A. Zuberer (Eds.), *Principles and Applications of Soil Microbiology (Third Edition)* (pp. 239–267). Elsevier. https://doi.org/10.1016/B978-0-12-820202-9.00010-1

Zheng, H., et al. (2005). Effects of regenerating forest cover on soil microbial communities: A case study in hilly red soil region, Southern China. *Forest Ecology and Management*, *217*(2), 244–254. https://doi.org/10.1016/j.foreco.2005.06.005

Bedaprana Roy*, Debapriya Maitra, Rajeshwari Podder,
Tannishtha Das, Snata Pandey, Jaydip Ghosh and Arup Kumar Mitra

Chapter 10
Etymology of microbial association in high-altitude permafrost regions of the world

Abstract: Permafrost (perennial frozen zones) are found in the northern hemisphere where the ground remains frozen throughout the year. About 85% of the regions of Canada, Siberia, Alaska, Greenland, and the Arctic are covered with permafrost. The microbial population that inhabits these zones are subjected to subzero temperatures, minimal availability of water, and severe lack of nutrition (oligotrophic conditions). In spite of such limitations a varied plethora of microbial communities that are essentially extremophiles are found in these regions thus making them a promising sector for metagenomic analysis. Metagenomic studies of these regions reveal presence of ancient microbial classes that can not only survive these extremes but also conduct their metabolism in the same.

In this chapter, we will discuss in detail the metagenomic studies of certain selected permafrost regions of the world, such as Siberia, Alaska, and the Arctic,. Analysis of this will lead us to the identification of some keystone classes and species of bacteria that are capable of survival in such extreme conditions. The etymology of these keystone species will explain the evolutionary history of extremophily. In this chapter, we will also explore mechanisms of survival, genetic makeup, and pathways implemented by these microbial populations to sustain themselves in such environmental extremes of permafrost.

10.1 Introduction

Permafrost or perm(anent) frost, a term first coined by English-Russian geologist, Siemon W. Muller, indicates regions of subsurface soil that remain frozen throughout the year. The global distribution of these permafrost regions includes areas where the temperature rarely goes above the freezing point. Therefore, the major global permafrost belt includes major chunks of the regions of the Arctic Circle, major portions of

***Corresponding author: Bedaprana Roy,** Department of Microbiology, St. Xavier's College (Autonomous), Kolkata, West Bengal, India, e-mail: bedaprana@gmail.com
Debapriya Maitra, Snata Pandey, Jaydip Ghosh, Arup Kumar Mitra, Department of Microbiology, St. Xavier's College (Autonomous), Kolkata
Rajeshwari Podder, Tannishtha Das, Department of Biotechnology, St. Xavier's College (Autonomous), Kolkata

https://doi.org/10.1515/9783110793635-010

Alaska, the Siberian regions of Russia uplands, parts of China, as well as certain parts of Eastern Europe. According to geologists, the range of permafrost thickness can vary between 1 m and 1,000 m and covers around 8.8 million square miles across the Northern Hemisphere. These permanently frozen subsurface soil regions are hard without any energy flow, and are nutritionally depleted, yet these permafrost regions are known as the hub for microbial diversity. According to the research by Hultman et al. (2015), it was found that this microbial plethora not only performs their basal metabolic functions but also thrives and grows. This in turn indicates that the microbial colonies living in permafrost conditions are content with the suite of stressors associated with the subzero environment. Normally, at such a temperature range, the nucleic acid-based secondary structures get stabilized, which in turn greatly reduces the membrane fluidity of cells and the tertiary structural flexibility of proteins. According to De Maayer et al. (2014), at a temperature range of approximately −10 °C, the concentration of ROS and the extent of DNA, RNA, and protein damage increase considerably.

Therefore, the interest in microbial communities thriving in these regions has existed for decades; however, it was only after the intervention of high-throughput omic studies that scientists have been able to characterize and study the phylogenetic and functional diversity of these species. In this study, an attempt has been made to study the prevailing microbial diversity across the permafrost regions of the globe. An attempt has also been made to correlate various mechanisms of polyextremophily, such as biofilm acting as a protective matrix, two sensor-based systems, or novel membrane-stabilizing mechanisms, which play interesting roles in helping bacteria survive in the harsh environmental conditions of the permafrost.

10.2 Metagenomic studies of different permafrost regions of the world

10.2.1 Siberia

Siberian permafrost is one of the oldest and deepest permafrosts, frozen and undisturbed, dating back more than 30,000–1,000,000 years. With accelerated global warming, the rapid thawing of these ancient and unexplored layers of permafrost has compelled many researchers to begin an investigation of the obscured permafrost microbes residing in the depths of the permafrost before they vanish within the layers of melting ice. Several metagenome-based studies are being conducted today, aimed at inspecting and characterizing the permafrost microbes sooner or later to decipher the novel mechanisms that govern their survival in such frozen depths.

The Siberian permafrost consists of layers of ice frozen for over 1,000,000 years and serves as an important reservoir for ancient permafrost microbes. The Siberian

Kolyma-Indigirka Lowland region is an example of Siberian permafrost that is spread over an area of 100 km^2 and belongs to the late Pleistocene Era, offering the possibility of unearthing microbes, both living and fossilized, that have been preserved for more than 30,000 years. There have been many metagenomic analyses that have given promising results. In this segment, a discussion on the microbes isolated from the Siberian permafrost utilizing metagenome analysis has been done.

A metagenomic study of the Siberian Kolyma-Indigirka Lowland region, part of northeastern Siberia, was conducted by Sipes et al. (2021). They gathered samples from various depths and subjected the extracted DNA to metagenome analysis. When Sipes et al. cultured organisms from the Kolyma-Indigirka Lowland region, they were able to isolate microbes belonging to the bacterial genera Actinobacteria, Firmicutes, *Carnobacterium*, and Gammaproteobacteria, which are in agreement with previous data corresponding to 40,000–50,000 years old permafrost (Sipes et al., 2021). Nevertheless, culture-mediated analysis was not enough to inspect the majority of the permafrost microbial population; thus, they supplemented culturing data with extensive metagenome analysis.

Sipes et al. collected samples from the Kolyma-Indigirka Lowland region spread over seven different depths: 3.5, 7.2, 14.1, 14.8, 16.6, 17.2, and 20 m. Out of 33 metagenome-assembled genomes (MAGs) produced from the metagenomes collected from these samples, eight MAGs were of superior quality, with >80% genome completeness and <10% contamination. The eight MAGs were analyzed using the Genome Taxonomy Database Toolkit (GTDB-Tk), which placed them in five taxonomic groups: Aminicenantes, Actinobacteria, Atribacteria, and Chloroflexi, belonging to bacteria, and one group, Thermoprofundales, belonging to archaea (Sipes et al., 2021).

In reference to previous data, the archaeal order Thermoprofundales, belonging to the class Thermoplasmata, was predominantly found in the marine benthic region, and Atribacteria MAGs were dominant in non-permafrost locations like deep marine sediments and temperate soil. Chloroflexi were found to be abundant in a variety of locations like marine sediments, hypersaline environments, host-associated environments, and hot springs. Such unusual abodes suggest an idea of adaptation unique to these microbes that could explain their viability in subzero cryoenvironments.

In another metagenome analysis, Liang et al. coupled DNA repair strategies with a metagenomics approach to study the presence of both fossilized and living microbes in the permafrost. They included the extent of damage to the genomic DNA as one of their parameters for assessment, thus combining a metagenome study with the PreCR DNA repair protocol (Liang et al., 2021). The sample was collected from Cape Chukochii near the East Siberian Sea coast from three different depths, 3.4, 5.8, and 14 m, such that it spans three geological ages, including the mid-Holocene, late Pleistocene, and mid-Pleistocene.

Permafrost sediments from the three depths were analyzed and generated a total of 144 MAGs, out of which 52 MAGs characterized by >80% completeness and <10% contamination were selected for GTDB-Tk and phylogenetic analysis (Liang et al.,

2021). The 52 Mags were identified into eleven bacterial phyla – Candidatus Nomurabacteria, Firmicutes, Actinobacteria, Proteobacteria, Nitrospirae, Acidobacteria, Gemmatimonadetes, Chloroflexi, Bacteroidetes, and Spirochaetes, and three archaeal phyla – Euryarchaeota, Bathyarchaeota, and Heimdallarchaeota (Liang et al., 2021).

Additionally, Liang et al. presented a depth-wise distribution of the microbial population. MAGs from the Yedoma suite, reaching a depth of 3.4 m, were found to be abundant in Actinobacteria and had a lesser population of Bacteroidetes and Acidobacteria. MAGs obtained from a 5.8 m depth of the marine Kon'kovaya suite were dominated by microbes from the phyla Firmicutes, Actinobacteria, and Spirochaetes. A conglomerate of both marine and freshwater microbes predominated at a 14 m depth, including microbes from archaeal genera – Bathyarchaeota, Heimdallarchaeota, ANME-1, and Deltaproteobacteria, and two from bacterial genera – Chloroflexi and Gemmatimonas (Liang et al., 2021).

Another study involved samples IC4 and IC8 that were collected from the Kolyma–Indigirka lowland in Siberia and analyzed by Rivkina et al. (2016) who isolated and also characterized the permafrost-residing microbes into methanotrophic, nitrogen-fixing, and sulfate-reducing bacteria. Rivkina et al. (2016) collected the samples IC4 (depth 22.5 m) and IC8 (depth 16 m) from the floodplains of the Ambolikha River and Omolon River, respectively. The genomic DNA was aseptically isolated, and a metagenomic analysis was performed. Metagenome analysis showed that the most abundant microbes in the samples were bacteria, with Proteobacteria being the most abundant, followed by Actinobacteria, Firmicutes, Bacteroidetes, Chloroflexi, Planctomycetes, Acidobacteria, Cyanobacteria, and Verrucomicrobia (Rivkina et al., 2016). One archaeal phylum was identified in samples, represented by Euryarchaeota. Free-living eukaryotes like *Schizosaccharomyces pombe* were detected in both the samples IC4 and IC8, while the ciliate, *Paramecium tetraurelia*, was found only in sample IC4 (Rivkina et al., 2016).

Rivkina et al. isolated methane-oxidizing bacteria that comprised bacteria from both α-Proteobacteria and γ-Proteobacteria. The most abundant among the α-Proteobacteria were populations of *Rhizobiales, Methylocella, Methylosinus, Methylocystis,* and *Methylobacterium.* The γ-Proteobacteria were represented by the genera *Methylococcus, Methylobacter,* and *Methylophaga* (Rivkina et al., 2016). The abundance of the methane-oxidizing bacteria representing both groups was found to be higher in the IC4 sample compared to IC8.

Rivkina et al., in their analysis, further characterized certain sulfate-reducing and nitrogen-fixing bacteria associated with the permafrost samples IC4 and IC8. The sulfate-reducing bacteria included the bacterial genera *Desulfovibrio, Desulfatibacillum, Desulfococcus, Desulfobacterium,* and *Desulfomicrobium,* with a higher abundance in the IC4 sample (Rivkina et al., 2016). They isolated bacteria of the genera *Bradyrhizobium, Sinorhizobium, Rhizobium, Rhodospirillum,* and *Afipia* that could fix nitrogen. These nitrogen-fixing bacteria were found in a higher percentage in the sample IC4, while the symbiotic actinobacteria, *Frankia*, which is involved in root nodule-mediated symbiotic nitrogen-fixing in plants, dominated in the IC8 (Rivkina et al., 2016).

A metagenomic study of the subsurface of Siberian permafrost was conducted by Rigou et al. Results obtained from the metagenome analysis of the isolated DNA coincided with the former findings, revealing that 90% of the microbial population was bacterial, with Actinobacteria and Proteobacteria being the most abundant (Rigou et al., 2022). The archaeal population, dominated by *Nitrososphaeria*, *Bathyarchaeota*, and *Methanomicrobia*, showed an abundance of only 3.2% (Rigou et al., 2022). Analysis of the viral population revealed an extremely low abundance of 0.4%, dominated by groups of large DNA viruses, including Mimiviridae and Pithoviridae, which coincided with the previous permafrost analysis conducted by Legendre et al. (2014, 2015). Other viruses included Caudovirales, Phycodnaviridae, and Ascoviridae (Rigou et al., 2022).

These findings provide undeniable proof of the diversity and microbial richness that have somehow survived the harshness of the permafrost through all these years. Metagenomics has greatly aided in broadening the scope of identifying and characterizing microbes residing in these ancient cryoenvironments. Even under conditions of extremely low temperatures, water deficiency, and absence of an energy source, metagenome analysis of the samples has shown the plethora of microbial abundance that exists entrapped and frozen in time within the ancient Siberian permafrost. This is evident from the above list of microbes, accounting for not only fossilized microbes but a large population of living bacterial populations. The metagenomic analysis of the Siberian permafrost has unearthed some major bacterial genera dominating the permafrost of Siberia, along with a few representatives from the archaeal phylum. Although viral abundance in the Siberian permafrost was analyzed to be minimal, a few large DNA viruses seem to have made the list. The aim of the metagenomic analysis does not end with the identification of these microbes but to further our understanding of their mechanisms of survival, their metabolic processes under conditions of negligible nutrients, and their impact on today's climate.

10.2.2 Alaska

Almost 85% of Alaska is covered with permafrost. It is present almost everywhere, and it is thickest and most extensive, extending as far as 2,000 feet below the surface of the Arctic Coastal Plain in Arctic Alaska to the north of the Brooks Range. With the exception of sections of high-elevation alpine permafrost, it gets thinner and more discontinuous as it moves south from the Brooks Range, punctuated by areas of unfrozen ground known as taliks.

Widespread permafrost in the United States causes classes of geologic hazards that are mostly exclusive to Alaska. The thawing of permafrost, which is structurally significant to the soils of Alaska, results in landslides, ground subsidence, erosion, lake disappearances, the formation of new lakes, and other environmental problems.

The major reasons why Arctic soils have received so much attention recently are likely their high levels of organic carbon storage and the unusually quick warming of this region as a result of climate change (Steven et al., 2009).

Despite the permafrost soil's extended subzero temperatures and limited water supply, a variety of microorganisms live there. In permafrost soils, the bulk of the microorganisms are heterotrophs adapted to cold, belonging to the bacterial phyla Proteobacteria, Firmicutes, Bacteroidetes, Actinobacteria, Acidobacteria, and Chloroflexi. Other microorganisms are from the archaeal phyla Euryarchaeota, Crenarchaeota, and Thaumarchaeota. Recent research has shown that a variety of abiotic parameters, including soil pH, moisture, conductivity, the quality of the soil's organic matter, and the availability of nutrients, have an impact on the microbial community's composition along the permafrost thaw gradient (Tripathi et al., 2018).

A study conducted in four locations in Alaskan permafrost by Johnston et al. revealed that microbial communities in the Alaskan tundra are less diversified and more homogeneous than previously thought. There is extensive OTU-sharing among tundra areas. The population bins comprised several taxonomic groups, and their metabolic activities are geared toward sulfur cycling, methanotrophy, hydrogen metabolism, and organic matter oxidation. Together, they accounted for around 10% of the metagenomic datasets.

Specifically, methanogenic archaebacteria groups – Methanobacteria and Methanomicrobia – were detected in all the tundra locales. Class Ignavibacteria under Chlorobi, class MCG of Crenarchaeota, Nitrospirae family Thermodesulfovibrio (obligatory anaerobic sulfate reducer), class Dehalococcoides under Chloroflexi, and phylum Lentisparae are further groupings that are unique to tundra soils. They also observed that the CiPEHR soil communities were dominated by the phylum Acidobacteria.

Extended generation times, stringent metabolic conditions (e.g., oxygen-deficient conditions or lower temperatures), and/or a reduced diversity of organic carbon derived from plants may all contribute to the tundra's lower diversity (Johnston et al., 2016).

For another investigation, four sites along the Trans-Alaskan Pipeline were chosen. One of these locations was in the tundra close to Sag River (TS), two other sites were in the north and south of the Yukon River (NY and SY) in the black spruce boreal forest, and the final location was at Poker Flat Research Range (PF) in an interior black spruce forest with sphagnum moss.

Bacterial and archaeal abundances are present in the TS core, decreasing with depth and peaking in abundance at the surface soils. The profusion of microorganisms and previous investigations have already noted the presence of more archaea in top soils than in deeper soils. Bacterial and archaeal abundances present in the NY and SY cores were found to vary in abundance at different soil layers but were generally rather uniform throughout the soil layers' depths. Throughout the depth of the PF core, bacterial abundance remained largely stable, whereas archaeal abundance was maintained until a depth of 80 cm before it substantially decreased at 60 cm.

Actinobacteria (23.9% on average) were the bacterial phylum that was most prevalent across all samples, followed by Proteobacteria (15.3%), then Acidobacteria (14.7%), Chloroflexi (12%), Firmicutes (5%), Bacteroidetes (4.9%), Verrucomicrobia (3.6%), Gemmatimonadetes (2.9%), Planctomycetes (1.7%), Caldiserica (1.7%), and AD3 (0.9%).

As depth increased, the structure of the bacterial phylogenetic community became less clustered. As nutrient levels often decrease from the top soils to the deeper layers, it is possible that the decline in relative abundance with depth of Bacteroidetes and Proteobacteria is connected to their life strategies, which are mainly copiotrophic. It was also discovered earlier that Verrucomicrobia and Planctomycetes predominate in peat soils of the surface, in contrast to mineral soils, which are at a greater depth. The reduction in Acidobacterial relative abundance may be attributed to the rise in soil pH, as that increases with depth. Cyanobacteria, a group of photosynthetic bacteria, may have a declining relative abundance with depth due to less sunshine exposure.

Additionally, in line with other research, Actinobacteria and Firmicutes, rose in relative abundance with depth, as they may also be crucial players in the anaerobic decomposition of soil organic carbon at greater depths. Chloroflexi, a group which was previously demonstrated to predominate in the deeper layers of the soil, interestingly displayed a different trend with depth between samples. While members of the family Anaerolineaceae (which are found frequently in areas enriched in petroleum and can potentially degrade hydrocarbons in anoxic conditions) make up the majority of the Chloroflexi in the upper layers of the TS core, the deeper layers of the PF core were dominated by an unidentified Chloroflexi lineage (GQ396871 f).

Methane peaks were discovered at different depth levels in two soil cores, and there was a substantial correlation between the relative abundance of two taxa of methanogenic archaea already known (*Methanosarcina* and *Methanobacterium*) (Tripathi et al., 2018).

Additionally, the Matanuska Glacier basal ice layer (Kayani et al., 2018) and Byron Glacier (Choudhari et al., 2014) can be considered permafrost layers since they remain frozen, and studies have been conducted there too for microbial diversity.

Unexpectedly, an abundant microbial ecosystem with over 2,500 bacterial species and even Archaea, a group that had thus far eluded discovery in glaciers of the Northern Hemisphere, was found in the Byron Glacier. Bacteroidetes, Firmicutes, and Proteobacteria are the main contributors to the species diversity, according to the analysis of glacial ice taxonomic composition, which is based on the 16S rRNA gene. At the individual species level, *Cyanobacteria, Actinobacteria,* and *Planctomycetes* had the highest relative cellular abundance. This is probably due to the ecosystem's reliance on the photosynthetic energy produced and its close links to the soil microbial population (Choudhari et al., 2014).

According to the 16S rRNA sequencing results of the samples obtained from the basal ice layer of Matanuska Glacier, the families Comamonadaceae and Desulfobulbaceae had the highest abundance. Members of the Desulfobulbaceae and Syntrophaceae were concluded to be metabolically active. In keeping with the 16S rRNA data, the results of the metagenomic sequencing revealed that the microbial community was dominated by bacteria and that the families Hydrogenophilaceae, Comamonadaceae, and so on were highly abundant (Kayani et al., 2018).

The overall microbial diversity across Alaskan permafrost is similar according to different studies, and it varies to some extent from the glacial permafrost findings.

The thawing of permafrost causes the decomposition of both fresh and old soil carbon stocks, which releases carbon that had been stored for several hundreds to thousands of years into the atmosphere (Mackelprang et al., 2016). This is a serious concern and impacts microbial diversity, which in turn can add to the detrimental effects of permafrost thaw.

Younger, thawed permafrost has shown a much higher bacterial richness, with Firmicutes, Actinobacteria, Chloroflexi, Deltaproteobacteria, and Alphaproteobacteria showing the greatest increases in richness. As the permafrost thawed, the organization of the bacterial community gradually changed, and determinism contributed more. Permafrost thawing had a bigger impact on bacterial diversity and community structure than permafrost age. The findings suggest that permafrost microorganisms of different ages respond to thawing in various ways, which ultimately results in varied bacterial community compositions and various organic carbon breakdown mechanisms in Arctic permafrost (Ji et al., 2020).

10.2.3 The Arctic

Arctic permafrost has been surviving for 11,700 years (Mackelprang et al., 2017). The studies on the Arctic permafrost are limited compared to those of Siberian and Alaskan permafrost. However, its study is significant with rising global temperatures, as the ice thaws, enabling microbial degradation of the previously protected organic matter, releasing CO_2 and CH_4 into the atmosphere, further boosting climate change. Identifying the microbiology operating in the Arctic permafrost could help understand the impact of climate on the regional microbial community and vice versa.

10.2.3.1 Microbial communities in different Arctic regions

A common trend in archaeal and eubacterial communities has been observed after analyzing several independent studies conducted in different parts of the Arctic, along with the discovery of novel species. The DNA analyzed might be extracellular (preserved at subzero temperatures) or intracellular (from dormant or metabolically active bacteria).

From the Markham and Ward Hunt ice shelves in the Canadian high Arctic, the major phyla obtained were eubacterial Bacteroidetes, Proteobacteria, and Actinobacteria, and a novel phylotype of archaeal Euryarchaeota, whose metabolisms were active at −10 °C (Bottos et al., 2008). From a 9 m deep Eureka core, Eur1, in the Canadian high Arctic, 16 isolates belonged to the Firmicutes, with 3 genera: *Bacillus*, *Sporosarcina*, and *Paenibacillus*; other phyla were Actinobacteria (genera: *Arthrobacter*, *Micrococcus*, *Rhodococcus*), Proteobacteria (genus: *Pseudomonas*), Euryarchaeota (genera: *Halobaculum* and *Haloarcula*), and Crenarchaeota (Steven et al., 2007). Figure 10.1 shows the relative abundances of the major eubacterial (Chart A) and archaeal (Chart B) phyla prevalent at the Eur1 core.

A vertical section of permafrost core, Eur3, collected from the Canadian high Arctic, had four distinct horizons, where the concentration of DNA decreased with depth, limiting the study to a 7 m depth (Steven et al., 2008). The surface active layer (which seasonally freezes and thaws) primarily contained the phyla Actinobacteria and Betaproteobacteria, and the genera *Dietzia* and *Rhodococcus*; the permafrost table (1 m deep) mainly contained Actinobacteria, the Cytophaga-Flavobacterium-Bacteroides group, Alphaproteobacteria, and some anaerobic genera like *Geobacter*; the permafrost (2 m deep) majorly contained Betaproteobacteria, CFB, and Actinobacteria groups, with the most common genus being the anaerobic iron reducer *Rhodoferax*; all the depth-levels of massive ground ice showed a similar DGGE fingerprint with low biomass, and Firmicutes were predominant, followed by Bacillaceae and Actinobacteria; *Clostridium* was significant (Steven et al., 2008). Archaeal diversity was more prominent in the ground ice and less varied in the top horizons in contrast to eubacterial diversity – in the active layer and permafrost table, Crenarchaeota was predominant; the permafrost contained the Euryarchaeota genus *Halorubrum*, and the ground ice contained both phyla with the methanogen *Methanobrevibacter* (Steven et al., 2008). Methanogens and type 1 methylotrophs like Methanomicrobia and Methylococcaceae genera were most prevalent (Steven et al., 2008). Thus, eubacteria dominated up to 2 m in the permafrost with significant Actinobacteria, while archaea and fungi were detected in much lower abundance (~200–10,000 times less) (Yergeau et al., 2010). Figure 10.2 shows the relative abundances of the major eubacterial (Chart A) and archaeal (Chart B) phyla prevalent at the Eur3 core.

In the high Arctic permafrost of Spitsbergen, Norway, Actinomycetes was the predominant phylum, with *Arthrobacter* and *Cellulomonas* being the predominant genera, and very few spore-formers; 41% of the isolates were also present in other cryohabitats, including Antarctica (Hansen et al., 2007). Methanogenic Euryarchaeota, Methanomicrobiales, Methanobacteriaceae, Methanosaeta (capable of both hydrogenotrophic and acetoclastic pathways), and Group I.3b of the uncultured Crenarchaeota were detected at two wetlands in this region (Høj et al., 2005). Figure 10.3 represents the relative abundances of the various prevalent Eubacterial phyla at Spitsbergen, Northern Norway.

In the northeastern Arctic Tundra, the Archaea in the five Arctic methane-containing permafrost samples of different ages – one from the active layer, two from the transition zone, and two from permafrost depth – were denoted. With depth, the

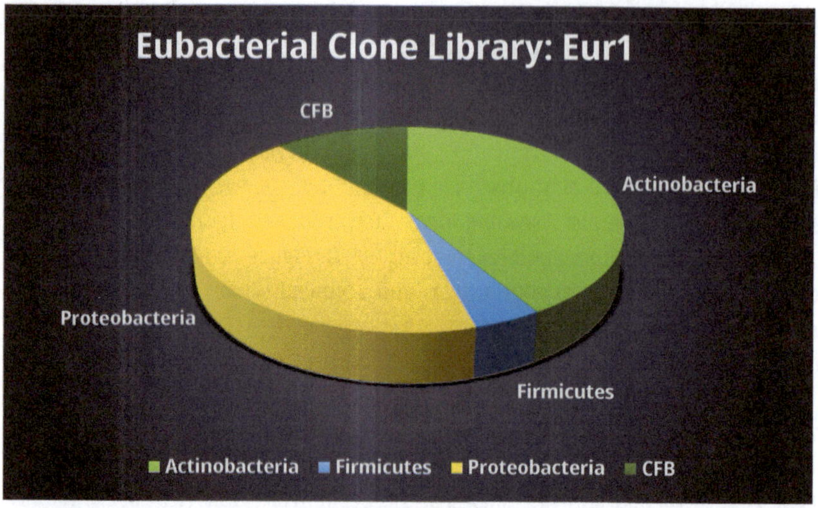

Chart (A)

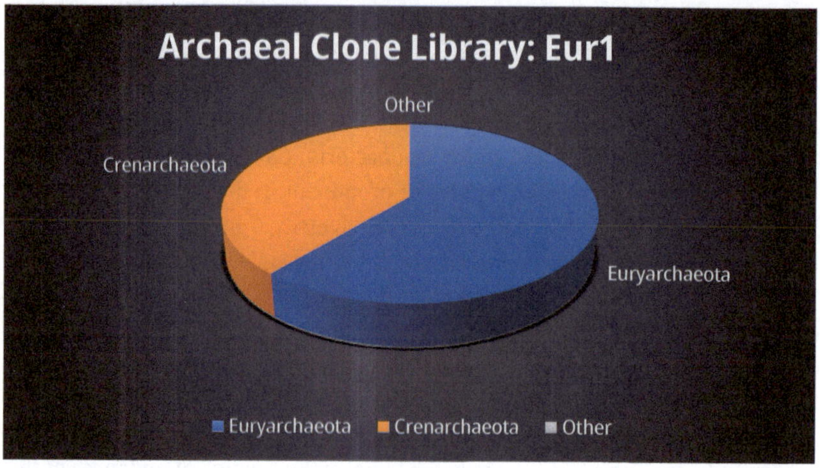

Chart (B)

Figure 10.1: The relative abundances of eubacterial (A) and archaeal (B) communities at the Eur1 core.

Euryarchaeota-related phylotypes decreased, and the Woesearchaeota diversity increased. Methanomicrobiales, Methanosarcinales, Methanocellales, and Methanobacteriales orders were found (Shcherbakova et al., 2016). Figure 10.4 represents the prevalent archaeal phyla in the studied samples of the northeastern Arctic Tundra.

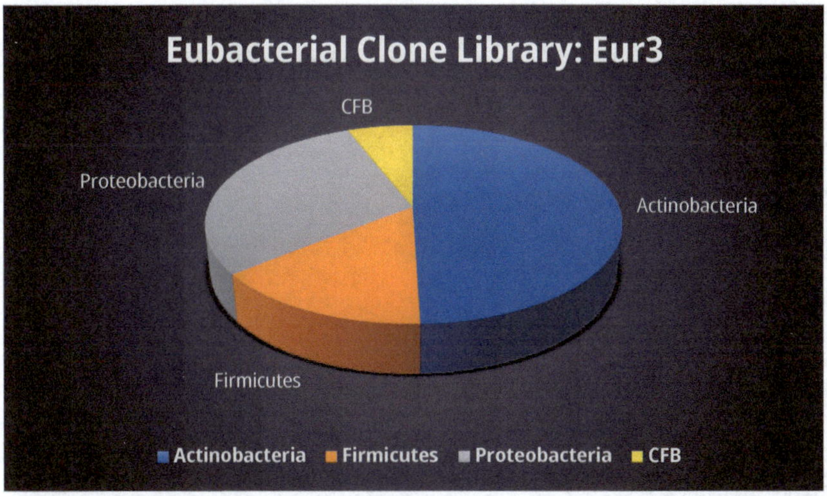

Chart (A)

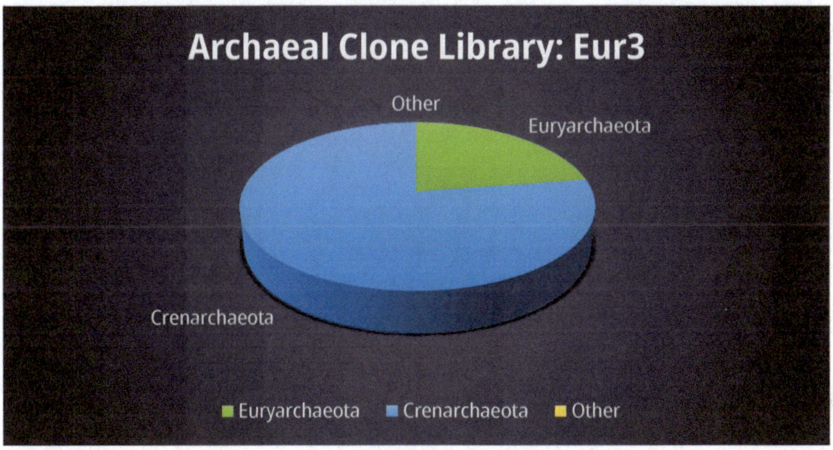

Chart (B)

Figure 10.2: The relative abundances of eubacterial (A) and archaeal (B) communities at the Eur3 core.

10.2.3.2 Microbial characterization

Isolates from the permafrost were mostly gram-positive, with some endospore-forming gram-negative bacteria. They were aerobic or anaerobic, cold-tolerant, and salt-tolerant. Of the isolates from the Eur1 and Eur3 cores, three to six grew at −5 °C. Permafrost selects its resident microbes (99.99% unculturable) based on its physical characteristics like salinity, permanently <0 °C, limited nutrition and water activity, and background radiation.

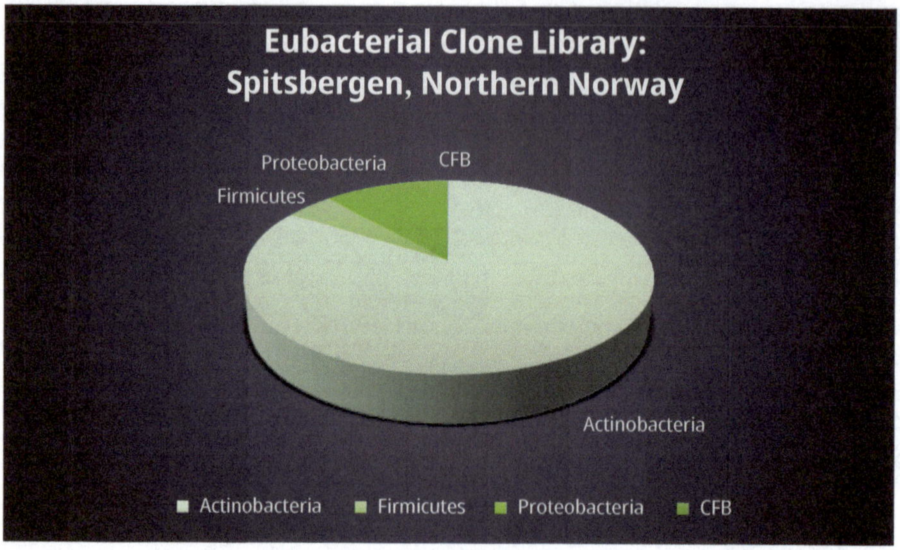

Figure 10.3: The relative abundance of eubacterial communities at Spitsbergen, Northern Norway.

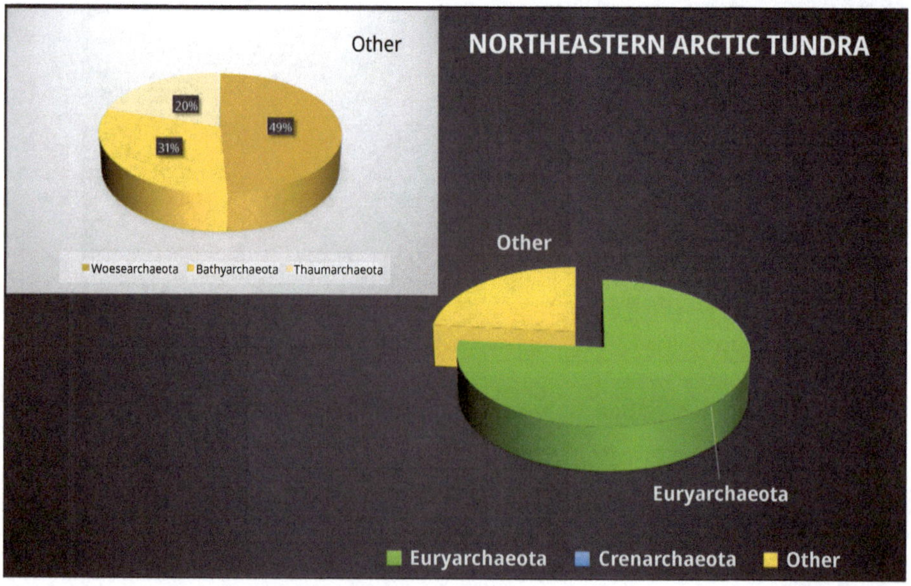

Figure 10.4: The relevant abundance of archaeal communities in the northeastern Arctic Tundra.

Active layer microbes performed aerobic respiration, methane, and ammonia oxidation, mainly up to 5 cm below the surface, while permafrost microbes were anaerobic heterotrophic or fermenting Bacteroidota, and many could degrade plant polysaccharides and metabolize nitrate (Wu et al., 2021). Some photosynthetic cyanobacteria and microalgae may be present in the Arctic tundra permafrost.

In the Eur3 core, the active layer communities could not grow in the presence of sucrose, although they could survive at −5 °C. In its permafrost layer, growth was possible at 5 °C in 7% (w/v) sucrose but not at −5 °C. This layer is highly saline, and so all its salt-tolerant isolates grew in 5% NaCl (Steven et al., 2008). Acetate and glucose were mineralized equally without preference, except in the Eur1 core where acetate was preferred at 5 °C. The Eur1 isolates mineralized them at high rates at 5 °C, but the rates dropped below 0 °C, while the Eur3 active layer and permafrost isolates mineralized them at −15 °C, and ground ice isolates mineralized at 5 °C only (Steven et al., 2008).

In the Eur3 core, the gene for methyl-coenzyme M reductase was absent, while genes for trehalase, chitinase, β-glucosidase, and β-galactosidase were detected in the permafrost and active layers. Organic matter-degrading Actinobacteria were crucial in the C cycle (Yergeau et al., 2009).

Mesorhizobium loti from the active layer or *Anabaena variabilis* from the permafrost showed the nifH gene, while ammonia-oxidizing *Nitrosomonas, Nitrosopumilus*, and so on from the two layers were derived, with the ammonia monooxygenase gene seen in the active layer (Yergeau et al., 2009).

At the physiological level, five representative Arctic bacteria showed cold shock proteins, RNA helicase and other enzymes, higher transposase copy numbers, and cold-adaptive protein flexibility (Raymond-Bouchard et al., 2018).

High-Arctic bacteria may have roles in hydrocarbon bioremediation, as their community structure, functions, and genome expressions are altered with bioremediation treatments after a lag phase, as well as with the effects of pollution and remediation depending on the localized site and treatment introduced (Yergeau et al., 2009).

10.3 Key stone classes and species of the microbiome inhabiting permafrost

10.3.1 The etymology

The percentage of different microbes, their variety, and the metabolism they follow in any permafrost will depend on factors like age, the layer (ground surface to 400 m deep cores), and physical and chemical parameters like salinity, temperature, and aeration of the permafrost being studied, and whether the permafrost is frozen or thawed. We will discuss the major psychrophilic and psychro tolerant eubacterial and archaeal communities, which are millions of years old, prevalent in the northern permafrost.

10.3.1.1 Actinobacteria (gram-positive, high GC content)

Actinobacteria is a predominant phylum in the northern permafrost, presenting 30 important permafrost-residing genera isolated from Canada, Siberia, and other places, including *Arthrobacter, Streptomyces, Nitrobacter, Rhodococcus, Microbacterium, Micrococcus,* and *Mycobacteria,* most of which are aerobic. Their count varies with temperature, oxygen content, and differences in carbon, phosphorus, and nitrogen levels, and their role in cellulose, chitin, and lignin degradation is crucial (Vašková, 2014). The exact molecular biology of actinobacterial survival in permafrost remains largely unknown, although it is known that the phylum produces cyst-like resting "arthrospores" (in *Arthrobacter* and *Micrococcus* isolated from permafrost) (Bottos et al., 2008). Two mechanisms for survival for the phylum include osmotolerance in the permafrost salinity and active metabolism for quick DNA repair.

Streptomyces are aerobic, filamentous spore-producers, producing bioactive antibacterials (chloramphenicol), antibiotics (amoxicillin), antifungals (nystatin), and so on, H^+ ions to buffer soils, and decomposing organic matter (Vašková, 2014). A psychrotolerant *Streptomyces avidinii* was further isolated from the Arctic Ocean, which produces the antimicrobial di-(2-ethylhexyl) phthalate (DEHP) (Lyutskanova, 2009).

Arthrobacter are facultatively anaerobic rods (exponential phase) or cocci (stationary phase) isolated from the Arctic permafrost, some of which could be helpful in chromium bioremediation; *Arthrobacter chlorophenolicus* degrades xenobiotics and pesticides (Vašková, 2014).

Cutibacterium, a common inhabitant of human skin, was also obtained – it's one complete nuclease gene had 100% similarity to *Cutibacterium* acnes, and the entire MAG had 99.72% average nucleotide identity with certain other human bacteria (Sipes et al., 2021).

10.3.1.2 *Firmicutes* (gram-positive, low GC content)

Firmicutes increased with the age of Alaskan permafrost, while *Actinobacteria* decreased (Mackelprang et al., 2017). The most important are the spore-forming classes of clostridia and bacilli. Many *Firmicutes* employ genes for osmoprotection and quick DNA repair for survival through geological times.

Clostridium tagluense sp. nov., isolated from the Canadian high Arctic, was anaerobic, psychrotolerant, and spore-forming, with a sequence similarity of 98.5% with *C. bowmanii.* Its physiology contained specific fatty acids and a 31.5 mol% GC content, and its metabolism included the utilization of specific carbohydrates (like glucose, fructose, trehalose, maltose, and starch) as its energy and carbon source, and gelatin hydrolysis (Suetin et al., 2009).

Desulfosporosinus was the only Firmicutes MAG isolated from Svalbard permafrost capable of fixing nitrogen (Xue, 2020).

10.3.1.3 Proteobacteria

Different permafrost microbial operators of the nitrogen cycle occur in the classes Alphaproteobacteria and Betaproteobacteria. The permafrost soils contain dissolved inorganic and organic nitrogen. The important isolates of the class Alphaproteobacteria are *Sphingomonas* and *Nitrobacter*, while those of the class Betaproteobacteria are *Alcaligenes, Gallionella, Nitrosomonas*, and *Nitrosospira.*

Nitrosospira contains a majority of nitrifiers (nitrifying and controlling the dissolved inorganic nitrogen levels in the permafrost in the form of ammonium and nitrate). It is hypothesized as a potent ammonia oxidizer as well (Sanders et al., 2019).

The Gallionella genus showed pathways for both assimilatory as well as dissimilatory sulfate reduction to H_2S (Xue, 2020). The most abundant genera in the class Gammaproteobacteria were *Psychrobacter, Xanthomonas, Pseudomonas, Acinetobacter*, and so on. *Pseudomonas* isolates (7 out of 10) showed the highest number of peptidase genes and the highest activity of leucine aminopeptidase (Sipes, 2022).

Psychrobacter arcticus strain 273-4 (from Kolyma, Siberia) genome sequence shows its employment of cold-survival strategies, including membrane alterations through saturation and desaturation of membrane lipids, use of acetate as a carbon source, and cold shock proteins, RNA and protein chaperones, reduced use of proline and arginine to ensure protein flexibility and hence growth at low temperatures, and so on (Ayala-del-Río et al., 2010). Moreover, a new multidrug-resistant plasmid, pKLH80, was isolated from *Psychrobacter maritimus* MR29-12 of the ancient permafrost with antibiotic resistance cassettes, including β-lactam resistance (Petrova et al., 2014).

Acinetobacter lwoffii strains collected from Kolyma permafrost contained seven plasmids bearing proposed cold-adaptive genes, such as resistance to heavy metals and organic peroxides, potassium and sulfate uptake, and so on (Mindlin et al., 2019).

10.3.1.4 Acidobacteriota

Several *Acidobacteria* were prevalent in the soil sample from a thawed and drained Alaskan coast, and sequences corresponding to *Solibacter usitatus, Terriglobus saanensis,* and *Geothrix fermentans* were isolated; these performed anaerobic metabolism, specifically iron reduction, in the reducing permafrost (Lipson et al., 2013).

Geothrix fermentans was characterized as an iron-reducer using specific decaheme cytochromes (Lipson et al., 2013). *Candidatus Koribacter usitatus* produced three sequences that hinted at the production of acidobacterial conductive biofilms in the soil (Lipson et al., 2013).

10.3.1.5 *Bacteroidota*

Bacteroidota showed potential for complete denitrification in the deep Svalbard permafrost. Important genera isolated from the Siberian permafrost included *Sphingobacterium* and *Flavobacterium.* Some can produce acetate in the environment from hydrogen and carbon dioxide for subsequent acetoclastic methanogenesis.

10.3.1.6 Phototrophs

From a sample collected from Siberia, *Oscillatoria* (non-heterocyst-forming cyanobacteria) and *Chlorella* (green algae) were isolated (Vishnivetskaya & Rivkina, 2021). Chloroflexi, showing different metabolisms including photoheterotrophy, isolated from Siberia, showed genes for trehalose and maltose hydrolase and the mevalonate pathway (Sipes et al., 2021). Isolates from Norway showed 14 enzymes, including D-xylose transport ATP-binding protein XylG and beta-hexosaminidase (EC 3.2.1.52) (Xue, 2020).

10.3.1.7 Archaeal communities

The salinity of permafrost supports halophilic and methanogenic archaea. *Thermoprofundales* was the only isolate from Siberia, with more genes for sugar and inorganic ion metabolism and transport, pyruvate kinase, cellobiose phosphorylase, and antiporters than in non-permafrost genomes (Sipes, 2021). Methane-trapped permafrosts of the Arctic Tundra showed methanogens like acetoclastic genera *Methanosarcina* and *Methanosaeta*, and hydrogenotrophic genera *Methanoregula* and *Methanocella* (Shcherbakova, 2016). Nonmethanogenic Bathyarchaeota were observed in all samples of the Arctic tundra, from the active layer to the deeper core (Shcherbakova, 2016).

10.3.2 Genetic makeup and diversity

Various metagenomic studies have managed to identify the genes that are dominant in the psychrophiles inhabiting the permafrost regions.

On a cellular or subcellular level, many permafrost microorganisms exhibit a variety of adaptations to subzero, oligotrophic, high salinity, and cumulative radiation environments, including the following: spore formation and dormancy, increased concentrations of intracellular compatibility, increased membrane fluidity, amino acid composition favoring protein flexibility, upregulation in the expression of stress-responsive genes, proteins, and transcripts involved in survival and maintenance, production of osmotic stress and cold stress proteins, temperature-related translation factors, increased dependence and inclination on horizontal gene transfer, environmental sensing and chemotaxis, and detrital biomass scavenging. Because of this increased genetic shift toward investment in stress response and maintenance, a large percentage of the species in the population are specialized for survival (Ernakovich et al., 2022)

Arctic soils have been revealed to have a variety of essential genes involved in carbon decomposition. For instance, Yergeau et al. (2010) discovered genes in the active layer as well as the permafrost that were involved in the breakdown of starch, lignocellulose, chitin, and trehalose. The potential in the genetic makeup to break down polymers from plants, which make up the majority of the carbon in peat, was thoroughly examined in the active layer study from two locations of Arctic peat. Researchers found the number of CAZy glycoside hydrolase families to be 73. (Endo hemi) cellulase families and debranching enzymes were also often seen, with oligosaccharide-degrading families being the most prevalent.

An Alaskan permafrost metagenome, which extensively represented genes implicated in the butanol pathway, revealed fermentation as a frequent microbial activity (Lipson et al., 2013). Alterations in the environment and parent soil chemistry can be linked to the quantity of genes involved in carbon metabolism in some instances.

Methanotrophic species were present because peptides and transcripts from the metaproteome and metatranscriptome emanated from genes crucially involved in the CH_4 oxidation pathway. Both the production of methane and the dynamics of consumption are anticipated to develop as the ground thaws as a result of interactions between microbial, vegetal, and hydrological elements (Mackelprang et al., 2016).

The active layer, as well as the permafrost, has high levels of a critical gene (nifH), which encodes a component of the nitrogenase complex that converts N_2 to ammonia. Additionally, it was shown that the prevalence of genes involved in nitrogen fixation was relatively high in frozen permafrost but drastically diminished following a brief thaw (Mackelprang et al., 2011).

Contrastingly, nitrate reduction, denitrification, and ammonification genes were more prevalent after thawing. The expression of narG (coding for nitrate reductase) was noted in permafrost. However, no proteins involved in denitrification were found, which is in keeping with the modest rates of denitrification that were observed. It is interesting to note that they discovered transcription and translation of the genes for glutamate synthase (gltS, gltB, and gltD) and glutamine synthetase (glnA). The findings collectively imply that permafrost cells were ready to absorb nitrogen (Mackelprang et al., 2016).

Microorganisms in mineral permafrost soils can use the high abundance of iron reduction genes to their advantage because it appears to be the anaerobic pathway that is primarily active in the permafrost and active layer (Lipson et al., 2013).

The presence of microorganisms using sulfate as a TEA was also discovered in the permafrost, thanks to the existence of genes related to sulfate reduction (Chauhan et al., 2014; Lipson et al., 2013).

This energy-limited, anoxic, and cold environment developed sugar metabolism that was deficient, and high levels of import, degradation, as well as salvage genes, indicating the recycling followed by utilization of detrital biomass as a carbon and nitrogen source across geologic timeframes, similar to deep seafloor deposits. Additionally, there were more aromatic hydrocarbon degradation genes in older permafrost (Mackelprang et al., 2017).

Chitin, cellulose, simple carbohydrates (average 81.2% for monosaccharides; 40.1% average for sugar acids; 68.6% average for sugar alcohols), and lipids like phospholipids and triglycerides were among the soil organic matter (SOM) degradation genes that the Arctic soil metagenomes were significantly abundant in (Johnston et al., 2016) These results were also in agreement with earlier research on Arctic tundra soils that discovered many sugar alcohol, chitinase, mono-, and disaccharide degradation genes in metagenomes and meta transcriptomes (Yergeau et al., 2010).

Genes involved in the catabolism of phenolic and lignin compounds, genes corresponding to the Wood-Ljungdahl carbon fixation pathway (carbon monoxide dehydrogenase/acetyl-CoA synthase subunit alpha, and 5-methyltetrahydrofolate: corrinoid/iron-sulfur protein co-methyltransferase proteins) are specifically used by acetate-producing bacteria and methanogens, and they were found to be abundant too.

The metagenomes' protein annotation revealed the presence of genes related to methanotrophy, such as those for the methanol and methylamine pathways (mttC1, mtbA, and mtaB). In addition, it had genes for the breakdown of a number of organic substances, such as D-glucuronate, L-arabinose, rhamnose, cellulose, xylan/xyloglucan, and ribose, as well as molecular H_2 metabolism (hupE, hndACD, hypABCDEF, hoxHY).

It also contained genes like sbp, cysATW, cysDGHIJK, and nodQ involved in sulfate reduction for assimilation and transport.

Some regions shared a large number of functional genes, including those needed for "normal prokaryotic TCA metabolism" and the TCA cycle or citric acid cycle with glyoxylate bypass.

They also contained genes for the utilization and degradation of trimethylamine (mttBC, etfABD, and mauG) as well as a transporter of ammonium ion on the same operon (nrgA). Additionally, both bins contained several genes involved in the degradation of many SOM constituents, including the degradation of fatty acids (fadADEN, lcf, and ech), rhamnose import and catabolism (rhaB, rhaT, and rhaM), genes for the breakdown of formate (fdhE, fdoGI, and fdnHG), succinate (sdhABC), and a number of other compounds, including chitin and *N*-acetylglucosamine (its derivative). Additionally, the genes involved in anaerobic methanotrophy, fermentative pathways, or sulfur reduction – B-coenzyme M heterodisulfide reductase (hdrABC), as well as additional methane-related coenzyme activities (cofDEGH) – were found. The genes hcrAB and bcrABCD had the ability to degrade aromatic chemicals both anaerobically and aerobically. There were genes for motility proteins (MotAB), bacterial chemotaxis (cheABRWY), and mutual gliding protein (MglA). The genome in another region contained genes necessary for flagella assembly as well as motility proteins (MotAB) (flhAB, flgBCL, and fliEGHKMOPQR). It also carried genes for polysulfide reduction utilizing molecular H_2 (shyACD), sulfate binding, import, and assimilation (sbp, cysATW, nodQ, cysH, and sir1), trimethylamine dehydrogenase (tmd), as well as genes for the breakdown of oligopeptides, fatty acids, lipopolysaccharides, mannose, chitin, galactose, pyruvate, succinate, and acetate (Johnston et al., 2016).

In comparison to their non-permafrost counterparts, the permafrost MAGs had higher proportions in the clusters of orthologous group energy generation and conversion categories and carbohydrate metabolism and transport. They also contained genes that were less common in genomes of nonpermafrost, such as those for mevalonate biosynthesis, trehalose synthesis, thymine metabolism, and cellulose breakdown. As a result of adaptation to the anoxic, highly ionic-strenuous, and freezing conditions of permafrost brine films, membrane stabilization and osmotic stress responses involve several of these genes. Our findings indicate that the DNA in this old permafrost is of a high enough quality to allow the assembly of MAGs from microbes that have been modified to survive in this harsh environment and tolerate long-term freezing.

As in Atribacteria and Thermoprofundales, several of the genes exclusive to the Siberian MAGs were involved in transport, osmoregulation, and carbohydrate use. The Chloroflexi MAG, Thermoprofundales MAG, and Atribacteria MAG all contain several genes and enzyme kinases for the mevalonate pathway. The manufacture of isoprenoids, which have been demonstrated to stabilize membranes and increase the organism's survival at low temperatures, occurs via the mevalonate pathway using acetic acid. Three Na1/H1 antiporters and osmoregulators (NhaD, MnhC, and MnhG), as well as a biotin transporter that functions in high-ionic-strength environments, were annotated similarly by all groups. Genes for cellobiose phosphorylase and carbamate kinase were found which help them survive in low-energy anoxic environments (Sipes et al., 2021).

Diamond BlastP against RefSeq identified 1,071 β-lactamase genes, which were then verified by a domain search. A different metagenomic study of the ARG distribution in 17 uncontaminated Antarctic surface soils gave an estimate of the relative frequency of ARGs to all genes in the range, which roughly equates to 17% of soil bacteria that encode at least one ARG. Additionally, this fraction comprises all ARGs, of which β-lactamases only make up a minor amount, including the aminocoumarin-resistant alanyl-tRNA synthetase, aminoglycoside acetyltransferase/nucleotidyl transferase, and the multidrug resistance efflux pump.

The highly extreme warming of the Arctic may increase the spread of bacterial antibiotic resistance, the present's main public health concern, in addition to established environmental hazards. Additionally, β-lactamases were frequently found in other kinds of soil, pointing to a broader function for these enzymes in controlling bacterial populations.

Although there is little chance that the ancient microbes preserved in the undisturbed permafrost strata investigated here will come into direct contact with people, there is still a chance that they could lead to new β-lactam resistances if they come into contact with modern bacterial species.

Their spread to modern harmful bacteria in a clinical setting could worsen the antibiotic resistance dilemma, which is currently regarded as one of the greatest public health problems of our day (Rigou et al., 2022).

Global warming, giving rise to rapid permafrost thaw, has become critically alarming at this point. In terms of projections of permafrost C breakdown, terrestrial feedbacks to the climate, the post-thaw microbiota, and greenhouse gas release represent a crucial unknown. When permafrost thaws, it has occasionally been observed that C degradation genes are rapidly upregulated, the methane flux in the soil fluctuates depending on the state of the microbial community, and the C:N ratio is disturbed, all of which can have long-term unfavorable effects in the future (Ernakovich et al., 2022).

10.3.3 Mechanisms and pathways of survival

Perennially frozen over millennia, the permafrost is marked by extreme subzero temperatures with little or no energy and water reserves, complemented by elevated salinity. It is hard to imagine life thriving within these layers of ancient ice, devoid of any luxuries of sustenance. Yet several decades of sampling and investigation have proven the existence of a diverse microbial population not just surviving the stressors but also growing and multiplying. The successful identification of several microbial communities within the permafrost around the world has set in motion the investigation of their survival strategies in such an oligotrophic cryoenvironment.

Permafrost microbes have equipped themselves with considerable mechanisms to survive the innumerable stressors that persist within the layers of permafrost. To survive in the frozen, oligotrophic, water-deficient, and highly saline environments, most microbes have adapted not just one but multiple strategies. Many of the microbial populations are therefore polyextremophiles, with intriguing adaptive strategies that could answer the questions of survival. Microbes thrive despite the permafrost's shortcomings thanks to extensive cellular and molecular modifications. A discussion of the possible survival mechanisms based on the various studies that have been conducted to date will be presented in the upcoming segment.

The freezing temperature of the permafrost, reaching an average of −10 °C, does not make it easy for the membrane to retain its fluidity, but rather causes a gradual transformation into a rigid gel-like state with the decreasing temperatures (De Maayer et al., 2014). Retention of membrane integrity is vital for proper communication with the surroundings and overall cell survival. Most studies have shown that the microbes accumulate membrane modifications involving an increase in unsaturation, particularly cis-unsaturation, and by replacing long-chain fatty acids with short-branched fatty acids. Investigations have identified many membrane-modifying genes and enzymes that help maintain membrane fluidity and intact stability. An extensive study of cell structure under the electron microscope from the samples collected from Siberian permafrost conducted by Soina et al. revealed myriads of membrane and cytoplasmic adaptations (Soina et al., 2004).

The gram-negative bacteria, *Sphingobacterium antarcticus*, and the gram-positive bacteria, *Micrococcus roseus*, isolated from Antarctica, were shown to have an increased accumulation of unsaturated fatty acids owing to an upregulation of the desaturase enzymes (Chattopadhyay et al., 2001). The same was observed in the Antarctic strain, *P. syringae*, where a decrease in membrane fluidity triggered a concurrent accumulation of unsaturated trans-fatty acids. The investigation of the *P. syringae* cti gene encoding the cis-trans-isomerase found it to be constitutively active irrespective of the environmental temperature, suggesting a post-transcriptional regulation strategy at play (Kiran et al., 2005). Mackelprang et al.'s (2017) investigation of the various metabolic pathways of the permafrost microbes revealed an upregulation of the membrane biosynthetic pathways, peptidoglycan, LPS, and fatty acid biosynthesis. They also detected an upregulation of the lpxD gene, which had been previously analyzed to be associated with temperature-regulated remodeling of the membrane (Li et al., 2012). Sipes et al.'s (2021) analysis of the Siberian permafrost MAGs revealed an abundance of enzyme kinases linked to the mevalonate pathway implicated in the synthesis of isoprenoids, which has a role in maintaining membrane fluidity and stability at subzero temperatures.

The scarcity of water activity in the permafrost can drive the majority of microbes to death or dormancy. Instead, evidence suggests the presence of growing and dividing cells. Many papers suggest that the scant amount of available liquid water associated with the saline brine veins or cryopegs could make them potential sites for active microbial communities (Gilichinsky et al., 2003, 2005). As the polar oceans regressed, the marine sediments were exposed and subsequently frozen, causing the formation of methane hydrates and supercooled NaCl lenses. These high solute brine lenses retain liquid water even in the face of subzero temperatures owing to the high salt content (170–300 g/L) of the lenses. Isolation of active motile microbes, sulfate reducers, anaerobic heterotrophs, psychrophiles, and methanogens from the cryopegs supports the belief that brine lenses could be a potential microhabitat for the permafrost microbes to thrive (Gilichinsky et al., 2003; Shcherbakova et al., 2005). For example, an obligate spore-forming anaerobe, *Clostridium algoriphilum*, was isolated from the brine films of Siberian permafrost by Shcherbakova et al. (2005). This was further supported by the findings of Gilichinsky et al. (2003), who were able to isolate psychrophilic microbes, *Clostridium* and *Psychrobacter*, in the Siberian permafrost.

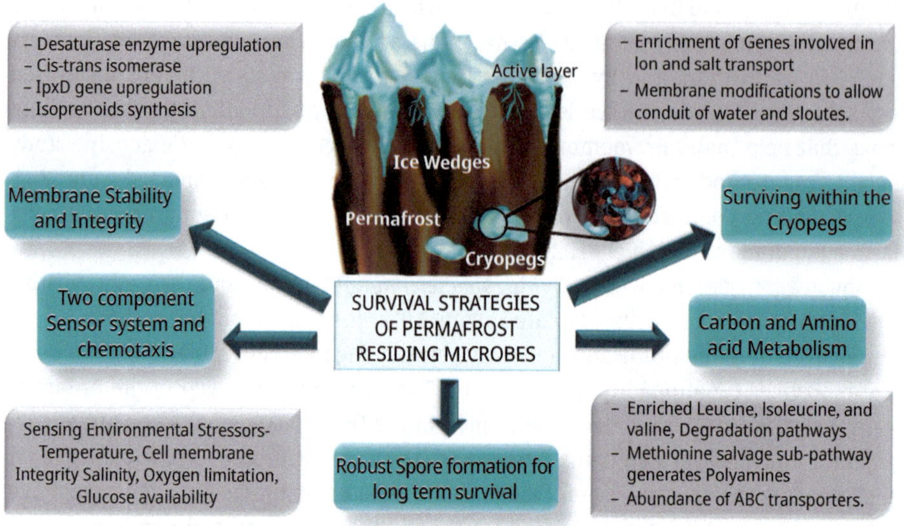

Figure 10.5: Survival strategies of bacteria residing in permafrost. Illustration courtesy of the author Bedaprana Roy.

Microbes isolated from Siberian permafrost, when grown in high saline conditions of 0.5–15% NaCl, were found to have optimal growth despite the high salinity, suggesting that they could survive within the brines. Analysis of the Siberian MAGs revealed an enrichment of genes involved in ion and salt transport, implying an adaptation to survive high saline environments like the cryopegs (Sipes et al., 2021). The uptake of glucose by the residents of the brine lenses, as studied by Gilichinsky et al. (2003), points to the fact that these microbes in the cryopeg eco-niche are indeed metabolically active. To allow efficient utilization of the resources offered by the cryopegs, proper membrane adaptations need to be incorporated for adequate siphoning of the water and solutes. The cell membrane needs to be in contact with brine films to allow a continuous flow of the solute and water. The presence of cells with unique membrane adaptations suited for such conduits across brine lenses was captured under an electron microscope from the Siberian permafrost samples by Soina et al. (2004).

Securing a source of decent water supply, as the brine theory promises, would not solve the problem of energy and nutrient deficiency, as the brines only promise a low-carbon diet. Studies to date have unraveled myriads of pathways that have become upregulated in these cryophilic microbial populations (Mackelprang et al., 2017; Sipes et al., 2021). Extensive analysis of the permafrost microbial communities along the differently aged permafrost by Mackelprang et al. (2017) revealed an abundance of at least 18 KEGG pathways. These included pathways ranging from cell envelope component synthesis, amino acid biosynthesis, carbohydrate metabolism, environmental sensing, chemotaxis, and membrane transport. MAGs from the Siberian permafrost, as analyzed by Sipes et al. (2021), revealed clusters of orthologous genes associated with carbohydrate metabolism and transport and energy production that were in greater abundance compared to the non-permafrost microbes.

An increase in the two-component sensor system was observed, as evident from the increased abundance of genes that were part of the two-component system. The two-component system is an essential attribute for sensing environmental stressors. Genes involved in sensing trace metals, temperature, cell membrane integrity, salinity, oxygen limitation, glucose availability, and nitrogen were shown to have increased in abundance Schematically represented in Figure 10.5 (Mackelprang et al., 2017). With an improved ability to sense finite resources, their chemotaxis pathways were concurrently upregulated to enable them to perceive and move toward the stimulus. Chemotaxis is an important adaptation in such a constrained environment, where localization to a site with a higher concentration of resources can be beneficial in the face of diminishing energy sources (Mackelprang et al., 2017). In particular, evidence for increased transporters, notably the ABC transporters that are involved in carbohydrate, amino acid transfer, inorganic ion transport, and trace metal intake, was found in increased abundance as supported by the works of Sipes et al. (2021) and Mackelprang et al. (2017).

KEGG analysis of the microbial population in these oligotrophic sites has revealed an increase in amino acid metabolic pathways, especially the cysteine and methionine metabolism pathways, along with leucine, isoleucine, and valine degradation pathways (Mackelprang et al., 2017). The methionine salvage sub-pathway is significant in combating stress as it generates polyamines like spermine and putrescine that are tagged as "stress molecules" owing to their diverse role in conferring increased survivability in response to cold and oxidative stress (Rhee et al., 2007). Enrichment of the aromatic hydrocarbon degradation pathways like toluene, dioxin, and xylene degradation pathways was observed. These pathways were found to be part of peripheral anaerobic catabolic pathways, yielding benzyl CoA, the starting point of the major carbohydrate central pathway. Three other carbohydrate metabolism pathways, including glyoxylate, butanoate, and galactose metabolism, were also found to be abundant in the samples from the 33 kyr (Mackelprang et al., 2017). The abundance of multiple catabolic pathways allows the utilization of alternate sources of energy and the recycling of detrital biomass, a beneficial adaptation against energy starvation in the oligotrophic permafrost habitat.

Analysis of the Siberian permafrost MAGs by Sipes et al. revealed enrichment of trehalose-6-phosphate synthase involved in the synthesis of trehalose (Sipes et al., 2021). Trehalose is a disaccharide that has been associated with cryoprotection by stabilizing DNA and membrane integrity under low-temperature conditions. It is also implicated in increasing cellular longevity by preventing damaged protein aggregation and diminishing oxidative damage. Two other enzymes detected in the Siberian MAGs, cellobiose phosphorylase and carbamate kinase, were thought to play an advantageous role in such an oligotrophic environment. Cellobiose phosphorylase catalyzes the conversion of cellobiose to glucose-1-phosphate, allowing slow degradation of the available carbon (Schimz et al., 1983). Detection of carbamate kinase suggests a way of generating ADP and carbamoyl phosphate from ammonia, carbon dioxide, and ATP. Additionally, the role of carbamate kinase in the nitrogen, glutamate, and purine pathways makes the enzyme beneficial for microbes under low-energy conditions (Yoon et al., 2017). A protein called carbon starvation protein, which is supposed to enhance peptide catabolism during carbon starvation, was identified in some of the Siberian MAGs (Borrel et al., 2012). These adaptations offer a chance of survival in the hostile, resource-limited, and energy-deficient conditions of the permafrost.

To survive, the microbes must adapt, but the subzero temperature of the permafrost has prolonged the generation time of the permafrost microbes, which drastically reduces the chance of quick adaptations. In this aspect, horizontal gene transfer offers an accelerated approach to acquiring favorable adaptive traits. For example, Mackelprang et al. (2017) observed an increased abundance of genes associated with the secretion systems, particularly the Type IV secretion system, which, along with multiple other functions, is responsible for the transfer of DNA between closely related species. This gene transfer between closely related microbes is essential for their survival by upgrading their fitness to face new stressors and successfully sustaining the cryophilic permafrost habitat.

Microbial survival of the majority depends on the upregulation of alternate pathways of energy production and carbon metabolism and a clever utilization of low water activity along with many structural modifications. However, one mechanism that has been identified in many cryophilic microbes, adapted by the microbes for long-term survival, is the production of robust spores. Metagenomic analysis has identified a substantial number of sporulation-related genes, supplemented by the presence of anaerobic spore-forming bacteria like *Clostridium* (Liang et al., 2019). Genes like spoVA, dpaA, and SASP were detected that are involved in the synthesis of dipicolinic acid (DPA) and small acid-soluble proteins (SASPs) in the spores, respectively (Liang et al., 2019). The synchronous complexation of DNA by the SASPs and the maintenance of a high concentration of DPA within the spore by the spoVA and dpaA is crucial for the long-term survival of the spores (Donnelly et al., 2016; Setlow et al., 2007). The predominance of both spore-forming microbes and sporulation-related genes in the permafrost sample suggests that attaining dormancy could be an essential mechanism of survival in the permafrost.

This discussion highlights the plethora of adaptations that accompany their survival in the permafrost. Such adaptations are not constant throughout the permafrost but have shown great variability among the communities at different depths. Such a concurrent relationship between adaptation and the chronosequence of the permafrost suggests that these microbes may not have started off with adaptive traits early on but have acquired gradual changes in their cellular and metabolic mechanisms to suit their respective chronosequences (Mackelprang et al., 2017). Although the discussion highlights many survival strategies, it is still only the tip of the iceberg. The investigation of the survival strategy contributes to a better understanding of their long-term survival in the cryoenvironment. The study demonstrates the possible niches within the permafrost that could sustain liquid water. The enrichment of various metabolic pathways provides crucial insight into their mechanisms of energy production and utilization under circumstances of energy starvation. The observed higher efficiency of gene transfer in many is critical for rapid adaptation to stressors, but an efficient horizontal transfer system also implies accelerated resistant gene transfer among these relatives. With the increment in permafrost thawing, more of these microbial communities are being exposed. Analysis of their metabolic pathways can help understand their impact on climate change. The road to understanding these microbes, their unique adaptations, and their concerning role in climate change has just begun, and further exploration into the unknown is guaranteed.

10.4 Conclusion

This chapter summarizes the plethora of microbial communities present in the different permafrost regions around the world, such as Siberia, Alaska, and the Arctic. Analysis of the various keystone species found in these regions has also been included to provide a profound understanding of the microbial kingdom that inhabits these uninhabitable areas of the Earth. The chapter also enlightens us about the different strategies these unique bacterial strains adopt to survive in these stringent conditions. The origins of these species, such as the Archaebacteria, Firmicutes, and *Arthrobacter*, have also been discussed in the etymology. Though a lot of research is currently in progress, there is much more left to explore in this field. The mechanisms of survival can also shed light on the concepts of polyextremophily, where certain bacterial strains can tolerate more than one such extreme condition. The area is still quite unexplored and holds a lot of potential for further research exploration.

References

Ayala-del-Río, H. L., Chain, P. S., Grzymski, J. J., Ponder, M. A., Ivanova, N., Bergholz, P. W., Di Bartolo, G., Hauser, L., Land, M., Bakermans, C., Rodrigues, D., Klappenbach, J., Zarka, D., Larimer, F., Richardson, P., Murray, A., Thomashow, M., & Tiedje, J. M. (2010). The genome sequence of Psychrobacter arcticus 273–4, a Psychroactive Siberian Permafrost Bacterium, reveals mechanisms for adaptation to low-temperature growth. *ASM Journals, Applied and Environmental Microbiology, 76* (7). https://doi.org/10.1128/AEM.02101-09

Borrel, G., Lehours, A.-C., Crouzet, O., Jézéquel, D., Rockne, K., Kulczak, A., Duffaud, E., Joblin, K., & Fonty, G. (2012). Stratification of archaea in the deep sediments of a freshwater meromictic lake: Vertical shift from methanogenic to uncultured archaeal lineages. *PLoS One, 7*, e43346. https://doi.org/10.1371/journal.pone.0043346

Bottos, E. M., Vincent, W. F., Greer, C. W., & Whyte, L. G. (2008). Prokaryotic diversity of arctic ice shelf microbial mats. *Environmental Microbiology, 10*, 950–966. https://doi.org/10.1111/j.1462-2920.2007.01516.x

Chattopadhyay, M. K., & Jagannadham, M. V. (2001). Maintenance of membrane fluidity in Antarctic bacteria. *Polar Biology, 24*, 386–388.

Choudhari, S., Lohia, R., & Grigoriev, A. (2014). Comparative metagenome analysis of an Alaskan glacier. *Journal of Bioinformatics and Computational Biology*, *12*, 1441003. 10.1142/S0219720014410030

De Maayer, P., Anderson, D., Cary, C., & Cowan, D. A. (2014). Some like it cold: Understanding the survival strategies of psychrophiles. *EMBO Reports*, *15*, 508–517.

Donnelly, M. L., Fimlaid, K. A., & Shen, A. (2016). Characterization of Clostridium difficile spores lacking either SpoVAC or DPA synthetase. *Journal of Bacteriology*, *198*, 1694–1707. https://doi.org/10.1128/JB.00986-15

Ernakovich, J. G., Barbato, R. A., Rich, V. I., Schädel, C., Hewitt, R. E., Doherty, S. J., Whalen, E. D., Abbott, B. W., Barta, J., Biasi, C., Chabot, C. L., Hultman, J., Knoblauch, C., Vetter, M. C. Y., Leewis, M.-C., Liebner, S., Mackelprang, R., Onstott, T. C., Richter, A., & Winkel, M. (2022). Microbiome assembly in thawing permafrost and its feedbacks to climate. *Global Change Biology*, 1–20. https://doi.org/10.1111/gcb.16231

Etienne, Y., Mélanie, A., Roland, B., David, J., Lawrence, J. R., Luke, M., Whyte, L. G., & Greer, C. W. (2009). Microarray and real-time PCR analyses of the responses of high-Arctic soil bacteria to hydrocarbon pollution and bioremediation treatments. https://doi.org/10.1128/AEM.01029-09

Gilichinsky, D., et al. (2003). Supercooled water brines within permafrost-an unknown ecological niche for microorganisms: A model for astrobiology. *Astrobiology*, *3*(2), 331–341. 10.1089/153110703769016424

Gilichinsky, D., et al. (2005). Biodiversity of cryopegs in permafrost. *FEMS Microbiology Ecology*, *53*(1), 117–128. 10.1016/j.femsec.2005.02.003

Hansen, A. A., Herbert, R. A., Mikkelsen, K., Jensen, L. L., Kristoffersen, T., Tiedje, J. M., Lomstein, B. A., & Finster, K. W. (2007). Viability, diversity and composition of the bacterial community in a high Arctic permafrost soil from Spitsbergen, Northern Norway. *Environmental Microbiology*, *9*, 2870–2884. https://doi.org/10.1111/j.1462-2920.2007.01403.x

Høj, L., Olsen, R. A., & Torsvik, V. L. (2005, June). Archaeal communities in High Arctic wetlands at Spitsbergen, Norway (78°N) as characterized by 16S rRNA gene fingerprinting. *FEMS Microbiology Ecology*, *53*(1), 89–101. https://doi.org/10.1016/j.femsec.2005.01.004

Hultman, J., Waldrop, M. P., Mackelprang, R., David, M. M., McFarland, J., Blazewicz, S. J., et al. (2015). Multi-omics of permafrost, active layer and thermokarst bog soil microbiomes. *Nature*, *521*, 208–212.

Ji, M., Kong, W., Liang, C., Zhou, T., Jia, H., & Dong, X. (2020). Permafrost thawing exhibits a greater influence on bacterial richness and community structure than permafrost age in Arctic permafrost soils. *The Cryosphere*, *14*, 3907–3916. https://doi.org/10.5194/tc-14-3907-2020

Johnston, E. R., Rodriguez-R, L. M., Luo, C., Yuan, M. M., Wu, L., He, Z., Schuur, E. A. G., Luo, Y., Tiedje, J. M., Zhou, J., & Konstantinidis, K. T. (2016). Metagenomics reveals pervasive bacterial populations and reduced community diversity across the Alaska Tundra ecosystem. *Frontiers in Microbiology*, *7*, 579. 10.3389/fmicb.2016.00579

Kayani, M., Doyle, S. M., Sangwan, N., et al. (2018). Metagenomic analysis of basal ice from an Alaskan glacier. *Microbiome*, *6*, 123. https://doi.org/10.1186/s40168-018-0505-5

Kiran, M. D., Annapoorni, S., Suzuki, I., Murata, N., & Shivaji, S. (2005). Cis-trans isomerase gene in psychrophilic Pseudomonas syringae is constitutively expressed during growth and under conditions of temperature and solvent stress. *Extremophiles*, *9*, 117–125.

Legendre, M., et al. (2014). Thirty-thousand-year-old distant relative of giant icosahedral DNA viruses with a pandoravirus morphology. *Proceedings of the National Academy of Sciences of the United States of America*, *111*(11), 4274–4279. 10.1073/pnas.1320670111

Legendre, M., et al. (2015). In-depth study of Mollivirus sibericum, a new 30,000-y-old giant virus infecting Acanthamoeba. *Proceedings of the National Academy of Sciences of the United States of America*, *112*(38), E5327–35. 10.1073/pnas.1510795112

Li, Y., et al. (2012). LPS remodeling is an evolved survival strategy for bacteria. *Proceedings of the National Academy of Sciences of the United States of America*, *109*(22), 8716–8721. 10.1073/pnas.1202908109

Liang, R., Lau, M., Vishnivetskaya, T., Lloyd, K., Wang, W., Wiggins, J., Miller, J., Pfiffner, S., Rivkina, E., & Onstott, T. (2019). Predominance of anaerobic, spore-forming bacteria in metabolically active microbial communities from ancient Siberian permafrost. *Applied and Environmental Microbiology, 85*. 10.1128/AEM.00560-19

Liang, R., Li, Z., Lau Vetter, M. C. Y., et al. (2021). Genomic reconstruction of fossil and living microorganisms in ancient Siberian permafrost. *Microbiome, 9*, 110. https://doi.org/10.1186/s40168-021-01057-2

Lipson, D. A., Haggerty, J. M., Srinivas, A., Raab, T. K., Sathe, S., et al. (2013). Metagenomic insights into anaerobic metabolism along an Arctic peat soil profile. *Plos One, 8*(5), e64659. https://doi.org/10.1371/journal.pone.0064659

Lyutskanova, D., Ivanova, V., Stoilova-Disheva, M., Kolarova, M., Aleksieva, K., & Peltekova, V. (2009). Isolation and characterization of a psychrotolerant streptomyces strain from permafrost soil in spitsbergen, producing phthalic acid ester. *Biotechnology & Biotechnological Equipment, 23*(2), 1220–1224. 10.1080/13102818.2009.10817642

Mackelprang, R., Burkert, A., Haw, M., Mahendrarajah, T., Conaway, C. H., Douglas, T. A., & Waldrop, M. P. (2017, Oct). Microbial survival strategies in ancient permafrost: Insights from metagenomics. *ISME Journal, 11*(10), 2305–2318. 10.1038/ismej.2017.93. Epub 2017 Jul 11. PMID: 28696425; PMCID: PMC5607373.

Mackelprang, R., Saleska, S., Jacobsen, C., Jansson, J., & Taş, N. (2016). Permafrost meta-omics and climate change. *Annual Review of Earth and Planetary Sciences, 44*. 10.1146/annurev-earth-060614-105126

Mackelprang, R., Waldrop, M. P., DeAngelis, K. M., David, M. M., Chavarria, K. L., et al. (2011). Metagenomic analysis of a permafrost microbial community reveals a rapid response to thaw. *Nature, 480*(7377), 368–371.

Petrova, M., Kurakov, A., Shcherbatova, N., & Mindlin, S. (2014). Genetic structure and biological properties of the first ancient multiresistance plasmid pKLH80 isolated from a permafrost bacterium. *Microbiology, 160*, 2253–2263. 10.1099/mic.0.079335-0

Raymond-Bouchard, I., et al. (2018, Apr.). Conserved genomic and amino acid traits of cold adaptation in subzero-growing Arctic permafrost bacteria. *FEMS Microbiology Ecology, 94*(4), 1e+. *Gale Academic OneFile*, link.gale.com/apps/doc/A690688861/AONE?u=anon~ea4d7f5c&sid=googleScholar&xid=fdd5a3c6

Rhee, H. J., Kim, E. J., & Lee, J. K. (2007). Physiological polyamines: Simple primordial stress molecules. *Journal of Cellular and Molecular Medicine, 11*, 685–703.

Rigou, S., Christo-Foroux, E., Santini, S., Goncharov, A., Strauss, J., Grosse, G., Fedorov, A. N., Labadie, K., Abergel, C., & Claverie, J.-M. (2022). Metagenomic survey of the microbiome of ancient Siberian permafrost and modern Kamchatkan cryosols. *MicroLife, 3*, uqac003. https://doi.org/10.1093/femsml/uqac003

Rivkina, E., Petrovskaya, L., Vishnivetskaya, T., Krivushin, K., Shmakova, L., Tutukina, M., Meyers, A., & Kondrashov, F. (2016). Metagenomic analyses of the late Pleistocene permafrost – Additional tools for reconstruction of environmental conditions. *Biogeosciences, 13*, 2207–2219. 10.5194/bg-13-2207-2016

Sanders, T., Fiencke, C., Hüpeden, J., Pfeiffer, E. M., & Spieck, E. (2019). Cold adapted Nitrosospira sp.: A potential crucial contributor of ammonia oxidation in cryosols of permafrost-affected landscapes in Northeast Siberia. *Microorganisms, 7*(12), 699. https://doi.org/10.3390/microorganisms7120699

Schimz, K. L., Broll, B., & John, B. (1983). Cellobiose phosphorylase (EC 2.4.1.20) of Cellulomonas: Occurrence, induction, and its role in cellobiose metabolism. *Archives of Microbiology, 135*, 241–249. https://doi.org/10.1007/BF0041347

Setlow, P. (2007). I will survive: DNA protection in bacterial spores. *Trends in Microbiology, 15*, 172–180. https://doi.org/10.1016/j.tim.2007.02.004

Shcherbakova, V., Yoshimura, Y., Ryzhmanova, Y., Taguchi, Y., Segawa, T., Oshurkova, V., & Rivkina, E. (2016, October). Archaeal communities of Arctic methane-containing permafrost. *FEMS Microbiology Ecology, 92*(10), fiw135. https://doi.org/10.1093/femsec/fiw135

Shcherbakova, V. A., et al. (2005). Novel psychrophilic anaerobic spore-forming bacterium from the overcooled water brine in permafrost: Description Clostridium algoriphilum sp. Nov. *Extremophiles: Life under Extreme Conditions*, *9*(3), 239–246. 10.1007/s00792-005-0438-3

Sipes, K., Almatari, A., Eddie, A., Williams, D., Spirina, E., Rivkina, E., Liang, R., Onstott, T. C., Vishnivetskaya, T. A., & Lloyd, K. G. (2021). Eight metagenome-assembled genomes provide evidence for microbial adaptation in 20,000 to 1,000,000-year-old Siberian permafrost. *ASM Journals, Applied and Environmental Microbiology*, *87*(19). https://doi.org/10.1128/AEM.00972-21

Sipes, K., Paul, R., Onstott, T. C., Vishnivetskaya, T. A., & Lloyd, K. G. (2022). Draft genome sequences of 10 Pseudomonas sp. isolates from the active layer of permafrost in Ny Ålesund, Svalbard, Norway. *ASM Journals, Microbiology Resource Announcements*, *11*(6). https://doi.org/10.1128/mra.00201-22

Sofia, M., Alexey, B., Andrey, M., & Mayya, P. (2019). Adaptive dif modules in permafrost strains of acinetobacter lwoffii and their distribution and abundance among present day acinetobacter strains. *Frontiers in Microbiology*, *10*. 10.3389/fmicb.2019.00632

Soina, V. S., Mulyukin, A. L., Demkina, E. V., Vorobyova, E. A., & El-Registan, G. I. (2004). The structure of resting bacterial populations in soil and subsoil permafrost. *Astrobiology*, *4*, 345–358. https://doi.org/10.1089/ast.2004.4.345

Steven, B., Briggs, G., McKay, C. P., Pollard, W. H., Greer, C. W., & Whyte, L. G. (2007, February). Characterization of the microbial diversity in a permafrost sample from the Canadian high Arctic using culture-dependent and culture-independent methods. *FEMS Microbiology Ecology*, *59*(2), 513–523. https://doi.org/10.1111/j.1574-6941.2006.00247.x

Steven, B., Niederberger, T. D., & Whyte, L. G. (2009). Bacterial and Archaeal Diversity in Permafrost. In R. Margesin (Ed.), *Permafrost Soils. Soil Biology*, vol. 16. Berlin, Heidelberg: Springer. https://doi.org/10.1007/978-3-540-69371-0_5

Steven, B., Pollard, W. H., Greer, C. W., & Whyte, L. G. (2008). Microbial diversity and activity through a permafrost/ground ice core profile from the Canadian high Arctic. *Environmental Microbiology*, *10*, 3388–3403. https://doi.org/10.1111/j.1462-2920.2008.01746.x

Suetin, S. V., Shcherbakova, V. A., Chuvilskaya, N. A., Rivkina, E. M., Suzina, N. E., Lysenko, A. M., & Gilichinsky, D. A. (2009, Jun). Clostridium tagluense sp. Nov., a psychrotolerant, anaerobic, spore-forming bacterium from permafrost. *International Journal of Systematic and Evolutionary Microbiology*, *59*(Pt 6), 1421–1426. 10.1099/ijs.0.002295-0. PMID: 19502327.

Tripathi, B. M., Kim, M., Kim, Y., et al. (2018). Variations in bacterial and archaeal communities along depth profiles of Alaskan soil cores. *Scientific Reports*, *8*, 504. https://doi.org/10.1038/s41598-017-18777-x

Vašková, D. (2014). *The Importance of Actinobacteria in Arctic Soil. Bc. Thesis, in English* (p. 21). Czech republic: Faculty of Science, University of South Bohemia, České Budějovice.

Wu, X., Chauhan, A., Layton, A. C., Lau Vetter, M. C. Y., Stackhouse, B. T., Williams, D. E., Whyte, L., Pfiffner, S. M., Onstott, T. C., & Vishnivetskaya, T. A. (2021). *Environmental Science & Technology*, *55*(18), 12683–12693. 10.1021/acs.est.1c00802. Active layer had more biodiversity which decreased with increasing depth.

Xue, Y., Jonassen, I., Øvreås, L., & Taş, N. (2020, May). Metagenome-assembled genome distribution and key functionality highlight importance of aerobic metabolism in Svalbard permafrost. *FEMS Microbiology Ecology*, *96*(5), fiaa057. https://doi.org/10.1093/femsec/fiaa057

Yergeau, E., Hogues, H. E., Whyte, L. G., & Greer, C. W. (2010). The functional potential of high Arctic permafrost revealed by metagenomic sequencing, qPCR and microarray analyses. *ISME Journal*, *4*(9), 1206–1214.

Yoon, S.-H., Ha, S., Lim, J., Kwon, S., & Chun, J. (2017). A large-scale evaluation of algorithms to calculate average nucleotide identity. *Antonie Van Leeuwenhoek*, *110*, 1281–1286. https://doi.org/10.1007/s10482-017-0844

Sanjana Ghosh*, Bidisha Chatterjee, Debapriya Maitra, Bedaprana Roy, Sudeshna Shyam Choudhury, Ayan Chandra and Arup Kumar Mitra

Chapter 11 An insight to microbial metagenome in wetland ecosystem

Abstract: Wetlands are areas having distinct types of ecological diversity. Agricultural practices are performed on certain reclaimed wetlands. Analysis of the physicochemical properties of soil, like pH, organic carbon, and soil texture, showed variation in different climates, suggesting heterogeneity in soil fertility and also a paradigm shift in species diversity. On the basis of the varied types of geomorphology and hydrological conditions, alterations have been observed in the microbial community of wetlands, and therefore fluctuations in the nutrient levels of plants have been observed. A combined knowledge of the physicochemical properties of the wetland soil, crop ecology, and microbial diversity is essential for understanding variations in crop fertility. Analysis of the genetics of the soil microbial community can be done using metagenomics, whereby data of detailed taxonomic stratification of the microbial flora gives an understanding of the species diversity. Data integration is usually performed before the data analysis, whereby extraction of data and their matching with heterogeneous data build a frame of reference for the genome, which otherwise becomes difficult in establishing a relation between the data received and its interpretation. Therefore, in this chapter, we shall understand the heterogeneity amongst the bacterial community in the agricultural wetlands and their implications on agricultural production for the growing population. This will also emphasize the role of keystone species in the productivity of certain crops and the importance of data integration in the metagenomics project.

11.1 Introduction

Wetlands are the transitional zones between two different ecosystems, the aquatic ecosystem and the terrestrial ecosystem, whereby the level of water is more or less adjacent to the land surface, and sometimes there is a constant layer of water over

*Corresponding author: Sanjana Ghosh, Department of Microbiology, St. Xavier's College (Autonomous), Kolkata, West Bengal, India, e-mail: sanjana.ghosh2018@gmail.com
Bidisha Chatterjee, Debapriya Maitra, Bedaprana Roy, Sudeshna Shyam Choudhury, Arup Kumar Mitra, Department of Microbiology, St. Xavier's College (Autonomous), Kolkata
Ayan Chandra, Department of Statistics, St. Xavier's College (Autonomous), Kolkata

https://doi.org/10.1515/9783110793635-011

the land mass (Prasad et al., 2002). It is a crucial terrestrial ecosystem. Wetlands are found on every continent except Antarctica. They cover almost 5–8% of the total Earth's surface (Mitsch et al., 2009). However, they are one of the most sensitive in terms of ecology and have a very adaptive ecosystem (Turner et al., 2000). The geographical variation has resulted in diverse soil, sediment, and aquatic ecosystems, which gave us a diverse microbial community. Wetlands are more popularly known as the "Kidneys of the landscape" because of their importance in water purification, processing of minerals like carbon, global emission of greenhouse gases, flood control, shoreline stabilization, and support of flora and fauna (Lv et al., 2014). This frequent flooding of the wetlands has led to the formation of anoxic and oxic conditions. This has resulted in the infestation of anaerobic along with aerobic microbial populations. Wetlands harbor a diverse microbial community, which leads to the rapid cycling of minerals and nutrients. Thus, they are known as the "most productive ecosystem." They are also considered the "biological supermarkets" due to their rich biodiversity, unique habitats, and extensive food habitat (De Mandal et al., 2020).

With the passage of time, wetlands are gaining importance all over the world because of their contribution to maintaining a healthy environment. One of the important properties of wetlands is that they retain water to maintain the groundwater level during the summer months. Streams and water channels flowing through the wetlands into lakes contain a reduced quantity of suspended solids. Wetlands also contribute to flora and fauna including microbial diversity (Prasad et al., 2002). They also provide social benefits such as a habitat and food for wildlife and fish including endangered and threatened species, flood control, water quality upgradation, shoreline erosion check, water storage, and economically favorable biological products for human utilization as well as opportunities for research, education, recreation, and esthetic appreciation. They are included in the highest yielding and biologically varied ecosystems globally, which can be compared to coral reefs and rainforests. They act as a reservoir of considerable biodiversity that supports all living organisms from microbes to mammals (Dorney et al., 2018).

Wetlands are of various types. They include high-altitude wetlands, saline, interim wetlands from the arid and semiarid regions, marine wetlands, mangrove swamps, salt marshes, and coastal wetlands like estuaries, lagoons, backwaters, coral reefs, and wetlands located across the flood plains, bogs, and fens. Hence, the water may be freshwater, marine, or brackish (Prasad et al., 2002). However, almost 70% of the wetlands have been lost since 1900. India has lost almost one-third of its wetlands. One of the main reasons for the reduction of wetland ecosystems is the presence of profound anthropogenic activities. The Ramsar Convention on Wetlands was signed in Ramsar, Iran in 1971. It was a global treaty signed to protect the wetlands. Now, almost 170 nations are involved in the treaty. The definition proposed by the Ramsar Convention to define wetlands stated "areas of marsh, fen, peat, and or water, whether natural or artificial, permanent or temporary, with water that is static or flowing, fresh, brackish or salt, including areas of marine water the depth of which at low tide does not exceed six meters." However, different countries define wetlands in different ways due to certain legal issues. The Ramsar Convention in India harbors 42 Ramsar sites that have international importance. According to the WorldWide Fund (WWF), the wetland ecosystem is the most endangered ecosystem in India. Various man-made causes like water pollution, overexploitation of land, invasive species, and excessive inundation have put the wetlands of India at high risk (Balwan et al., 2021).

Wetlands are one of the most fertile ecosystems on the Earth, equivalent to coral reefs or rain forests. They harbor a wide variety of flora and fauna, which are significantly dependent on the geographical and climatic conditions of the ecosystem. Wetlands provide a diverse variety of food for the population and thus can be termed "Biological Supermarkets" (Richardson et al., 2001). This amalgamation of primary productivity, shallow water, and high nutrient levels can support the growth of species ideal for the food web base formation and provide food for a large population of animals. The microbial species inhabiting the wetland are a major part of different biogeochemical cycles like nitrogen and sulfur along with climatic cycles like the water cycle. Another function of this ecosystem is the maintenance of atmospheric carbon dioxide. As we know, the wetland ecosystem has the ability to trap carbon in the plants and also in the soil instead of liberating it back into the atmosphere as methane or carbon dioxide. Thus, wetlands usually have a crucial role in modulating the climatic conditions on the Earth and may work as a crucial factor resulting in climate change (Balwan et al., 2021).

The actual assessment of wetland production cannot be obtained through the estimation of plant cover and animals. Therefore, a study is being undertaken to evaluate the edifice and heterogeneity of the microbial population because of their noteworthy association with the trophic status. However, it was observed that the microbial community present in the sediment gave a more accurate environmental estimation as a result of higher taxonomical diversity and plethora than a water body. This was mainly led by the development of a multiplex grid, which provides better support for the growth of microbes. The microbial population at a specific place includes a wide plethora based on the principle of environmental discretion. This microbial plethora is mainly responsible for the biogeochemical cycles and climate modulation. They also have the potential to alter the physicochemical characteristics of the wetland, which may have a positive or negative impact on the wetland ecosystem (Iliev et al., 2019). The identification of these nonculturable microbes in the wetland ecosystem is carried out through metagenomic analysis. The study of wetland metagenomics helps in two ways. First, it will provide us with a thorough knowledge about the wide diversity of the microbial population coexisting in the wetland ecosystem. Investigations show that metagenomics can be used to study the microbial flora of various environments including soil, water, and extreme environments. One such report stated that the soil microbial plethora is a consortium of various antimicrobial agents using metagenomic analysis (Gillespie et al., 2002). It is a known fact that around 70% of the Earth's surface is enveloped with water, so they present a wide variety of aquatic microbes. These microbes produce secondary metabolites using various metabolic pathways. With the help of metagenomics, it was reported that inland water systems like lakes and rivers harbor a high amount of antibiotic-resistant genes such as the methicillin-resistant gene and vancomycin-resistant gene (Nnadozie et al., 2019). Microbes growing in extreme conditions produce unique enzymes representing different metabolic pathways and regulations. These can be identified and studied only using metagenomics since culturing these microbes in their natural condition is not a feasible idea. Hence, we can observe the sheer diversity and outreach of metagenomic analysis, which can be easily applied in the wetland ecosystem (Zhang et al., 2007). Molecular procedures involved in revealing bacterial diversity are represented in Figure 11.1.

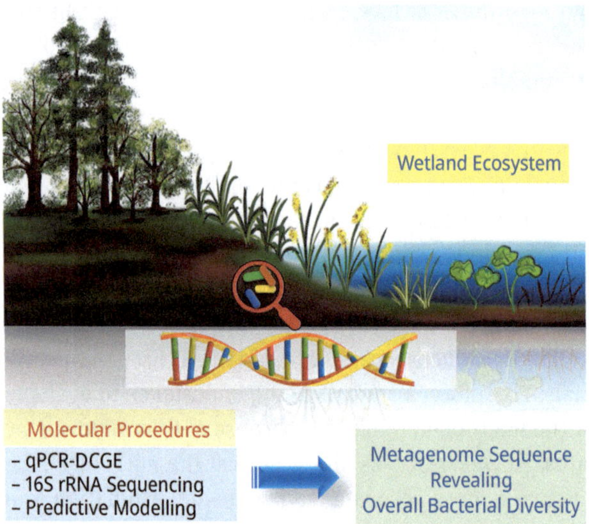

Figure 11.1: A schematic diagram representing molecular procedures like qPCR-DGGE and 16S rRNA sequencing are used for revealing overall bacterial diversity in wetland ecosystems.

Once the microbial flora is known, it can be used for various purposes like bioremediation of the wetland ecosystem. Bioremediation is a process in which the natural microflora of that ecosystem is utilized to maintain and purify the environment. It can be carried out by degrading the pollutants completely or rendering them ineffective. However, certain criteria have to be fulfilled for bioremediation to take place, like pollutants have to be organic in nature, the microbes used for this purpose have to produce the requisite number of desired metabolites, and the environment has to support the degradation. The study of metagenomics will help us to screen new genes producing desired metabolites (Zhang et al., 2021). In this chapter, we will discuss wetlands and their various types. Then, a discussion about the metagenomics of wetlands will be done, which has found a widespread use in modern research and has become important for understanding the microbial diversity of a wetland ecosystem. This will be explained through various reports of metagenomic studies on wetlands that were carried out in the recent past. Along with this, the chapter will give us insight into how certain microbial keystone species identified through metagenomes have proved their importance in maintaining the environment of the corresponding wetland.

11.2 Classification of wetlands

Wetlands were subdivided into marine, estuarine, lacustrine, riverine, and palustrine, which include coastal wetlands, deltaic tidal marshes, lakes, rivers, and marshy areas, respectively. This classification was made on the basis of their hydrometrical, ecological, and geomorphological characteristics. An international treaty was signed in the year 1971 on wetlands, referred to as the Ramsar Convention. According to this convention, certain national forces and cooperation from international agencies were impacted for the conservation of wetlands, ensuring the sustainable use of these wetlands. On an overall basis, there are around 1,052 wetland sites in Europe, followed by 289 sites in Asia, 175 areas in South America, and 211 places in North America. Africa has around 359 sites of wetlands, followed by around 79 sites in Oceania. The wetlands are divided into six crucial types, which include areas of swampy, marshy, fen, bogs, wet meadow, and shallow water wetlands (aquatic). Different amalgamations of events like the occurrence of floods, climatic changes, and variation in soil nutrients are accountable for the generation or evolution of the above types of wetlands. Wetlands having a high content of salinity are categorized into salty marshes and mangrove regions, which are categorized as a different wetland. Wetlands having high saline content are usually found to be present along the coastal areas. Some types of mangrove wetlands are also found to be present along the noncoastal regions, where the intensity of evaporation transcends the amount of rainfall in these areas. These areas usually include the arid regions of western parts of North America, northern regions of Africa, and portions of central regions of Eurasia (Sheikh et al., 2021).

Wetland floristics and vegetation are greatly influenced by the presence of telluric water, that is, the water that has transuded through the bedrocks or the soils heavily laden with minerals. Depending on the presence of telluric water, the wetlands can be subdivided into ombrotrophic and minerotrophic. Ombrotrophic regions have very low pH, calcium, and minerals. Their occurrence is correlated with the presence of dust fall and high precipitation. A lesser concentration of metal ions is found to be deposited within the peats. However, in the case of minerotrophic wetlands, a minor amount of telluric water infiltrates. Therefore, the value of pH, peat deposition, and metal ions depends upon the extent of intrusion of telluric water (Schwintzer, 1981).

The wetlands can be further classified as follows:
Floodplains: The wetlands that are present along the rivers are usually inundated with water. This usually results in the deposition of sediments and nutrients that revitalize the growth of plants by the annual pulsation of water (e.g., Riparian Wetlands). These pulses usually aid in the deposition of thick layers of sediment or dissolved nutrients that vitalize plant growth. These are widespread along the northeastern regions, followed by the eastern regions of India, encompassing the edges of the Ganga-Brahmaputra rivers. Degradation of floodplains is usually closely associated with the degradation of freshwater biodiversity. Invasion of species, pollution, and alteration of habitats are the main factors associated with the degradation of wetlands (Erwin et al., 2009). Some examples of these floodplains include the Deepor Beel located in Assam and the East Kolkata Wetlands located in West Bengal along the lower Bhagirathi basin (ENVIS, 2019; Mukherjee et al., 2021).

Tidal (coastal) marshes: These are present along the shorelines of oceans, tidal rivers, estuaries, and coastal areas. These areas are highly influenced by the incidence of tides along the coast, which are in turn regulated by the factors that affect the tides, for example, the position of the Sun and Moon in accordance with the surface of the Earth. The freshwater available from the regions of runoff, groundwater, or from rivers also has its influence upon these areas. One of the very common types of tidal marshes is the salt marsh, characterized by highly saline-tolerant plants. These salty marshes are usually inundated by nutrients from either the surface water or water from the tidal coasts. Estuaries are situated downstream of these freshwater tidal marshes. The absence of high levels of salinity and salt stress usually allows the flourishing growth of plants. These regions have a high ability for carbon sequestration (Tiner, 2018; Yang et al., 2022).

Nontidal (inland) marshes: These areas are characterized by the incidence of floods. They harbor varied types of plants, animals, microorganisms, as well as soil. Plants with soft or hollow stems are often found in these regions, which usually allow the movement of oxygen into the roots. These marshes are usually found in regions that have floodplains and shallow areas with water around rivers and lakes. The predominant properties of these freshwater marshes are regions of shallow water, which greatly impact their locations of existence. Most of the water for these inland marshes usually arrives from surface waters, runoff, and regions inundated with flood water. Marshes have different kinds of microorganisms that have the ability to metamorphose the available nitrogen, phosphorus, as well as sulfur into different chemical states (Keddy, 2008). The ground water has a profound role in the formation of these inland marshes. Great Lakes coastal marshes, regions of the Florida Everglades, and the pothole regions of prairies in the United States are some examples of inland marshes (Winter et al., 1988).

Swamps are usually the regions that have wetlands with a profound dominance of trees as well as woody plants. These are usually found along freshwater regions or areas heavily inundated with saltwater. The general characteristics of these areas are the presence of wet soils and the perpetual presence of water during a major part of the year. These salt marshes are severely affected by climate change. A rise in sea level usually results in the aversion of marsh vegetation. Swamps can be further compartmentalized into mangroves, forests, or shrubs. Forest areas with swamps are usually found in the flooded plains of the United States, especially in the northeastern, subcentral, and southeastern regions. These areas receive water from nearby rivers and lakes. Shrub swamps are almost homogeneous to forested swamps, except that there is a greater dominance of shrubs. Mangrove swamps have saline-tolerant plants and are usually found along the regions of coastal wetlands, with influence from the saline tidal waters (Erwin et al., 2009).

Bogs are referred to as areas of freshwater wetlands that have deposits of peat. Broadly, they are referred to as wetlands that are made up of organic soils. They are characterized by the presence of a water table and an overlay of *Sphagnum* sp. moss that encrusts the carpet on the surface. A bog is a peatland with a high-water table and a surface carpet of mosses, chiefly *Sphagnum* sp. They have peat layers that are usually ombrotrophic or sometimes may be minerotrophic; as a result, they are usually very strongly acidic in nature (Schwintzer, 1981). Evergreen trees and ericaceous shrubs, followed by a thick cover of *Sphagnum* sp. moss, are found in these wetlands. These wetland areas are dependent on rainwater and are sometimes found in the glacial lakes of the United States (Gorham et al., 2003).

Fens are referred to as wetlands that are usually characterized by the deposition of peat or minerotrophic peatland. These wetlands are usually covered by a bounty of varied types of grasses, wildflowers, reeds, and sedges. They have different kinds of bryophytic plants in these areas. Fens were formed when the glaciers retreated. The source of water for fens is usually streams and groundwater. Furthermore, precipitation is one of the sources of the formation of fens. They are freshwater peat-forming wetlands covered mostly by grasses, sedges, reeds, and wildflowers. Compared to bogs, fens usually have an alkaline pH or less acidic and have a high concentration of nutrients (Schwintzer, 1981; Gorham et al., 2003).

11.3 Significance of metagenomic analysis in understanding the microbial species diversity of wetlands

In earlier days, classical microbiology was greatly dependent on lab-based culturing techniques of microorganisms. However, this had a major setback since not all microbes can be cultured in a laboratory environment. This has led to the inception of a new technique called metagenomics. It is also defined as "the genome of the total microbiota found in nature." It comprises both bacterial and fungal genomes present in the environmental sample. Thus, metagenomics is the study of the genetic material of microorganisms from a mixed community collected from a specific environment using screening techniques for functional gene identification and analysis of sequenced data. The main focus areas of metagenomic research include population structure, functional activities, microbial diversity, and various relationships like genetic, cooperative, evolutionary, and, lastly, environmental (Zhang et al., 2021).

Metagenomics is a relatively new area that has contributed largely to solving different problems related to the discovery of new organisms and their interactions with the environment. The change from traditional microbiology to metagenomic studies has led to the development of various specializations like modern sequencing technologies (Escobar-Zepeda et al., 2015). Sequencing techniques have evolved into next-generation sequencing (NGS) and third-generation sequencing (TGS) with high throughput (Zhang et al., 2021). This has also led to solutions for computational problems that arise with DNA sequencing. Modern computational techniques are continuously being developed for processing and extracting necessary biological information from various databases and samples. However, metagenomics requires the amalgamation of these methods and techniques. Therefore, it is mandatory that scientists have a clear understanding of these techniques, which would help them in their experimental design to analyze and present the metagenome information (Escobar-Zepeda et al., 2015).

The metagenome of a sample is determined through a series of procedures starting with sample processing, followed by sequencing techniques, assembly, binning, annotation, experimental design, statistical analysis, data storage, and finally, data sharing (Thomas et al., 2012).

11.3.1 Sample processing

The most important step in metagenomics is sample processing. The DNA isolated from the sample needs to represent all types of cells present. This is followed by library production and sequencing, for which nucleic acids of a standard quality are required. Standardized protocols for processing individual types of samples are required followed by DNA extraction methods. Nowadays, research is being carried out to standardize a single protocol for various types of DNA extraction (Delmont et al., 2011). Host (plant or other invertebrates) DNA is removed from the sample using selective lysis or fractionation to reduce the chance of contamination. This concept is also applied if the target of analysis is only a part of the community (Burke, 2009). Sometimes cells have to be isolated physically (especially in the case of soil samples) to improve the quality of DNA extraction, which can otherwise be hindered due to the presence of certain enzymatic inhibitors like humic acid (Delmont et al., 2011).

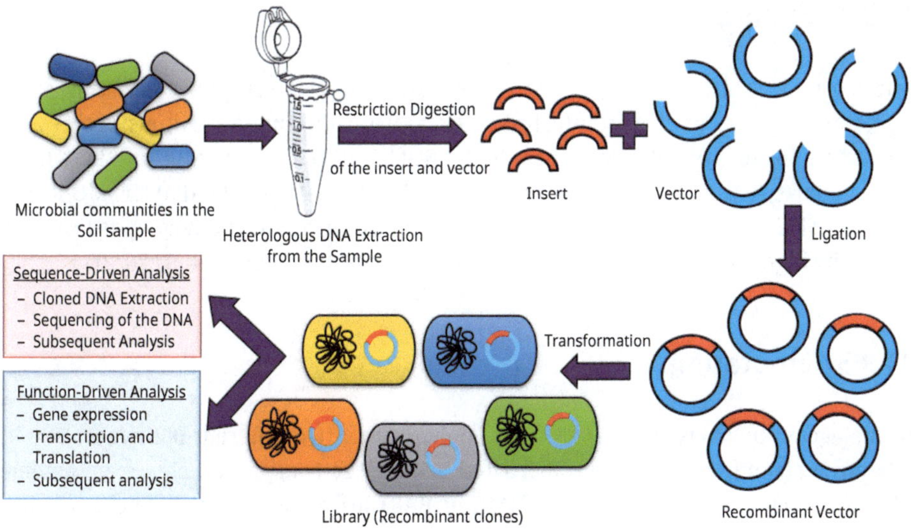

Figure 11.2: Schematic diagram representing the different steps involved in forming clone libraries that help in the analysis of microbial diversity.

11.3.2 Sequencing techniques

In the case of metagenomics, the sequencing technique has undergone a transition from traditional Sanger sequencing to NGS. The most commonly used sequencing techniques are the 454/Roche and the Illumina/Solexa System. Apart from these, there are some other upcoming sequencing techniques that might be useful in the near future like the SOLiD sequencer and Ion Torrent. TGS has two popular methods: single-molecule fluorescence sequencing and single-molecule real-time sequencing. The sequencing technique should be based on certain factors such as read length, error rate, speed, methylation, and cost. Although NGS is more widely used, TGS is an upcoming alternative with the advantage of long reads (Zhang et al., 2021).

11.3.3 Assembly

Since the main purpose of metagenomics is to obtain information on uncultured organisms for further characterization, shorter read fragments are assembled to produce longer contigs. There are mainly two different strategies. Firstly, reference-based strategies using software such as MIRA, ROCHE, and AMOS (Chevreux, 1999) which have high-speed efficient algorithms. Secondly, de novo assembly, which most commonly uses de Bruijn assemblers, has a higher run time and requires huge memory (Pevzner et al., 2001).

11.3.4 Binning

Binning is the method of segregating DNA sequences into categories denoting a genome or its group of closely interlinked organisms. Binning algorithms are of two types: composition-based, which includes S-GSOM, TACAO, Phylopythia, and PCAHIER, and similarity-based, which includes MEGAN, IMG/M, CARMA, MetaPhyler, MG/RAST, and Sort-ITEMS. However, there are some algorithms, including both composition and similarity algorithms such as MetaCluster and PhymmBL (Thomas et al., 2012).

11.3.5 Annotation

In metagenomic analysis, annotation is applied using two different pathways. One is when the genome has to be reconstructed using large contigs available from assembly, for which existing pipelines like IMG or RAST are used. The other one is applied to the community as a whole, and it is completely dependent on the short contigs (Thomas et al., 2012).

11.3.6 Statistical analysis

The vast amount of data obtained from metagenomics is usually much higher than the sample number. As a result, hypothesis correction is applied (Bonferroni correction is an example in the case of *t*-test analyses). One of the well-known tools, the Primer-E package, is readily applicable for various multivariate statistical analyses such as generating multidimensional scaling plots, ANOSIM (analysis of similarity), and SIMPER.

That helps to identify the functions and the various species that contribute to the differences present between two samples (Thomas et al., 2012).

11.3.7 Data storage

Data obtained from metagenomics requires a high level of organization and collaborative work to obtain metadata along with the sharing of computational results and data. The Minimum Information about any Sequence checklist (MIxS) is a commonly used language currently used for metadata. The storage of metagenomic data has resulted in the question of centralized and decentralized storage, depending on who is paying for it. Although NCBI stores all the data, the amount of metagenomic data being generated has led to an urgency in generating proper ways to store this data (Thomas et al., 2012).

11.3.8 Significance

The diversity of the microbial plethora nurtured by the wetland ecosystem is a key factor in the biogeochemical cycles (including phosphorus, carbon, iron, sulfur, and nitrogen cycles), nutrient recycling, and is pivotal in the numerous roles of wetlands (Bodelier et al., 2013). These biogeochemical cycles are carried out by the microbial community in the wetlands along with other ecosystem services. They are easily affected by anthropogenic activities (Iliev et al., 2019). However, eutrophication can be averted in a wetland ecosystem by keeping a check on the phosphate and nitrate concentrations in freshwater and saline water bodies. The change in parameters can be directly correlated with the microbial diversity. Another important factor is that the soil microbial community differs from that of the wetlands due to the restricted diffusion of oxygen and altered soil pH. So the study of the microbial population in a wetland ecosystem is crucial for elucidating the working principles of wetlands such as modulating the cycle, retention, and the liberation of minerals and soil carbon. But the microbial population of wetland ecosystems is uncharted with respect to aquatic and terrestrial ecosystems (De Mandal et al., 2020). So metagenomic study is the best possible solution in today's research field for obtaining a thorough knowledge of the bacterial community in a wetland. This metagenomic data can then be analyzed and used to design future experiments (Iliev et al., 2019).

As discussed earlier, the microbial genome obtained from environmental samples is analyzed using metagenomics tools. It usually lacks background information on the preexisting microbial population (Borrel et al., 2013). Moreover, the information obtained is only on the amplified genes and not of the total genome. The study results in the evaluation of the microbial population present in varied ecosystems. Thus, a complete understanding of the assortment and phylogeny of the microbial population is procured (Riesenfeld et al., 2004). Useful information on microbial antibiotics, materials with metabolic activities, enzymes, and their description of the function in their biotechnological applications, production, and medicinal processes can be retrieved from metagenomic libraries (De Mandal et al., 2020).

11.4 Metagenomic dossier of some wetlands across the globe

Wetlands have different varieties of microbiomes. These microbiomes have different biochemical roles that have significant impacts on the wetlands. Microbes perform different kinds of microbial activities that are crucial for the proper functioning of the wetlands. The potential of the microbes and their diversity can be deduced from their microbial resources. Microbes within the wetlands perform various kinds of activities that are essential for the well-being of the soil as well as the plants growing on that soil. According to the literature, different kinds of bacteria and archaea have been studied with reference to their role in biodiversity in natural wetlands as well as those that have been constructed. Various techniques in molecular biology have suggested that variations in the constitutions of microbial groups are associated with the activities in soil. 16SrRNA gene sequencing, gel electrophoresis, terminal-RFLP, and quantitative PCR are some of the molecular biology techniques that are used for understanding the molecular activity of the microbes. Identification of unknown microbes is usually done with the help of 16SrRNA sequencing. Clone libraries having greater than 200 sequences are used for the analysis of microbial diversity (Lv et al., 2014). The steps for the formation of clone libraries, essential for the analysis of microbial diversity, are presented in Figure 11.2.

The diversity of the microbial flora of natural wetlands has been studied using various advanced techniques of metagenomics. The diversity of various types of natural wetland areas has been studied, like that wetland in the northern parts of Russia, which is abundant in species like *Sphagnum*. Other wetlands include Salar de Huasco, Chile, a high elevation-based wetland area, marshy lands in the eastern regions of Massachusetts, Great Lakes coastal wetlands, spread across large areas of North America, peatlands present in a glacial lake called Agassiz in the central North American region, and Sundarbans mangrove, India. The metagenomic analysis using high-throughput sequencing technologies showed an abundance of mainly Proteobacteria. In aquatic wetlands, an abundance of bacterial phyla like Actinobacteria, followed by Bacteroidetes and certain types of Cyanobacteria, is usually found, whereas in soil, dominance of bacterial phyla like Acidobacteria, with paramount preponderance of Bacteroidetes, and different types of Actinobacteria are seen. In the mangrove areas of Sundarbans, an abundance of Deltaproteobacteria, with varied types of Alphaproteobacteria, and diverse forms of Gammaproteobacteria were observed, showing a distinct variation in the microbial flora from the normal bacterial phyla that are usually found in the soil and aquatic wetlands. Apart from Proteobacteria, other kinds of phyla that also show dominance usually include Actinobacteria and Acidobacteria, with a preponderance of Bacteroidetes, Chloroflexi, and Cyanobacteria. Diverse types of Firmicutes, *Nitrospira*, and other prevalent types of Planctomycetes were found in the sediments of Sundarbans in different seasons. Lonar soda lake in Maharashtra, India, is a wetland region. The different bacterial phyla that are present in these areas usually include *Bacillus*, *Paenibacillus*, and *Alkalibacillus*, and most importantly, some other associated species like *Exiguobacterium*, *Planococcus*, and *Enterococcus*. Although at relatively lower levels, species like *Vagococcus*, *Halomonas*, and *Stenotrophomonas* are present, yet they have a very dominant role in these regions. Certain types of phyla like *Providencia*, *Cellulosimicrobium*, and *Dietzia* are also present in these areas. Species like *Arthrobacter* and *Micrococcus* are found predominantly in these areas as well. Both anaerobic and facultative bacteria have been found in the Salar de Huasco, Chile, with the predominance of certain species like *Sulfuricurvum*, *Sulfurimonas*, *Rheinheimera*, *Thioalkalivibrio*, Epsilonproteobacteria, and Gammaproteobacteria. In the region of the Qinghai-Tibetan Plateau, three wetland areas, namely Dangxiong, Hongyuan, and Maduo, were studied. Apart from Proteobacteria and Actinobacteria, dominance of Gammaproteobacteria, belonging to the order of Xanthomonadales, was detected in places of Dangxiong and wetlands of Maduo. Different types of methanogens representing mostly the genera of *Methanobacterium*, types of *Methanosarcina*, and certain forms of *Methanosaeta*, with additional methanotrophic organisms mainly representing *Methylocystis*, were found to be present in wetlands of Hongyuan. Bacterial isolates having greater than 98% similarity with bacterial species like *Psychrobacter*, *Halomonas*, and *Aeromonas*, and some other species like *Pseudomonas*, *Paracoccus*, *Brevundimonas*, and *Rhizobium* with great congruence from *Exiguobacterium*, *Bacillus*, *Planococcus*, *Dietzia*, and *Chryseobacterium* were found. These bacteria have the ability to withstand extremophilic conditions such as extreme high altitude and low temperature (De Mandal et al., 2020). Analysis of the bacterial isolates from the mangrove soil areas of the Mahanadi River delta, region of Odisha, in India exhibited 30 types of bacteria that had properties of sulfur oxidation, which were found to be isolated from varied areas of the mangrove soil (Behera et al., 2014).

The microorganisms present in natural wetlands have the ability to eliminate virulent pollutants from the soil or water bodies using their own biological as well as chemical mechanisms (Zedler et al., 2005). These microorganisms perform a variety of functions that help in alleviating the contaminants that are present (Abou-Elela, 2019).

Certain types of bacteria in the wetlands have abilities related to nutrient cycling and phosphate solubilization (Abed et al., 2018). The properties of these PGPR are represented in Figure 11.3.

Certain wetlands, like the Pallikaranai marshland in Chennai, are also referred to as bogs. These areas are highly acidic and are usually deprived of nutrients due to the presence of peat and other decayed plants (Surya, 2016; Jose et al., 2018). The bacteria present in these wetlands are responsible for degrading xenobiotic compounds. With the use of metagenomics, different microbial flora present in these areas can be identified. Soil metagenomics of this area was done using kits like DNeasy PowerMax Soil Kit, MG-RAST (i.e., Metagenome Rapid Annotation using Subsystem Technology server), was used for the classification of the taxonomic data and interpretation of the role of bacteria present in these marshy wetlands (Gregoracci et al., 2012). For the degradation of xenobiotic compounds, novel genes related to pathways can be identified. Analysis of the taxonomic orders of microbes was done by balancing the reads of MG-RAST and SSU-SILVA (unwanted database) (Yurgel et al., 2019). The metagenomic analysis of the area showed that it had various types of bacteria belonging to different phyla, such as Planctomycetaceae, Flavobacteriaceae, Burkholderiaceae, Flammeovirgaceae, Rhodospirillaceae, Nitrospiraceae, Verrucomicrobiaceae, Clostridiaceae, and Xanthomonadaceae (Khairun et al., 2021).

Diversity of bacteria in two varied types of wetlands of the Maritsa River Basin, Bulgaria, that is, the Zlato Pole wetland, followed by paddy fields of rice. The soil samples were collected, and during metagenomic analysis, the hypervariable region, that

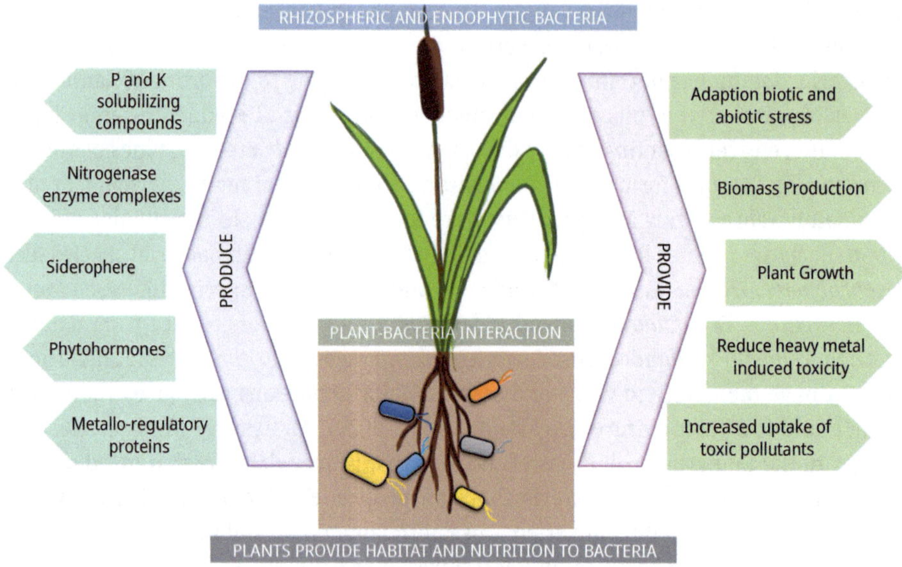

Figure 11.3: Plant growth-promoting properties of rhizospheric and endophytic bacteria.

is, V3–V5 of the 16S rRNA gene, was amplified. The sequencing was done using the MiSeq Illumina platform. FastQC was used for checking the raw data quality. Only sequences of high quality are usually preserved. Low-quality bases are usually removed. The quality of the raw data was checked using FastQC (Bolger et al., 2014). Operational taxonomic units were identified using the QIIME script, and the taxonomy to the phyla was assigned using SILVA (Quast et al., 2012). KRONA charts are usually generated for visualizing the diversity of the taxonomic groups of the samples (Ondov et al., 2011). The taxonomic composition of the studied samples showed the dominance of certain major phyla like Proteobacteria, Actinobacteria, Acidobacteria, and Bacteroidetes. Proteobacteria was the most dominant phylum that was found in the varied wetlands of the Maritsa River Basin. Proteobacteria has a profound role in the cycling of nutrients as well as carbon, nitrogen, and sulfur cycling. The presence of Acidobacteria is usually known to be an extremely sensitive index of the changes in the environment due to the impact of activities related to agriculture. In the Zlato Pole wetland sample, dominance of Firmicutes and Bacteroidetes was seen over Acidobacteria. In the rice paddy fields of these regions, a plenitude of Chloroflexi and Gemmatimonadetes was found as compared to the pole wetland sample. The variety of the dominant flora obtained in these areas is closely associated with the contents of nitrogen, organic matter, and ammonium nitrogen. The foundation of the varied flora found in these areas is based on the type of soil present in these wetlands as well as the kinds of management practices followed in this area (Iliev et al., 2019).

11.5 Microbial keystone species and their significance in wetland ecosystem

The significance of wetlands in maintaining the ecological balance of the environment, as well as their economic impact, is far and wide. Due to the virtue of being in the transition zone of terrestrial and aquatic ecosystems, these regions act as a buffer zone between oxic and anoxic microenvironments (Bodelier, 2013). It is the rich diversity of microorganisms like bacteria, archaea, and fungi in the sediments of wetland regions that makes the biggest impact in creating these specific microenvironments. The dynamics between the microbes and their spatial ecology is one of the major causative factors behind carbon, nitrogen, sulfur, and other biogeochemical cycles. The rising concern about wetlands and exploring their microflora gained popularity in the last few decades. However, it was in recent times that high-throughput NGS mechanisms facilitated their understanding (Brown et al., 2011). A common example in this regard is to explore the tag-based pyroquenching of the conserved 16s rRNA sequence of bacteria to understand species diversity (Deng et al., 2015). A deeper look into further classes of bacteria will show that Proteobacterial species like the nitrate reductase and denitrifying bacteria *Nitrospira* sp., the anammox-based *Nitrosomonas* sp., contaminant-degrading species of *Pseudomonas*, sulfate reducers like *Desulfovibrio* sp. or *Geobacter* sp. are prevalent chemotrophs or autotrophs of these wetland ecosystems. The general abundance of Proteobacteria ranges somewhere around 37.5% in any wetland ecosystem and can be easily tagged as keystone species for a variety of essential functions like nitrification, ammonium oxidation or anammox, denitrification, iron, and sulfur reduction. Following the prevalence of bacterial genera, the *Actinomycetes* or *Actinobacteria* come next with around 17.3% relative abundance. Although it can be termed as a relatively lower abundance due to the slow decomposition of some of the major species in this genus, they play an important role in bioprospecting and bioreclamation. Some examples of species in this genus include *Streptomycetes*, a very common resistant type of bacteria, and toxin-degrading *Arthrobacter* sp. The third category of bacteria are the Firmicutes and the Bacteroidetes, which comprise even less than 11% of relative abundance in wetland ecology. This group includes facultative aerobic species of *Bacillus* sp. and spore-forming pathogenic *Clostridium* sp. Apart from these important ones, the prevalence of cyanobacteria and other phototrophic bacteria showed the bioprospecting nature of wetland soils (Deng et al., 2015).

Archaea are a group of microorganisms that are found vastly in any wetland ecosystem, playing a major role in different grades of anaerobic reduction processes. These processes include anaerobic reduction of sulfur, ammonia, chemolithotrophism, and so on. According to Head et al. (1998), such organisms were classified into nitrifiers and methanogenic bacteria like Euryarchaeota, Methanobacteria, *Methanosarcina*, and Crenarchaeota. Another important class of organisms that have a major role to play in wetland ecosystems are the algal, fungal, and other higher-order organisms. According to Singh et al., 2018, algal species along with other eukaryotes like *Daphnia* sp. are responsible for

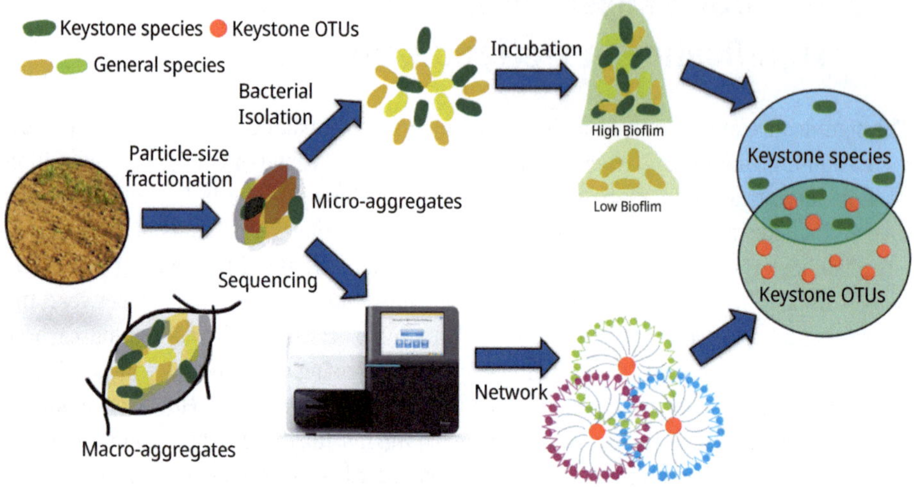

Figure 11.4: Identification of keystone species and their operational units from a microaggregation setup.

maintaining the energy production in wetland ecosystems by photosynthesis. On the other hand, the presence of fungal populations in wetland ecosystems is low to moderate due to the extreme anoxic environment and lower decomposition rates of fungi (Lodge et al., 1995). According to theories proposed by Wu et al. (2013), the diversification of fungal colonies found across various wetlands can be attributed to different interactions of rivers, riverine currents, and sediments. This can help in devising a true relationship between geographical locations and fungal keystone species across corresponding wetlands. Approximately 74,000 different species of fungi have been identified in wetlands across the globe, predominantly belonging to different genera like *Penicillium*, *Trichoderma*, and *Aspergillus* (Kennedy & Clipson, 2003; Wu et al., 2013). A detailed study on the fungal diversity of East Kolkata Wetlands was conducted by Nath and Kalam (2014), where they identified around 300 different fungi including species like *Penicillium*, *Aspergillus*, *Fusarium*, *Cylindrocladium*, and *Umbelopsis*. Another interesting observation was made about the presence of endophytic fungi living in association with the leaves and roots of wetland plants that are quite abundant in nature and endure a symbiotic relationship with the wetland flora. Some extensive researches made by Waller et al. (2005) and You et al. (2012) showed that the endophytic fungi help in an array of plant growth-promoting activities as well as induce systemic resistance. In the mangrove wetlands of Goa, India, *Glomus* sp. was noted as the most predominant keystone species of endophytic fungi, closely followed by other species like *Racocetra*, *Rhizophagus*, and *Funneliformis* (D'Souza & Rodrigues, 2013). Meanwhile, in Deepar Beel, a very famous Ramsar site of Assam showed *Glomus*, *Gigaspora*, *Scutellospora*, and so on as the predominant keystone genera of the region (Choudhury et al., 2010). Apart from bacterial, fungal, and algal populations of wetlands, another class of organisms which is found predominantly in wetland ecosystems is the archaeal population. Although the archaeal population is profoundly studied in the wetlands of extensive cold climatic regions like the Tundra, the methanogenic family is found quite frequently in local wetland ecosystems as well. An overall understanding of these methanogenic bacterial and archaeal strains becomes important to estimate and understand global methane emissions. Approximately 50 different species of archaea were isolated and grouped based on their morphological diversity and the presence of two very unique cofactors (F 420 and 2-mercaptoethanesulfonic acid, or commonly known as cofactor-M or CoM). The process of identification of keystone species and their operational units are represented in Figure 11.4. Some of the important keystone species with respect to archaeal populations include *Methanobacterium*, *Methanobrevibacter*, *Methanothermococcus*, *Methanofollis*, and *Methanoignis* (De Mandal et al., 2020). Next to methylation and demethylation, the sulfate reducers play a pivotal role in wetland ecosystems. According to studies done by Pester et al., 2012, around 36–50% of the total Earth's anaerobic carbon mineralization occurs in wetland ecosystems, and this dissimilatory sulfate reduction system can be considered one of the oldest metabolic processes occurring. Major sulfate-reducing organisms acting as keystone species in wetland ecosystems include *Desulfosporosinus*, *Desulfovibrio*, and *Desulfobulbus*. Another major category of bacteria found in wetland ecosystems is the class of phosphate solubilizers. These are the class of organisms that survive by hydrolyzing ester-phosphatase bonds of inorganic phosphorus, making it available and soluble. This class of microbes belongs to both aerobic and anaerobic classes. Major strains predominantly found in wetland ecosystems include *Bacillus*, *Burkholderia*, *Micrococcus*, and *Rhizobium* sp.. Apart from these major groups of microbial species, denitrifying microbes and microbes with metal-reducing or fermentation abilities play a pivotal role in maintaining the balance of ionic flow in a wetland ecosystem (De Mandal et al., 2020).

11.6 Conclusion

In this chapter, we deduce the concept that wetlands function as an interface between aquatic and terrestrial ecosystems. The geographical diversity of wetlands is a key indicator of the variation of microorganisms found in these wetlands. The wide variety of wetlands usually includes floodplains, tidal and intertidal marshes, swamps, bogs, and fens, which are characterized depending upon the presence of telluric water that has infiltrated the bedrocks of the wetland soil. The extrapolation of metagenomic analysis from this chapter highlights the fact that metagenomic analysis is essential for understanding the microbial diversity of wetlands. It involves the process of sample processing followed by sequencing, assembly, binning, annotation, experimental design, statistical analysis, data storage, and data sharing. Beneficial microorganisms in the soil have a crucial role in the promotion of plant growth and are therefore referred to as PGPR. It can be inferred that the keystone species of wetlands have a vital role in the oxidation of ammonia, phosphate solubilization, nitrification, and so on, whereby they have a pivotal role in regulating the functioning of wetlands. Wetlands are a rich source of biodiversity; therefore, it is essential to protect the wetlands to ensure the well-being of organisms on the planet and the planet itself.

References

Abed, R. M. M., Al-Kharusi, S., Gkorezis, P., Prigent, S., & Headley, T. (2018). Bacterial communities in the rhizosphere of *Phragmites australis* from an oil-polluted wetland. *Archives of Agronomy and Soil Science, 64*(3), 360–370. 10.1080/03650340.2017.1352087

Abou-Elela, S. I. (2019). Constructed wetlands: The green technology for municipal wastewater treatment and reuse in agriculture. *Handbook of Environment Chemistry, 75*, 189–239. https://doi.org/10.1007/698_2017_69

Balwan, W. K., & Kour, S. (2021). Wetland- an ecological boon for the environment. *East African Scholars Journal of Agriculture and Life Science, 4*(3), 38–48. 10.36349/easjals.2021.v04i03.001

Behera, B., Patra, M., Dutta, S., & Thatoi, H. (2014). Isolation and characterisation of Sulphur oxidising bacteria from mangrove soil of Mahanadi river delta and their Sulphur oxidising ability. *Journal of Applied & Environmental Microbiology, 2*(1), 1–5.

Bodelier, P. L., & Dedysh, S. N. (2013, Apr 5). Microbiology of wetlands. *Frontiers in Microbiology, 4*, 79. 10.3389/fmicb.2013.00079. PMID: 23577010; PMCID: PMC3617397.

Bolger, A. M., Lohse, M., & Usadel, B. (2014). Trimmomatic: A flexible trimmer for Illumina sequence data. *Bioinformatics, 30*(15), 2114–2120.

Borrel, G., O'Toole, P. W., Harris, H. M., Peyret, P., Brugere, J.-F., & Gribaldo, S. (2013). Phylogenomic data support a seventh order of methylotrophic methanogens and provide insights into the evolution of methanogenesis. *Genome Biology and Evolution, 5*(10), 1769–1780.

Bowen, J. L., et al. (2011). Microbial community composition in sediments resists perturbation by nutrient enrichment. *ISME Journal, 5*, 1540–1548.

Burke, C., Kjelleberg, S., & Thomas, T. (2009). Selective extraction of bacterial DNA from the surfaces of macroalgae. *Applied Environmental Microbiology, 75*(1), 252–256.

Chevreux, B., Wetter, T., & Suhai, S. (1999). Genome sequence assembly using trace signals and additional sequence information computer science and biology. *Proceedings of the German Conference on Bioinformatics, 99*, 45–56.

Choudhury, B., Kalita, M., & Azad, P. (2010). Distribution of arbuscular mycorrhizal fungi in marshy and shoreline vegetation of Deepar Beel Ramsar Site of Assam, India. *World Journal of Microbiology and Biotechnology, 26*(11), 1965–1971.

D'Souza, J., & Rodrigues, B. F. (2013). Seasonal diversity of arbuscular mycorrhizal fungi in mangroves of Goa, India. *International Journal of Biodiversity, 2013*.

De Mandal, S., Laskar, F., Panda, A. K., & Mishra, R. (2020). Chapter 12 – Microbial Diversity and Functional Potential in Wetland Ecosystems. In S. De Mandal, & P. Bhatt (Eds.), *Recent Advancements in Microbial Diversity* (pp. 289–314). Academic Press, ISBN: 9780128212653, https://doi.org/10.1016/B978-0-12-821265-3.00012-8

Delmont, T. O., Robe, P., Clark, I., Simonet, P., & Vogel, T. M. (2011). Metagenomic comparison of direct and indirect soil DNA extraction approaches. *Journal of Microbiological Methods, 86*(3), 397–400.

Deng, X., Peng, X. J., Jin, B. S., Xiao, M., Chen, J. K., Li, B., Fang, C. M., & Nie, M. (2015). Spatial distributionof bacterial communities driven by multiple environmental factors in a beach wetland of thelargest freshwater lake in China. *Frontiers in Microbiology, 6*, 129.

Dorney, J., Savage, R., Adamus, P., & Tiner, R. (2018). *Wetland and Stream Rapid Assessments: Development, Validation, and Application*. London, San Diego, CA: Academic Press.

Erwin, K. L. (2009). Wetlands and global climate change: The role of wetland restoration in a changing world. *Wetlands Ecology and Management, 17*, 71–84. https://doi.org/10.1007/s11273-008-9119-1

Escobar-Zepeda, A., Vera-Ponce de León, A., & Sanchez-Flores, A. (2015). The road to metagenomics: From microbiology to DNA sequencing technologies and bioinformatics. *Frontiers in Genetics, 6*(348). 10.3389/fgene.2015.00348

Gillespie, D. E., Brady, S. F., Bettermann, A. D., Cianciotto, N. P., Liles, M. R., Rondon, M. R., et al. (2002). Isolation of antibiotics turbomycin A and B from a metagenomic library of soil microbial DNA. *Applied and Environmental Microbiology, 68*, 4301–4306. 10.1128/AEM.68.9.4301-4306.2002

Gorham, E., & Rochefort, L. (2003). Peatland restoration: A brief assessment with special reference to Sphagnum bogs. *Wetlands Ecology and Management, 11*, 109–119. https://doi.org/10.1023/A:1022065723511

Gregoracci, G. B., Nascimento, J. R., Cabral, A. S., et al. (2012). Structuring of bacterioplankton diversity in a large tropical bay. *PLoS One, 7*. https://doi.org/10.1371/journal.pone.0031408

Head, I. M., Saunders, J. R., & Pickup, R. W. (1998). Microbial evolution, diversity, and ecology: A decade ofribosomal RNA analysis of uncultivated microorganisms. *Microbial Ecology, 35*(1), 1–21.

Iliev, I., Marhova, M., Kostadinova, S., Gochev, V., Tsankova, M., Ivanova, A., Yahubyan, G., & Baev, V. (2019). Metagenomic analysis of the microbial community structure in protected wetlands in the Maritza River Basin. *Biotechnology & Biotechnological Equipment, 33*(1), 1721–1732. 10.1080/13102818.2019.1697364

Jose, J. M., & Milton, J. (2018). Current status of Pallikaranai wetland: A review full length review article.

Keddy, P. (2008). Freshwater Marshes. In S. E. Jørgensen, & D. Brian (Eds.), *Fath, Encyclopedia of Ecology* (pp. 1690–1697). Academic Press, ISBN: 9780080454054, https://doi.org/10.1016/B978-008045405-4.00338-4

Kennedy, N., & Clipson, N. (2003). Fingerprinting the fungal community. *Mycologist, 17*(4), 158–164.

Lv, X., Yu, J., Fu, Y., Ma, B., Qu, F., Ning, K., et al. (2014). A meta-analysis of the bacterial and archaeal diversity observed in wetland soils. *The Scientific World Journal, 2014*.

Mitsch, W. J., Gosselink, J. G., Zhang, L., & Anderson, C. J. (2009). *Wetland Ecosystems*. John Wiley & Sons.

Mukherjee, K., & Pal, S. (2021). Hydrological and landscape dynamics of floodplain wetlands of the Diara region, Eastern India. *Ecological Indicators, 121*, 106961, ISSN 1470-160X, https://doi.org/10.1016/j.ecolind.2020.106961

Nath, S., & Kalam, A. (2014). Study of fungal diversity of some selected natural spot of east Kolkata wetland. *Indian Journal of Microbiology Research, 1*(1), 60–71.

Nisha, K., Sunny, J. S., Natrajan, A., et al. (2021, 06 July). Metagenomic analysis of microbial communities from Acidic Marshland Habitat of India. PREPRINT (Version 1) available at Research Square. https://doi.org/10.21203/rs.3.rs-652700/v1

Nnadozie, C. F., & Odume, O. N. (2019). Freshwater environments as reservoirs of antibiotic resistant bacteria and their role in the dissemination of antibiotic resistance genes. *Environmental Pollution, 254*, 113067. 10.1016/j.envpol.2019.113067

Ondov, B. D., Bergman, N. H., & Phillippy, A. M. (2011). Interactive metagenomic visualization in a Web browser. *BMC Bioinformatics, 12*(1), 385.

Pester, M., Knorr, K.-H., Friedrich, M. W., Wagner, M., & Loy, A. (2012). Sulfate-reducing microorganisms in wetlands–fameless actors in carbon cycling and climate change. *Frontiers in Microbiology, 3*, 72.

Pevzner, P. A., Tang, H., & Waterman, M. S. (2001). An Eulerian path approach to DNA fragment assembly. *Proceedings of the National Academy of Science USA, 98*(17), 9748–9753.

Prasad, S. N., Ramachandra, T. V., Ahalya, N., Sengupta, T., Kumar, A., Tiwari, A. K., Vijayan, V. S., & Vijayan, L. (2002). Conservation of wetlands of India – A review. *Tropical Ecology, 43*(1), 173–186.

Quast, C., Pruesse, E., Yilmaz, P., et al. (2012). The SILVA ribosomal RNA gene database project: Improved data processing and web-based tools. *Nucleic Acids Research, 41*(D1), D590–D596.

Richardson, J. L., Arndt, J. L., & Montgomery, J. A. (2001). Hydrology of Wetland and Related Soils. In J. L. Richardson, & M. J. Vepraskas (Eds.), *Wetland Soils*. Boca Raton, FL: Lewis Publishers.

Riesenfeld, C. S., Schloss, P. D., & Handelsman, J. (2004). Metagenomics: Genomic analysis of microbial communities. *Annual Review of Genetics*, *38*, 525–552.

Schwintzer, C. R. (1981). Vegetation and nutrient status of northern Michigan bogs and conifer swamps with a comparison to fens. *Canadian Journal of Botany*, *59*(5), 842–853. 10.1139/b81-118

Sheikh, A. (2021). *Human Impacts on Wetland Ecosystems: Case Study of Ramsar Site of Jammu Region*. Jammu and Kashmir, India.

Singh, R., Upadhyay, A. K., Chandra, P., & Singh, D. P. (2018). Sodium chloride incites reactive oxygenspecies in green algae Chlorococcum humicola and Chlorella vulgaris: Implication on lipidsynthesis, mineral nutrients and antioxidant system. *Bioresource Technology*, *270*, 489–4.

Surya, S. (2016). Landscape ecological urbanism for restoration of Pallikaranai marsh land, Chennai, Tamil Nadu. *Procedia Technology*, *24*, 1819–1826. https://doi.org/10.1016/j.protcy.2016.05.227

Thomas, T., Gilbert, J., & Meyer, F. (2012). Metagenomics – A guide from sampling to data analysis. *Microbial Informatics and Experimentation*, *2*(3).

Tiner, R. W. (2018). Hydrology of Coastal Wetlands. In *The Wetland Book*. Dordrecht: Springer. https://doi.org/10.1007/978-90-481-9659-3_63

Turner, R. K., Van Den Bergh, J. C., Söderqvist, T., Barendregt, A., Van Der Straaten, J., Maltby, E., et al. (2000). Ecological-economic analysis of wetlands: Scientific integration for management and policy. *Ecological Economics*, *35*(1), 7–23.

Waller, F., Achatz, B., Baltruschat, H., Fodor, J., Becker, K., Fischer, M., et al. (2005). The endophytic fungus *Piriformospora indica* reprograms barley to salt-stress tolerance, disease resistance, and higher yield. *Proceedings of the National Academy of Sciences*, *102*(38), 13386–13391.

Winter, T. C. (1988). A conceptual framework for assessing cumulative impacts on the hydrology of nontidal wetlands. *Environmental Management*, *12*, 605–620. https://doi.org/10.1007/BF01867539

Wu, B., Tian, J., Bai, C., et al. (2013). The biogeography of fungal communities in wetland sediments along the Changjiang River and other sites in China. *ISME Journal*, *7*, 1299–1309. https://doi.org/10.1038/ismej.2013.29

Yang, X., Zhu, Z., Qiu, S., Kroeger, K. D., & Zhu, Z. (2022). Scott Covington, Detection and characterization of coastal tidal wetland change in the northeastern US using Landsat time series. *Remote Sensing of Environment*, *276*, 113047, ISSN 0034-4257, https://doi.org/10.1016/j.rse.2022.113047

You, Y.-H., Yoon, H., Kang, S.-M., Shin, J.-H., Choo, Y.-S., Lee, I.-J., et al. (2012). Fungal diversity and plant growth promotion of endophytic fungi from six halophytes in Suncheon Bay.

Yurgel, S. N., Nearing, J. T., Douglas, G. M., & Langille, M. G. I. (2019). Metagenomic functional shifts to plant induced environmental changes. *Frontiers in Microbiology*, *10*, 1–12. https://doi.org/10.3389/fmicb.2019.01682

Zedler, J. B., & Kercher, S. (2005). Wetland resources: Status, trends, ecosystem services, and restorability. *Annual Review of Environment and Resources*, *30*, 39–74. https://doi.org/10.1146/annurev.energy.30.050504.144248

Zhang, J., Lin, S., & Zeng, R. (2007). Cloning, expression, and characterization of a cold-adapted lipase gene from an antarctic deep-sea psychrotrophic bacterium, Psychrobacter sp. 7195. *Journal of Microbiology and Biotechnology*, *17*, 604–610.

Zhang, L., Chen, F., Zeng, Z., Xu, M., Sun, F., Yang, L., Bi, X., Lin, Y., Gao, Y., Hao, H., Yi, W., Li, M., & Xie, Y. (2021). Advances in metagenomics and its application in environmental microorganisms. *Frontiers in Microbiology*, *12*(766364). 10.3389/fmicb.2021.766364

Sophiya Alexander, Siddant Ratanpal, Nilesh S. Wagh* and Jaya Lakkakula*

Chapter 12 Metagenomics profiling of microbes for remediation process

Abstract: The presence of pollutants in environmental niches has proven to be detrimental to the overall functioning of natural processes and the industrial sector's production pipeline, resulting in a great deal of economic and ecological loss. The presence of heavy pollutants in soil and aquatic habitats has mitigated the overall diversity of natural biotas. The use of microbe-assisted techniques to eliminate pollutants has proven to be a cost-efficient and eco-friendly method to restore various contaminated ecological niches. Major advancements in the field of omics and high-throughput sequencing technologies, like shotgun sequencing and next-generation sequencing, in addition to the aid of bioinformatics, have provided assistance in the taxonomic profiling of species in microbial communities. Metagenomic profiling studies include diversity and physiological processes that govern the indigenous microorganisms present in the environment using 16S rRNA metagenomic sequencing for taxonomic screening and the construction of libraries for functional screening of genes. The DNA sequences are classified and characterized, and their taxonomic composition is determined. These serve as a tool, providing in-depth analysis of microbiota – various detoxifying microbes, novel functional genes, metabolites, active metabolic pathways, and bioactive substances have been identified. This has allowed us to improve upon the methods of bioremediation by tapping into the metabolism, enzymatic activity, and genetic makeup of microorganisms. This metagenomic approach has also helped in the environmental monitoring of microorganisms that cannot be cultured under laboratory conditions. The construction of metagenomic libraries and screening of improved strains is exploited to discover various degrading enzymes and other biomolecules that can assist in various bioremediation strategies. Data obtained from profiling also show the functional capacities of different microbial communities to degrade various pollutants present in wastewaters and can be applied

***Corresponding author: Nilesh S. Wagh,** Amity Institute of Biotechnology, Amity University Maharashtra, Mumbai-Pune Expressway, Bhatan, Panvel, Mumbai 410206, Maharashtra, India, e-mail: waghnil@gmail.com
***Corresponding author: Jaya Lakkakula,** Amity Institute of Biotechnology, Amity University Maharashtra, Mumbai-Pune Expressway, Bhatan, Panvel, Mumbai 410206, Maharashtra, India; Centre for Computational Biology and Translational Research, Amity Institute of Biotechnology, Amity University Maharashtra, Mumbai-Pune Expressway, Bhatan, Panvel, Mumbai 410206, Maharashtra, India, e-mail: spencerjaya@gmail.com
Sophiya Alexander, Siddant Ratanpal, Amity Institute of Biotechnology, Amity University Maharashtra, Mumbai-Pune Expressway, Bhatan, Panvel, Mumbai 410206, Maharashtra, India

https://doi.org/10.1515/9783110793635-012

to treat industrial effluents to eliminate their toxic nature. This chapter provides data and functionality of microbial species in different scenarios, such as eliminating hydrocarbons from water reservoirs contaminated with petroleum or metal residues, which can be used to redesign bioremediation strategies to increase the overall efficiency of the remediation process.

12.1 Introduction

The constant outpour of effluents and wastewater from industries containing xenobiotic compounds, especially in urban cities and developing countries, has led to pollution in water bodies worldwide and is disastrous for humans, aquatic life, plant life, microorganisms, and the overall ecosystem. This constant addition of toxic xenobiotic compounds like hydrocarbons, aromatic compounds, and dyes has caused their accumulation in water sources and the entire aquatic landscape (Kachienga et al., 2018). This calls for methods to completely remove or reduce the amounts of toxic wastes to permissible levels from the industrial wastewaters before discharging them into the aquatic systems. Conventional methods used for the elimination of these compounds are generally expensive and produce large quantities of various toxic compounds in the sludge, which, when discharged, can cause detrimental effects on living organisms Mishra et al., 2021). Moreover, the maintenance and sterilization of effluents on a large scale is not economically feasible. Hence, biological treatment methods have been used for this purpose due to their eco-sustainability and the ability to remove most of the toxic compounds from the soil and water bodies in a cost-effective manner.

Bioremediation is the most effective biological removal process known to remove toxic pollutants from wastewaters, break them down, and convert them into substances that can easily enter the biogeochemical cycles (Rathour et al., 2021). This technique has been used for decades for the removal of various toxic compounds from industrial effluents. The only limiting factor of this process is that it is a labor-intensive process, owing to the slow biological metabolism of the microbiome, its complexity, and insufficient knowledge of the mechanism of this process for large quantities of effluents. Secondly, microorganisms present in aquatic systems play a pivotal role by mediating the biogeochemical cycles and have much influence on the constant functioning of almost all ecosystems. Microbial communities are sensitive, and shifts in these communities can lead to an imbalance causing a multitude of problems (Wu et al., 2019). Hence it is important to study the microbial community extensively to understand their mechanism, thereby optimizing the techniques to fully utilize the microbial potential for an effective outcome of the process (Pandit et al., 2021). Bioremediation techniques have been applied to industrial wastewaters for the removal of organic and inorganic recalcitrant compounds, allowing the safe discharge of the wastewater into water bodies. The microbial community introduced into the bioreactors biodegrades these compounds and converts them into less harmful/nontoxic compounds that can further be discharged out in the open (Barak et al., 2020). Pollutants like heavy metals are toxic and inhibit various natural biological processes (Mishra et al., 2021). To understand the complexity and functioning of the microbiome present in the wastewater effluents, next-generation sequencing (NGS) and high-throughput techniques have been implemented to study and explore the diversity of microorganisms at the genomic level, to ascertain their interaction, various metabolic pathways, and the presence of significant genes that are responsible for the breakdown of the toxic compounds in a pool of microorganisms.

Although there have been several methods in the past for water remediation, like microbe-assisted phytoremediation, which have shown results in previous studies (Chaudhary et al., 2019), none of them have been as effective as integrating metagenomics into bioremediation. This technology helps us tap into the potential of microbes, which can enhance the process of bioremediation. Recent breakthroughs in profiling genetic material with 16S ribosomal RNA or rRNA are possible by incorporating high-throughput sequencing, which eludes the process of specific microbial isolation or cultivation. Many studies have used the 16S rRNA gene sequencing or whole genomic sequencing technique to study metagenomes found in various aquatic ecosystems (Zhang et al., 2017). Bioinformatics plays a huge role in metagenomic profiling. One such tool of bioinformatics is Phylogenetic Investigation of Communities by Reconstruction of Unobserved States (PICRUSt), which predicts gene families, their copy numbers, and constructs a reference tree with respective operational taxonomic units (Lee et al., 2019). The end goal of metagenomics is to perceive the bacterial interactions, understand the communication between them, and identify their part in the biodegradation and decontamination of xenobiotic pollutants present in the environment (Krishnaswamy et al., 2020). This will further help us improve our methods for biodegrading industrial wastewaters by developing new and advanced tools and strategies based on our understanding to cut down the toxic effects of xenobiotic compounds and for the safe disposal of treated industrial effluents into aquatic systems. For this purpose, amplicon-sequencing based on 16S rRNA gene is the most widely used technique to determine the taxonomy of the microbial species present dominantly in the sample. This technique is cost-effective, involves high-throughput multiplexing, and is capable of capturing rare biospheres (Lukhele et al., 2021). The results at different taxonomic levels help us understand the relationship between bacteria in a community and indicate that the dominant phyla could be responsible for the breakdown of the persistent toxic compounds. This information is inadequate to analyze the variations in the bacterial community; hence, the relative abundance of the metagenomes at genus levels can also be identified from the sample to study the characteristics of the bacterial community (Kim et al., 2020).

The objective of this chapter is to study the types of wastes that are present in industrial wastewaters, and the taxonomic and functional annotations of the dominant metagenomes present in the sample, which are involved in the biodegradation of toxic recalcitrant compounds by the microbial community present in the soil. These studies help us delve deeper into the understanding of the mechanism of breakdown of various pollutants including heavy metals, dyes, polyaromatic hydrocarbons (PAHs), and other organic compounds that are actively responsible for the increasing environmental pollution.

12.2 Metagenomic profiling of bacterial community for bioremediation of various contaminants

12.2.1 Metagenomic profiling of microbes involved in bioremediation of heavy metals

The presence of heavy metals in wastewater effluents has caused significant environmental and health problems due to their toxicity and bioaccumulation in living organisms. Biological treatments have gained significant attention in recent years as a substitute for chemical treatment methods for the reduction of concentrations of heavy metals present in industrial effluents or for scaling down the toxicity of the generated sludge containing metallic pollutants.

One of the main sources of heavy metal pollution is mining wastewater, which contributes to health problems due to its carcinogenic and/or mutagenic effects. These industries produce large amounts of acidic water from sulfur-bearing minerals, resulting in acid mine drainage (AMD). There are several other methods of treating AMD, which are considered unsustainable, so sulfate-reducing bacteria is generally employed for their degradation, with the limitation that it is highly sensitive toward heavy metals, hence limited to AMD with low heavy metal concentration. This called for studies related to the group of microorganisms present in the wastewater that are responsible for heavy metal degradation and their effectiveness in remediating AMD wastewater. In a study, Akinpelu et al. (2021) performed remediation of AMD containing high amounts of sulfur and heavy metals. The sample was poured into a reactor where it was stirred continuously for 7 days and kept in static batch mode for 14 days. After the adaptation period, the redox potential was observed to be low while sulfate reduction commenced. The pH was increasing, whereas concentrations of the metals were decreasing due to the action of the bacterial community present in the wastewater. BLAST analysis of the amplified 16S rRNA genes was performed to examine the microbes dominant in the wastewater, and the results were as follows: Firmicutes were predominant at the phylum level (39%), Bacilli (40.80%, also the most dominant microbial community at the class level), and Proteobacteria (15.5%) in the raw S_a sample. Although in the treated sample (S_b), class Actinobacteria (16.7%) was prevalent. Shannon-Chao indices were higher for the S_a sample compared to S_b, which indicated a decreasing species richness as a result of the remediation of the sample. The highest removal percentage of the metals present in the sample was Al^{3+} (97%), As^{3+} (65%), Cd^{2+} (98%), Cu^{2+} (91%), Cr^{3+} (69%), Fe^{2+} (88%), Mg^{2+} (55%), Mn^{2+} (94%), Ni^{2+} (78%), Pb^{2+} (94%), Sr^{2+} (94%), and Zn^{2+} (58%). Interestingly, Cd, Cu, Mn, Pb, and Sr were almost removed completely. The performance of the microbial diversity was attributed to the action of metallo-animoproteases in the presence of divalent ions. The presence of *Bacillus cereus* in the bacterial consortium facilitated efficient bioremediation of the heavy metals. The results suggested that the presence of *Bacillus* spp. in the AMD wastewater and their adaptability in extreme environments make them suitable agents in the bioremediation of the metal-laden AMD wastewater (Akinpelu et al., 2021).

A study conducted by Tripathi et al. (2021) provided insight into the bacterial diversity that facilitates the bioremediation of sugarcane molasses-based distillery wastewater containing heavy metal and organic pollutants. Samples were collected from three different sites: D1 (near industrial effluent discharge), D2 (the point where industrial and sewage wastewater gets mixed), and D3 (2 km away from the industry). High pH and electric conductivity (indicating high total concentration of Na^+, Cl^-, SO_4^{2-}, PO_4^{3-}) and heavy metal concentrations of Mn, Fe, Ni, Zn, Cu, Cd, and Cr (due to corrosion of equipment) were detected above permissible levels. GC-MS analysis revealed the presence of hazardous organic compounds in the samples. For taxonomic studies, 16S rRNA regions were amplified, and quantification of the prepared library was performed using qPCR. QIIME software was employed to reveal diversity indices, calculating the Shannon, Chao1, Simpson, and Observed-species indices of samples (0.0013, 10.18, 36,706.55, and 45,653.84, respectively), indicating high bacterial diversity and richness at the D1 site compared to D2 and D3. Predominant phyla present in all three sites were Proteobacteria (~39%), Firmicutes (~28%), Actinobacteria (~10%), Bacteroidetes (~10%, with hydrolytic capacities), Tenericutes (~9%), and Patescibacteria (~2%). Spirochaetes (~5%), Synergistetes (~2%), and Cloacimonetes (~1%) were only detected in D1 samples. At the class level, Clostridia, Deltaproteobacteria, Betaproteobacteria, Erysipelotrichia, and Spirochaetia were only detected in D1. Results showed that there was higher bacterial community diversity in D1, as it was closest to the discharge site. Bacteria present in D1 could be potential candidates to detoxify and biodegrade persistent organic and metal pollutants and improve the chemical quality of discharge (Tripathi et al., 2022).

In 2017, Selvarajan et al. conducted an experiment where five samples from different industrial wastewater sites located in Gauteng province (South Africa) were collected, namely Dixon (from automotive batteries), CWI (high-tensile fencing and barbed wire), Capegate (steel-related products), Ford (zinc-coating steel), and Chemreem (petrol wash and chemical tank trucks), all containing high amounts of toxic metal wastes and organic matter. High concentrations of heavy metals like Mg, Al, Si, and Zn and lower concentrations of B, Pb, and Sr were found in the samples. Bacterial community analysis was performed on the Illumina Mi-Seq platform, which classified bacterial diversity under 41 phyla, 111 classes, and so on. The sequenced datasets were processed using MOTHUR software, and the abundance of the bacteria was Dixon (27.01%), Chemreem (23.08%), Ford (22.10%), Capegate (18.04%), and CWI (9.76%). The Chao1 index showed that OTU richness was highest in Dixon (4,552 OTUs) and lowest in Capegate wastewater (974 OTUs). The Shannon-H index showed the highest diversity index in Dixon (6.04) and the lowest in CWI (4.15). Phylogenetic classification revealed 41 phyla distributed across the samples, the most dominant ones being Proteobacteria: 75.87% (Dixon) – 17.14% (Capegate), Actinobacteria: 39.66% (Capegate) – 1.95% (Dixon), Firmicutes: 34.74% (Capegate) – 2.05% (Dixon), Planctomycetes: 25.21% (Chemreem) – 1.06% (CWI), Verrucomicrobia (5.09% to 0.16%; found in all samples except Ford), and Cyanobacteria: 4.9% to 0.01%. Among these, the distribution of Proteobacteria, Cyanobacteria, and Verrucomicrobia was correlated to NO_3, Zn, Pb, PO_4, S, and Mg concentrations in Dixon and Ford. In Capegate, Actinobacteria and Firmicutes correlated to Al, and Bacteroidetes correlated to Na, Fe, and P. Planctomycetes and Chlamydiae correlated to chemical oxygen demand (COD), salinity, K, Ca, and Mn. Members of these phyla have been reported in other works of literature as well and suggest that these play a huge role in the removal of toxicants. Redundancy analysis (RDA) was generated to map the relationship between physicochemical parameters and bacterial diversity, which confirmed that the heavy metals (Mn, Pb, Zn, and Sr) were correlated to the microbiome present in the samples (Selvarajan et al., 2018).

In an experiment, Sharma et al. (2021) conducted a metagenomic analysis on wastewater from the pulp and paper industry. The samples were collected, and their heavy metal concentration was determined by atomic absorption spectroscopy (AAS), which identified high concentrations of metals as follows: Fe (75.29 mg/L), Zn (53.40 mg/L), Mg (18.57 mg/L), Ni (8.26 mg/L), Cu (8.01 mg/L), Cd (6.37 mg/L), and salts and ions like K^+, Na^+, and Cl were beyond the permissible limits. They amplified the 16S rRNA gene, constructed a library, and analyzed bacterial diversity using Illumina Miseq sequencing, wherein the taxonomic annotation revealed an abundance of Christensenellaceae group (19.8%), *Anaerolinea* (15.7%), *Desulfovibrio* (1.6%), *Candidatus-Solibacter* (1.1%), *Leptolinea* (1.1%), and others (53.2%). The *Proteobacteria* phylum was predominant in the wastewater sample, followed by *Bacteroidia*. The metagenomic analysis highly indicated that Proteobacteria may be responsible for the degradation of organics and detoxification of pollutants containing nitrogen like ammonia and other phenolic compounds and heavy metals. Results showed that microbial communities that facilitate the remediation of metal pollutants can change with various stresses in the environment, such as metals, which can be assessed by metagenomic profiling (Sharma et al., 2021).

A study was conducted in Uttar Pradesh, India, at a site near distillery sludge where *Saccharum arundinaceum* harbors rhizospheric bacterial communities that grow in an organometallic-rich environment. Two samples – rhizospheric (R_1) and non-rhizospheric (R_2) distillery sludge – were analyzed to identify the microbiome responsible for eliminating organic and inorganic pollutants. Heavy metals were present in high amounts, including Fe (~1,512 mg/kg), Zn (~94.25 mg/kg), Cu (~61.55 mg/kg), Cr (~17.524 mg/kg), Cd (~1.012 mg/kg), Ni (~10.115 mg/kg), and Mn (~95.15 mg/kg), which were found in the R_2 sample. The Shannon and Chao1 indices and further principal component analysis revealed that the bacterial community grown in R_1 regions is diverse and rich compared to R_2 regions (76.50% and 23.50% difference in their relative abundance). Results obtained from sequencing also confirmed the presence of *Alcanivorax* (1.07% reads), which indicated that these genera could adapt and grow in highly stressed conditions. Unique OTUs were more present in R_1 compared to the R_2 sample. Proteobacteria (predominant in both samples) and its classes including Alphaproteobacteria, Betaproteobacteria, Gammaproteobacteria, Deltaproteobacteria, and Gemmatimonadetes showed a significant increase of 50% in the R_1 sample. The findings, therefore, indicate that Proteobacteria play a fundamental role in the degradation of metallic contaminants in the soil (Kumar & Chandra, 2020).

In another study conducted on the mangrove ecosystem, which is vulnerable to the bioaccumulation of heavy metals and harbors a heavy metal resistome bacterial community, two sites (BTMV and BTPV) were analyzed for heavy metal concentration. The abundance of heavy metals was in the following concentrations (mg/kg): Cd (0.73), As (12.03), Pb (27.91), Cu (33.70), Ni (41.34), Cr (104.0), and Zn (139.15) in BTMV; and Cd (1.60), As (5.61), Pb (15.93), Ni (28.11), Cu (45.49), Cr (93.65), and Zn (295.48) in BTPV, where Cd (74% and 89%, respectively) and Zn showed the highest values of total potential ecological risk. Whole metagenome shotgun sequencing and taxonomic profiling revealed an abundance of Proteobacteria (53.593–75.99%, with the majority classes being Deltaproteobacteria and Gammaproteobacteria), Bacteroidetes, Firmicutes, Acidobacteria, and Actinobacteria. Comparative studies of taxonomy showed an abundance of *Anaeromyxobacter*, *Geobacter*, *Candidatus Solibacter*, *Pseudomonas*, and others. MRGs analysis from these regions revealed the prevalence of Cu-Zn-Cd resistance genes (>45%), Cu homeostasis, resistance to Zn and Ar (~90%), and the predominance of Proteobacteria (specifically Gammaproteobacteria) in samples. Results proved that exoelectrogenic bacteria like *Geobacter*, *Rhodopseudomonas*, and *Pseudomonas* can be employed in constructing MFCs for in situ bioremediation of heavy metals in soil. Future studies focus on identifying novel bacterial taxa to add momentum to current research in this field (Puthusseri et al., 2021).

In 2017, Bhakta et al. identified a bacterial consortium resistant to toxic heavy metals like As, Cd, and Hg through metagenomic-DNA fingerprinting using PCR-DGGE (polymerase chain reaction denaturing gradient gel electrophoresis) of the samples. Four samples were collected randomly from different polluted sites (S1, S2, S3, S4), and their As, Cd, and Hg concentrations were measured using ICP-AES and AAS. The presence of heavy metals was detected in high concentrations: As (1,938 to 283 µg/kg), Cd (1,407 to 108 µg/kg), and Hg (207 to 23 µg/kg). PCR-DGGE analysis showed that the majority of bacteria in the sample were resistant to As, which indicated that Hg and Cd have a higher toxic impact on the microbes than As. Some of them even showed multi-heavy metal resistance. The Cd-resistant consortium showed the highest removal efficiency, followed by the Hg and As-resistant consortium. Samples S1 and S4 showed higher bacterial diversity than S2 and S3. Results showed that bacterial consortia that are resistant to heavy metals can survive under heavy metal stress and can reduce the contamination by uptake and resistance capacity, which can help in the bioremediation of heavy metals in the environment (Bhakta et al., 2018).

Prakash et al. (2021) studied the bioreduction of Cr(VI) to Cr(III) from tannery effluent contaminated soil, and remediation of the sample was done using the bioelectrokinetic process wherein indigenous microbes were induced (Figure 12.1). An electric field (2 V cm^{-1}) was applied for 5 days, during which the microbes facilitated the conversion of sulfate ions into sulfide when K_2SO_4 was added as an electrolyte. This led to the formation of FeS particles that further reduced toxic Cr(VI) into nontoxic Cr(III) (as shown in Figure 12.1). Metagenomic analysis revealed that at the genus level, 177 bacterial genera were detected including Acidobacteria (48.47%), Deltaproteobacteria (9.70%), Alphaproteobacteria (6.80%), Gammaproteobacteria (5.70%), and others (4.80%). The phyla Proteobacteria, Firmicutes, and Bacteroidetes were present abundantly in the sample. Gammaproteobacteria proved to play an important role in the oxidation of the sulfides. Alphaproteobacteria and Deltaproteobacteria were highly resistant to heavy metals and were responsible for bioremediating Cr through the process of intracellular accumulation, bioabsorption, enzyme-catalyzed transformation, and other mechanisms (Prakash et al., 2021).

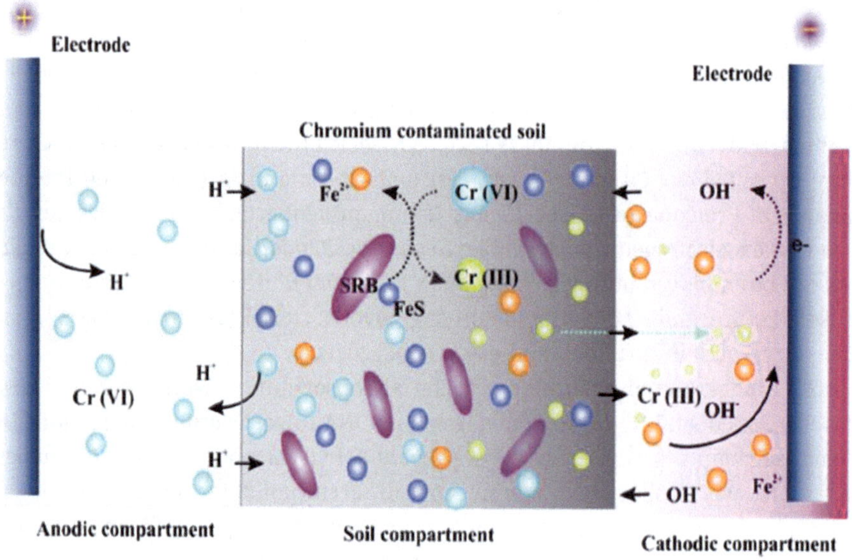

Figure 12.1: Schematic representation of bioelectrokinetic enhanced remediation of chromium-contaminated soil. Reprinted with permission from Prakash et al. (2021).

In one of the experiments, Tripathi et al. (2021) studied the degradation and detoxification of organometallic pollutants present in distillery wastewater using biostimulation and bioaugmentation of the samples. The wastewater contained heavy metals in high concentrations (mg/L) including Fe (2,403), Zn (210.15), Cr (22.825), Cu (73.62), and other heavy metals. A bacterial consortium containing two strains of *Klebsiella pneumoniae* (AS7 and AS1) in the stimulated sample and *Enterobacter cloacae* as the bioaugmented strain was introduced into the sample. The autochthonous bacterial strains were isolated from the biostimulated sample using 16S rRNA sequencing, which revealed that the bacterial consortium mediated the biotransformation and biodegradation of pollutants present in the sample. After biostimulation, there was a slight reduction in all the physicochemical parameters, but after the further bioaugmentation process, the degradation activity of the consortium improved significantly. SEM analysis showed an increase in bacterial diversity and population and confirmed that the organic compounds present in the wastewater are being used as nutrients by the cells for their growth and development through co-metabolism. The metal concentrations decreased significantly: Fe (852.529, by 39%), Cu (11.526, 81%), Cr (11.526, 51%), Cd (0.94, 19%), and Zn (21.463, 77%) (mg/L). The results indicate that *Klebsiella pneumoniae* and *Enterobacter cloacae* use a broad range of chemical compounds as their carbon source and hence can act as a niche for these indigenous microbes to degrade the pollutants in distillery wastewater (Tripathi et al., 2021).

A study was performed to investigate the area upstream of Miyun Reservoir (Beijing), close to a mining field with high amounts of metallic tailings, to study the microbial community harboring its sediments and their relationship with the geochemical properties. Cd concentration was found to be approximately 3.43 to 10.9 times higher than the risk screening value. Zn (89.4 mg/kg), Ni (31.8 mg/kg), Pb (18.1 mg/kg), and As (2.87 mg/kg) were also measured. RDA and multivariate regression tree analysis was performed, which showed that As, Zn, Pb, OM, available P, pH, and available K were the factors influencing the distribution of bacterial communities. Collected samples were analyzed using the Illumina MiSeq platform targeting the 16S rRNA gene. Bacterial diversity indices (Shannon index) were measured using MOTHUR software, which identified 51 phyla, 121 class, and so on. The predominant phyla identified were Proteobacteria (32.2%), followed by Actinobacteria (33.3 to 14.3%), Chloroflexi (11.7 to 7.5%), Acidobacteria (16.2 to 3.6%), Bacteroidetes (8.1 to 2.1%), and Firmicutes (6.8 to 1.3%). Functional annotation analyzed using KEGG (Kyoto Encyclopedia of Genes and Genomes) and eggNOG databases revealed that Proteobacteria (specifically class Betaproteobacteria) harbored the highest variety and abundance of metal resistance genes, followed by Bacteroidetes, Actinobacteria, and Firmicutes. Genera that possessed a great number of trace metal resistance genes include Pseudomonas, Flavobacterium, Geobacter, and Herminiimonas. These bacteria also manifested strong DNA repair capacity, as under heavy metal stress, they can be damaged and can produce an oxidative stress response. The study provided a broad understanding of the distribution and resistance mechanisms of the bacterial community in this reservoir. These can successfully be implemented as a bioremediation tool for heavy metal toxicity (Yan et al., 2020).

Verma and Sharma (2020) conducted a study on tannery waste dumpsites to identify potential candidates that are helpful in bioremediating heavy metal-polluted tannery waste. These contained heavy metals in high amounts including Cu, Mn, Cr, Zn, and Fe. Taxonomic annotation was done using the Illumina HiSeq Platform, and the ORFs were predicted as a part of functional profiling. KEGG analysis revealed that the serine protease and metalloproteases were predominant (~89%) in the two samples. Metalloproteases like rseP (sigma E protease regulator), pepA (leucyl aminopeptidase), pepN (aminopeptidase N), and mcp (metallocarboxypeptidases) were identified. Among these, ORFs belonging to metallo-proteases were assigned to Flavobacteriaceae, Bacillaceae, Xanthomonadaceae, Chitinophagaceae, and Clostridiaceae. KEGG and SEED, therefore, revealed the microbial community that carries potential genes responsible for the bioremediation of the heavy metal pollutants present in the soil. The genera identified in this study are also commonly found in TWW, which may give us an understanding of the role of the microbial community in a given environment (Verma & Sharma, 2020).

12.2.2 Metagenomic profiling of microbes involved in bioremediation of dyes

Dye contamination is a major concern in modern times. The majority of dyes that are discarded into the sea are mainly from textile industries. Apart from dyes, large quantities of xenobiotic compounds are also released by textile industries through effluents into water bodies, which degrade overall water quality and deprive aquatic life of their habitat. Hence, there is an urgent need for the remediation of aquatic ecosystems. Therefore, the use of metagenomics and isolating the 16S rRNA gene from a consortium of microbes present in the aquatic ecosystem can help us enhance the process of bioremediation.

A study conducted by Patil et al. (2020) focused on the degradation of a mixture of dyes from textile industry effluent performed with the help of dye-degrading bacteria and the process of bioaugmentation. Dye-degrading bacteria were collected from soils contaminated with textile plant effluent, and then the soils were inoculated in a nutrient broth and grown at 37 °C overnight. A mixture of dyes was then added to the broth-containing bacteria such as green HE 4B, red HE 3B, red HE 8B, scarlet RR, and scarlet GDR. To study the mixture of the bacterial consortium, its influence on the mixture of dyes, textile effluent, and the incorporation of bacterial species for bioaugmentation, six sets of microcosms were created: soil mixture (SM), soil with textile effluent (SME), soil dye mixture, augmented soil (ASM), augmented soil with dye mixture (ASMD), and augmented soil with textile effluent (ASME). Each microbial consortium was investigated using 16S rRNA amplicon sequencing and was placed in reactors along with dye mixtures for 30 days. The results of metagenomic analysis showed that Proteobacteria, Firmicutes, Acidobacteria, Actinobacteria, and Chloroflexi were the most dominant and comprised more than 79% of the total microbiome of soil from all microcosms. The effects of microcosms against the dyes were reported, and it was observed that decolorization, COD, and biological oxygen demand (BOD) removal gradually increased with time and were highest on the 30th day. KEGG pathway analysis was also performed to observe metabolic pathways for xenobiotic degradation. The results of KEGG pathway analysis showed that the SMD, SME, ASMD, ASM, and ASME were enriched with xenobiotic metabolic pathways for the degradation of various xenobiotic compounds like aminobenzoate, bisphenol, chlorobenzoate, methylnaphthalene, dichlorodiphenyltrichloroethane (DDT), xylene, and toluene, and were observed the most in ASMD and ASME microcosms, which were bioaugmented. This study showed that the integration of augmented bacteria, microflora from soil, and degradation of xenobiotics can lead to the bioremediation of contaminated soils (Patil et al., 2020).

In a study conducted by Rathour et al. (2021), bioremediation of textile industry effluent was performed to degrade xenobiotic compounds present in the effluent. The process of biodegradation was performed on an up-flow fixed-film microaerophilic bioreactor (UFMB); also, the microbial characteristics and their dynamics were studied at different hydraulic retention times (HRT) and organic loading rates. The bioreactor consisted of bacterial community DR4 and unsterile raw textile effluent. Every day for 5 days, HRT was measured, and the reactor was allowed to run for three to four cycles after changes in HRT to maintain a steady state. The steady state of the bioreactor was studied by measuring the removal of dye and reduction in COD, BOD, and other contaminants like phenols, sulfates, and nitrates. The biofilm present in the reactor was developed on various packaging materials, and these materials were observed under a scanning electron microscope. At each HRT, biofilm samples were collected, and metagenomic analysis was performed using metagenomic DNA as a template. The resulting amplicons were further sequenced on the Illumina MiSeq platform. Results of the steady state showed that with each measurement of HRT, the amounts of dyes, COD, BOD, phenols, sulfates, and nitrates gradually decreased. Results of the metagenomic analysis showed that Proteobacteria and Bacteroidetes were the most dominant, depending on the upper and lower zones of the UFMB, the packaging material, and the HRT measurement day. Hence, it was concluded that wood charcoal was the best packaging material for bacterial biofilm formation and establishment in textile effluent. The results revealed that the sequential microaerophilic-aerobic treatment process was efficient in removing 98% of dyes, 98.6% of COD, and 92% of BOD from the textile effluent (Rathour et al., 2021).

Another study conducted by Krishnaswamy et al. (2020) examined potential microbial pathways for the degradation of dyes and chemicals present in textile industry effluent. Samples were collected from a textile effluent in Chennai and were then sent for treatment in a laboratory-scale anaerobic sequential batch reactor. Further PCR amplification was performed using nanopore sequencing, and 16S rRNA primers were added. Then the PCR product was barcoded, and a sequenced DNA library was prepared. Among the 30,700 read counts of the sample, 28,245 reads were classified, and the rest remained unclassified. The abundance of read count percentage was obtained from *Burkholderia* taxa, and *Burkholderia pseudomallei* MSHR3960 had the highest count of 13.79% of all classified reads, followed by *Burkholderia* sp. JKS000303 with a read count of six and 10.34% of classified reads. Taxonomical results included Proteobacteria as the most abundant organism, and further classification of Proteobacteria revealed the family of *Burkholderia* to be a significant part of the organism with its 53% abundance in all of Proteobacteria. Pathway annotation was performed on the microbial community in the activated sludge, and the degradation of amines, carbohydrates, aromatic compounds, carboxylates, cyclohexane, and chlorinated compounds was found to be the most dominant. The xenobiotic degradation pathway showed that the degradation of toluene, naphthalene, xylenes, nitrotoluene, and chlorocyclohexane was predominant. Degradation pathways of atrazine, dioxin, phenylalanine, and aminobenzoate were dominant in *Burkholderia pseudomallei.* Interaction of *Burkholderia* was observed with more than 60 taxa, including humans, and was found to be pathogenic to all of them. Hence this study showed the interaction of the microbial community and its degradation pathways for organic and inorganic substances. Also, it was proven that *Burkholderia* is a pathogen and hence should be treated carefully in further studies (Krishnaswamy et al., 2020).

A study conducted by Krishnamoorthy et al. (2021) examined the exposure of azo dyes from textile industrial wastewater to surrounding water and its effect on the microbial community. The samples were collected from the waters and soils within the vicinity of Orathupalayam Dam, as the textile industry using azo dyes was nearby. The soil enzyme activity was measured using dehydrogenase as the enzyme and assaying 2,3,5-triphenyl tetrazolium chloride (TTC), with the assay defined by the production of triphenylformazan. Further, urease activity was measured by the amount of ammonium released after soil sample incubation for 2 h. DNA was isolated from the soil samples; PCR amplification was performed by amplifying 16S rRNA gene using PCR primers. After amplification, a phylogenetic library was prepared using raw beads, and sequence reads were generated. Surrounding waters from the dam showed significantly lower levels of dehydrogenase compared to nearby soils. The reason for this can be the high accumulation of azo dyes in water, which might have lowered the microbial population compared to soil samples. Metagenomic analysis revealed a total of 50 and 54 phyla for water and soil environments, respectively. In the case of water samples, Actinobacteria (20.02%), Proteobacteria (15.49%), Acidobacteria (4.16%), and Chloroflexi (3.31%) were the most prominent, while for the soil environment, Bacteroidetes (3.48%), Chloroflexi (6.34%), Gemmatimonadetes (7.59%), Planctomycetes (8.4%), Acidobacteria (9.94%), Actinobacteria (10.03%), and Saccharibacteria (15.24%) were the dominant ones. Various genera of Proteobacteria were retrieved from textile effluent and were the most active in effluent sludge, whereas Saccharibacteria and Actinobacteria were the ones responsible for the degradation of azo dyes in water and soil samples. Therefore, the use of Saccharibacteria and Actinobacteria can be employed to devise a remediation technique for the degradation of azo dyes from industrial effluents (Krishnamoorthy et al., 2021). A similar study performed by Guo et al. (2021) presented the use of a halo-thermophilic bacterial consortium (HT1) to perform azo dye decolorization in textile wastewater. The use of a halo-thermophilic bacterial consortium is justified as textile wastewater is highly saline and has a relatively higher temperature, making halo-thermophilic bacteria a perfect candidate for the decolorization of azo dyes. A liquid basal medium was created, which comprised Na_2SO_4, $CaCl_2$, KH_2PO_4, KCl, $MgCl_2{\cdot}6H_2O$, NaCl, and yeast extract as a cosubstrate. Bacterial colonies were enriched from textile wastewater sludge, and metanil yellow G (MYG) dye was added externally. Bacterial DNA was retrieved, and the 16S rRNA gene was amplified using Illumina Mi seq high-throughput sequencing. Key enzyme activity of the bacterial consortium was performed to determine the effect of salinity on enzyme activity; for this, the cells were grown at different salinities and were further disrupted using sonication, and disrupted cells were removed using centrifugation. Results were recorded, and it was reported that, from metagenomic analysis, the most abundant family was *Bacillus* followed by *norank Bacillaceae* and *Piscibacillus*. The decolorization rate was influenced by the types of carbon sources used in the media. Yeast extract was the most efficient for decolorization, and lactose was the least efficient for the decolorization of dyes. The most efficient combination of factors for decolorization was the use of yeast extract at 50 °C, pH 7, for 36 h of inoculation, which resulted in the removal of 92% of MYG dye. In the case of results of enzyme activity, LiP, MnP, Lac, azoreductase, and NADH-DCIP reductase are the key enzymes in the decolorization process. It was observed that an increase in salinity decreases the enzymatic activity of the bacterial consortium significantly. The azoreductase and NADH-DCIP reductase activities were significantly higher in the present study. This study showed that the use of a halo-thermophilic bacterial consortium with yeast as a cosubstrate showed high decolorization of MYG at different ranges of pH, temperature, salinity, and so on (Guo et al., 2021).

A study conducted by Sánchez-Reyes et al. (2022) analyzed the biodegradable potential of microbes found in a river basin against industrial textile dye. The water samples containing microbial species were collected from the Apatlaco River Basin in Mexico. Four samples from the surface and sediment were taken and processed. A composite sample consisting of both surface water and sediment was processed against Deep Blue 35 dye, which is an anthraquinone dye, and incubated for 30 days in a batch reactor. From these samples, a Hi-C library was generated in order to extract genomes from metagenomes and to deconvolve microbes and identify functional-phenotypic determinants related to community performance. The microbial community composition was estimated by Hi-C reads against the respective clade-specific marker catalog contained in the MetaPhlAn2 tool. The textile dye build environment contained an abundance of Euryarchaeota followed by Firmicutes, Proteobacteria, and Actinobacteria. COD levels were measured from the composite sample, and it was observed that the initial COD was 265.80 ± 5.0 mg/L and the final COD, after 30-day incubation, was found to be 133.33 ± 3.6 mg/L, almost a 50% reduction. Further, the samples were recorded for discoloration, and the rate of discoloration was recorded to be 23% at a maximum visible wavelength of ~592–630 nm. Various studies have also proven that *Bacillus cereus* DC11 is the highest dye decolorizing strain and achieved 55% decolorization in Acid Blue 25 dye. The mixed or composite culture showed 23% decolorization without bio stimulation, which proves beneficial for biotechnological and bioremediation applications. Further links were established between dye biodegradation and taxa deconvoluted in textile dye by conducting KEGG analysis, which revealed that NADH-azoreductase, polyphenol oxidase, and catalase-peroxidase sequence elements were the most abundant per genome. This study showed that nutritional changes by xenobiotic substances, such as anthraquinone textile dye in this case, can help shape the microbial niche in order to achieve necessary changes at the community level (Sánchez-Reyes et al., 2022).

A study performed by Chaudhari et al. (2017) deals with the biodegradation of anthraquinone dye reactive blue 4 by using aerobic bacterial granules (ABG). ABG were isolated from dye-contaminated areas containing diverse bacterial cultures and then cultured in a column reactor along with supplement media containing yeast extract, peptone, and NaCl, which was introduced at regular intervals for the maturation of bacterial granules. To determine the metagenomic functions of bacterial communities, DNA was extracted from ABG using the phenol-chloroform method. Amplicons extracted were then sent for PCR amplification for 15 cycles, and sequences obtained were formed into a metagenomic library. Further PICRUSt-based imputed metagenomic analysis was done. From this analysis, it was found that xenobiotic degradation pathways like naphthalene, aminobenzoate, and benzoate were upregulated in suspended culture. Pseudomonadaceae was found to be the most dominant among all other bacterial communities. The taxonomic analysis revealed that Proteobacteria, Actinobacteria, Chloroflexi, Firmicutes, and Acidobacteria were in relative abundance in an unexposed granule (Figure 12.2). The efficiency of ABG was checked by the decolorization rate. The decolorization rate during the 1st cycle was 61%, whereas at the 9th

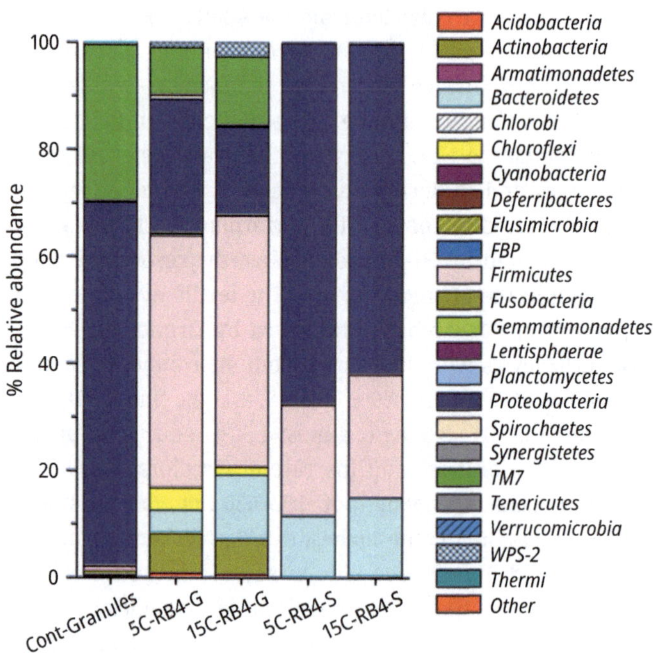

Figure 12.2: Phylum-level distribution of bacteria isolated from aerobic bacterial granules and supernatant after continuous exposure to reactive blue 4. Reprinted with permission from Chaudhari et al. (2017).

cycle, it was found to be at 86%, which was the peak, and then it declined to 55% on the 16th cycle. Few studies have reported that increasing the load of dye can lead to higher efficiency of decolorization by ABG. The above results of decolorization have potential application at the commercial level (Chaudhari et al., 2017).

A study performed by Santhanarajan et al. (2022) investigated biodegradative pathways for Acid Blue 113, which is an azo dye causing toxicity in textile industry effluent/wastewater. The researchers obtained the bacterial consortium commercially and isolated samples for dye wastewater from a wastewater treatment plant in South Korea. The consortium was introduced to wastewater and then placed onto a shaker at 120 rpm. It was found that two bacterial and two yeast cultures were isolated and identified by 16S (for bacterial cultures) or 26S rRNA (for yeast cultures) genes. Further amplicons were isolated by PCR, and bacterial strains were identified as *Mesorhizobium* sp. (B1) and *Sphingomonas melonis* (B2), whereas the yeast strains were identified as *Apiotrichum mycotoxinivarans* (Y1) and *Meyerozyma guilliermondii* (Y2). The dye, along with strains, was inoculated in MSM media and incubated at 27 °C on a shaker, and spectrophotometric readings were taken at 560 nm at regular intervals. It was observed that B1 and B2 achieved maximum decolorization of 67.8% and 75.4% at 50 mg/L of dye within 72 and 48 h of incubation, respectively. B1 was observed to have a slower growth rate, which can be the potential reason for lower degradation efficiency. On the other hand, Y1 showed peak decolorization of 77% after 9 h of incubation, and Y2 showed 60% decolorization after 48 h of incubation. The average decolorization of all four consortium members was found to be 80%. Degradation pathways such as benzoate and naphthalene degradation pathways significantly increased their activity after treatment with Acid Blue 113. This study shows that the use of AB113 can be done to monitor the dye wastewater treatment process (Santhanarajan et al., 2022).

Another study conducted by Kumar et al. focused on key microbial genes encoding enzymes for dye degradation. The researchers collected samples from three different wastewater systems: industry sites, a common effluent treatment plant (CETP), and an ETP, which involved a total of nine sites. DNA was isolated from the samples and was further sent for shotgun metagenome sequencing. Gene prediction was performed for homology against the KEGG database for functional annotations of genes involved in textile dye degradation. It was observed that overall, the most dominant phyla in industry and CETP sites were Bacteroidetes and Proteobacteria, while in ETP, it was found that Firmicutes, Cyanobacteria, and Proteobacteria were dominant. *Aeromonas caviae, Desulfovibrio, Klebsiella pneumoniae, Pseudomonas stutzeri, Shewanella decolorationis, Desulfuricans, Shewanella oneidensis, Shewanella putrefaciens,* and *Vibrio cholerae* were the most abundant species across all three sites. Shotgun metagenomics indicated a key role of biosynthetic cluster genes which were involved in chemical dye degradation. The results of the metagenome indicated the dominance of oxidoreductase and reductase classes of enzymes. More than 20 genes encoding different enzymes for bioremediation potential in the wastewater treatment process were found, and a higher amount of methyl-accepting chemotaxis proteins were also identified, which shows the role of bacterial consortiums in polluted environmental conditions. The result of this study shows that textile industry effluent influences the occurrence and abundance of xenobiotic compounds and genes degrading those compounds at the same time. Also, the identification of functions and roles of genes encoding enzymes can be beneficial in large-scale bioremediation projects (Kumar et al., 2021).

12.2.3 Metagenomic profiling of microbes involved in bioremediation of hydrocarbons

Hydrocarbons are usually generated from the burning of wood, charcoal, coal, gas, and so on, and are known to act as pollutants when ejected along with industrial effluents, which flow into seawater, hence degrading the quality of life in the aquatic ecosystem. Among the mentioned sources, coal is the largest source for the generation of PAHs. Microbial remediation has been considered the go-to method, but using the 16S rRNA gene has proven to be useful and much more efficient in the field of bioremediation.

In a study conducted by Wu et al. (2019), the degradation of PAHs like benzo[a]pyrene (BaP) was performed in activated sludge generated from coking wastewater using phenol as the substrate. A high-throughput sequencing approach was taken to examine the 16S rRNA gene in microbes. The degradation was performed under aerobic and anaerobic conditions separately. Some prominent genera involved in anaerobic degradation were *Tissierella_soehngenia, Diaphorobacter*, and *Geobacter*, while in aerobic degradation, *Rhodanobacter*, *Dyella*, and *Thauera* were dominant. The phylum Proteobacteria was the most dominant. The use of phenol as a substrate showed relatively more degradation of BaP than without phenol. BaP biodegradation efficiency increased from 78.6% to 140% when phenol was implemented as a substrate. BaP degradation with phenol as a substrate led to notable changes in the microbial diversity of wastewater-activated sludge samples, specifically in anaerobic conditioned bed reactors. The findings made in the research can contribute to the progress of practical in situ treatment of coking wastewater-activated sludge containing high contents of PAHs/BaP (Wu et al., 2019). Similarly, in another study performed by Muangchinda et al. (2018), a consortium of bacteria known as SWO was isolated from crude oil wastewater, which was able to degrade a mixture of PAHs, namely phenanthrene, anthracene, fluoranthene, and pyrene, approximately 73%, 69%, 52%, and 48%, respectively, within 21 days. Pseudomonas and Methylophaga were revealed to be the most abundant of the entire consortium with the help of the 16S rRNA sequencing technique. The consortium generated from the crude oil waste water showed resistance to a range of environmental conditions like high pH and temperature. Thus, the consortium SWO showed promising results and can be used for remediation (Muangchinda et al., 2018).

Another study by Xu et al. (2018) on the remediation of wastewater generated from coal gasification containing phenolic compounds and PAHs using methanol was performed. A pilot-scale anaerobic, anoxic, and oxic system was created to treat coal gasification in wastewater and methanol was introduced to the oxic system to eliminate refractory organic pollutants. By the addition of methanol in the oxic system, the average COD removal increased from 24.9% to 36.0%, and phenolic components and PAHs were decreased from 54.4 to 44.9 mg/L, and degradation intermediates were observed. Seventy bacterial communities were isolated from all three systems, and *Proteobacteria* was found to be the most dominant. The growth of phenolic and PAH-degrading bacteria in oxic sludge was enhanced with the introduction of methanol, and microbial degradation pathways for xenobiotics functioned efficiently in the presence of methanol. Consequently, adding methanol in the oxic segment could be a promising strategy for the bioaugmentation of coal gasification wastewater treatment (Xu et al., 2018).

Another study by Kim et al. (2020) examined the efficiency of the microbial community in a petroleum refining wastewater treatment plant, which contained carcinogenic agents and hydrocarbon contaminants such as benzene and phenol. From the samples obtained, the bacterial 16S rRNA was sequenced using the V4 hypervariable region, and genes were amplified using primers. The metagenomic analysis showed more than 30 genera, which included *Sulfuritalea* (11.9 ± 3.5%), *Ottowia* (4.3 ± 2.2%), *Thauera* (3.1 ± 7.2%), and *Hyphomicrobium* (1.3 ± 0.7%), as the most dominant. For the degradation of benzene and phenol, activated sludge process (ASP) and dissolved air flotation (DAF) methods were used. Among the two, DAF removed 8.5 ± 6.4% of benzene and phenol present, whereas ASP showed 100% removal. The results of the study showed that the process stability of the ASP method was significantly higher and can provide beneficial results for microbial activity functions (Kim et al., 2020).

In a study conducted by Muangchinda et al. (2018), the biodegradation of a mixture of PAHs which consists of fluorene, phenanthrene, pyrene, and acenaphthene obtained from river sediments, was analyzed through microcosm analysis. Gas chromatography-mass spectrometry (GC-MS) was performed for the detection of PAHs in sediment samples, and the above-mentioned hydrocarbons were the most dominant. The bacterial microcosm was identified by 16S rRNA gene sequencing, and it was found that actinobacteria became significantly more dominant and showed the highest percentage of PAH degradation (approximately 80% of phenanthrene and fluorene, and 40% of pyrene and acenaphthene were degraded) after a 14-day incubation of a 70-day incubation study. After the 70-day incubation period, approximately 50% of fluorene, 30% of phenanthrene, 60% of pyrene, and 70% of acenaphthylene were degraded. The findings from the study concluded that Actinobacteria have the potential to play a crucial role in the degradation of PAHs from river sediments (Muangchinda et al., 2017).

Another similar study by Mahajan et al. (2021) involved collecting pelagic samples from different sites in the Gulf of Gujarat to examine the fungal community. These samples were grown in basal MS media and supplemented with PAHs such as naphthalene, pyrene, fluoranthene, phenanthrene, and anthracene. Shotgun sequencing was used to identify fungal species and genes involved in the degradation of PAHs. GC-MS analyses revealed that *Aspergillus versicolor* and *Penicillium ilerdanum* were able to degrade the highest percentage of PAHs (approximately 80% of pyrene, fluoranthene, phenanthrene, and 90% of anthracene for *Aspergillus versicolor*, and 90% of phenanthrene, anthracene, and 80% of fluoranthene and pyrene for *Penicillium ilerdanum*). In the case of genes isolated, *Sarocladium implicatum* strain NPDF1239-K3-F20 was the most resistant against PAHs but had relatively less degradation efficiency, whereas *Penicillium ilerdanum* strain NPDF1239-K3-F21 and *Aspergillus versicolor* strain NPDF190-C1-26 showed maximum PAH degradation of more than 75% for all four PAHs (Figure 12.3). The study concluded that the use of marine-derived fungal isolates can be effective for degrading PAHs and can also exhibit high tolerance against some of them (Mahajan et al., 2021).

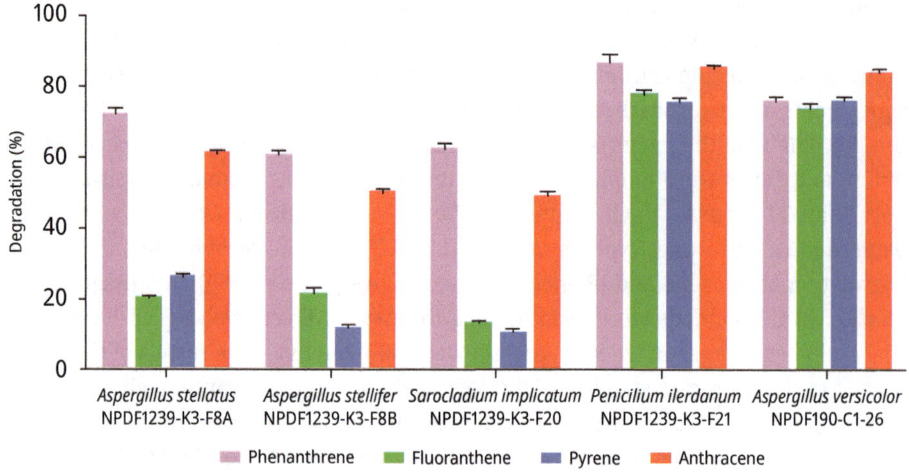

Figure 12.3: Percentage degradation of four PAHs by five fungal isolates as revealed by the GC-MS profile. (Naphthalene was not considered due to its high volatility.) Values are the mean of duplicates, and error bars represent SEM. Reprinted with permission from Mahajan et al. (2021).

The study conducted by Liu et al. (2021) examined the distribution of PAHs in sediments of two types of rivers: black odor rivers and those with relatively less black odor. Alongside PAHs, phthalates (PAEs) were also measured, and the linkage of the bacterial community was observed for both. It was revealed that black odor-containing rivers had significantly higher amounts of PAHs and PAEs (16 PAHs at 1405 µg/kg of dry weight and 6 PAEs at 7,120 µg/kg of dry weight) than rivers with less black odor (16 PAHs at 462 µg/kg of dry weight and 6 PAEs at 2,470 µg/kg of dry weight). Metagenomic analysis revealed the microbial community using 16S rRNA sequencing, which showed that at the class level, Gammaproteobacteria (35.6–40.3%), Anaerolineae (10.6–36.9%), Bacteroidia (9.7–24.0%), and Deltaproteobacteria (5.4–18.2%) were dominant. Further PCR quantification was done to check for functional genes, and it was found that 14 functional genes were identified, including three PAH degradation genes: nah, pdo1, and nidA, encoding the PAH dioxygenases, and three PAE degradation genes: cCEX (encoding carboxylesterase for PAE hydrolysis), phtAb, and phrC (encoding dioxygenases for benzene ring cleavage), along with sulfur, nitrogen, and phosphorus metabolism and reduction genes. The findings provided insight into the negative influence of PAHs and PAEs in urban rivers and how microbes can be used for their elimination and remediation of rivers (Liu et al., 2021).

In a study conducted by Han et al. (2021), samples were collected from the Three Gorges Reservoir in China to measure concentrations of PAHs. A total of 32 samples were isolated from 16 sites to study the distribution and bioremediation of PAHs from water level fluctuating regions. It was found that levels of PAH were 43.8–228.2 and 30.8–206.3 ng/g soil (dry weight) 19 ng/g soil under the high water level (HWL) and low water level (LWL) conditions, respectively. Five types of PAH were identified, which included 2, 3,4, 5, and 6 ring PAHs. Among these, 3 and 4 ring PAHs were significantly dominant, with percentages contributing 14.23 ± 6.81%, 9.88% ± 4.21%, and 13.15% ± 4.05% of the total PAHs under the HWL conditions and 12.43% ± 4.58%, 10.55% ± 4.19%, and 12.80% ± 3.34% under the LWL conditions, respectively. Genomic DNA extraction was performed, and further PCR amplification was performed on *bamA* and RHD genes, which are specific pathway biomarkers for PAHs degradation in anaerobic and aerobic pathways, respectively. Microcosm analysis revealed that *Proteobacteria* was the most dominant phylum found in both HWL and LWL conditions. Phylogenetic analysis based on the *bamA* gene revealed six genera including *Magnetospirillum, Rhodomicrobium, Azoarcus, Geobacter, Desulfobacterium*, and *Syntrophus*. The first three genera were able to degrade PAH in anaerobic conditions using nitrate as an electron 495 acceptor, and Geobacter has the ability to promote the biodegradation of PAHs by coupling iron reduction. Therefore, approximately 55.76% ± 497, 5.84%, and 67.27% ± 9.84% of functional bacteria had the ability to degrade PAHs by 498 coupling nitrate reduction under the HWL and LWL, respectively. The results of the study were able to provide us with an insight into PAH degradation and develop future strategies for efficient remediation of PAHs (Han et al., 2021).

In a study conducted by Zheng et al. (2021), a nano-Fe_3O_4-activated coke enhanced biosystem (FEBS) was applied for aromatic organics treatment in coal pyrolysis wastewater. Wastewater was synthetically created by phenolic products isolated from a coal chemical plant. Analysis of this wastewater revealed its constituents, which accounted for 80% organics along with PAH and aromatic organics. The COD and total phenols (Tph) of raw crude phenols were 2.6×10^3 and 1.4×10^3 g/L, respectively. Physiochemical analysis showed five major aromatic compounds present; these were indole, quinoline, 3,5-dimethylphenol, *p*-cresol, and phenol. Metagenomic analysis was performed with 16S rRNA sequencing, and it was found that *Alicycliphilus* (10% of all samples) and *Acidovorax* (5% of all samples) were the ones involved in the cleaving of aromatic compounds and naphthenic hydrocarbons. Incorporation of nano-Fe_3O_4-@activated coke in FEBS increased aromatic degradation efficiency along with *Alicycliphilus* and *Acidovorax*, which also assisted in aromatic degradation through an aerobic pathway (Zheng et al., 2021).

A study conducted by Zhang et al. (2017) created a system of biofilters for naphthenic acid (NA) degradation from oil sands process water (OSPW). The biofilter was designed so that OSPW was passed through the first biofilter, then through the ozone tank where it was ozonized, and finally through the second biofilter, which produced treated OSPW. The OSPW was passed through the filter directly or oxidized; oxidized NAs showed higher tolerance against the biofiltration process. A mild ozonation process increased the NA degradation efficiency to 84.8%. 16S rRNA sequencing revealed that Proteobacteria were most abundant (47.21%), followed by Actinobacteria and Bacteroidetes (13.24% and 12.59%, respectively). The presence of dominant bacterial phyla in the biofilter after the biofiltration-ozonation-biofiltration and ozonation-biofiltration of OSPW treatment was similar. Although Proteobacteria, Actinobacteria,, and Bacteroidetes were abundant, Rhodococcus had the highest contribution in the removal of NAs from OSPW (Zhang et al., 2019).

12.2.4 Metagenomic profiling of microbes involved in bioremediation of other organic pollutants

Xenobiotic compounds are chemical compounds produced by humans and are present in the environment unnaturally in high amounts. These jeopardize the environment, accounting for their long-term persistence in the soil and water bodies when dumped into the environment by industries through wastewaters in high amounts. They pose an existential danger to humans as they get added to the nutrient cycles and interfere in the process of decomposition, developing into recalcitrant for years. To reduce their impact and toxicity, biodegradation methods are applied and have been proven to be effective for the efficient breakdown of such compounds. A study on the effect of bioaugmentation vs. indigenous soil microcosms on the degradation of xenobiotic recalcitrant compounds in textile effluents was conducted by Patil et al (2020). Using MOTHUR and QIIME, taxonomic annotation was performed, and the dominant phyla were identified as follows: Proteobacteria (49–67%), Firmicutes (2–30%), Actinobacteria (9–24%), Acidobacteria (2–4%), and Chloroflexi (0.5–6%), which accounted for more than 79% of the total microbial diversity. Alpha diversity was calculated using Chao1 and Shannon indices, which revealed rich and even microbial diversity in the indigenous samples, which was increased in the bioaugmented microcosms. KEGG pathway analysis indicated that the indigenous microbial community was augmented with xenobiotic metabolism-associated pathways of various compounds like bisphenol, DDT, toluene, aminobenzoate, chlorobenzoate, methylnaphthalene, and xylene (Figure 12.4). Moreover, these pathways were overactivated in the bioaugmented microcosms, which enhanced the degradation of xenobiotic compounds in these samples. This eventually illustrated that enriching the soil bacterial consortium with biodegrading bacteria enhanced its biodegrading capacity. For instance, metagenes that encode enzymes responsible for the degradation of benzoate

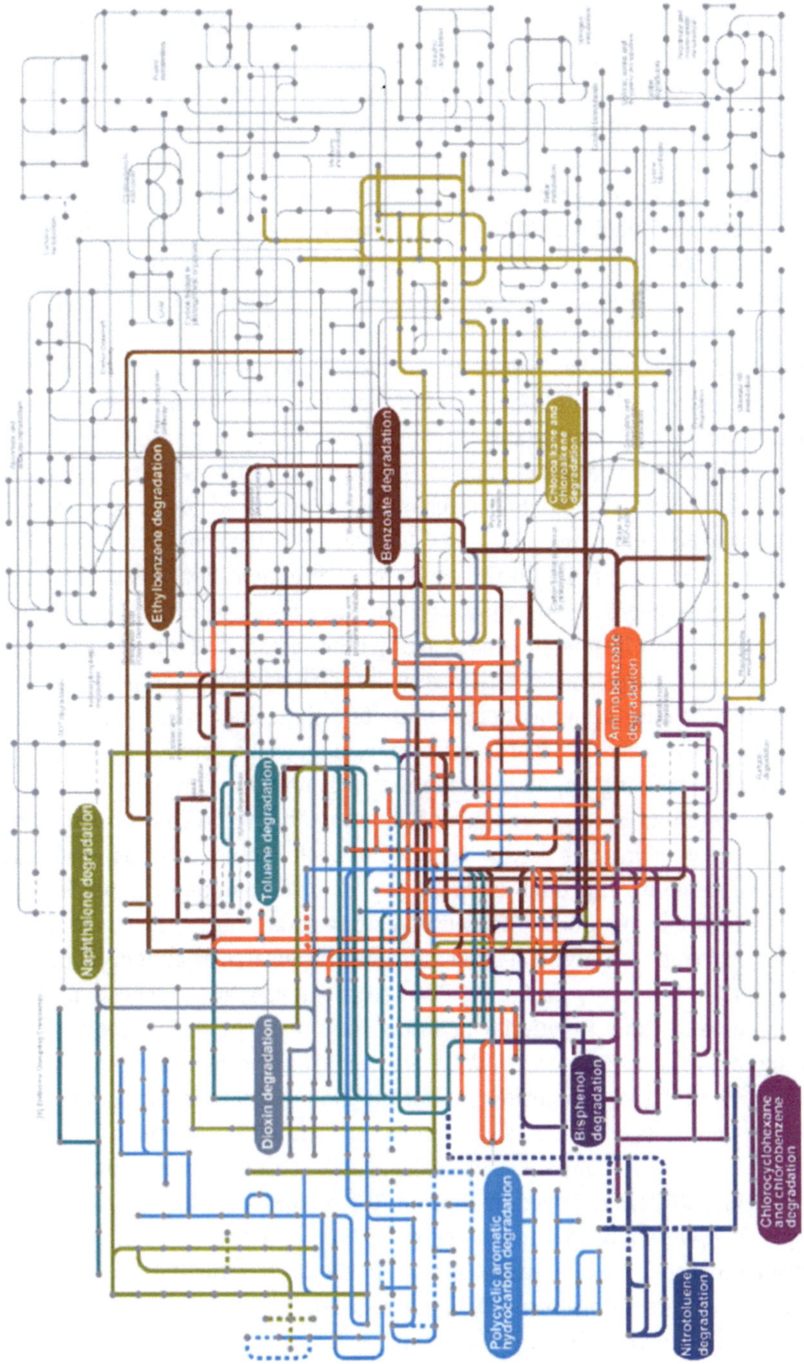

Figure 12.4: KEGG map of xenobiotic degradation pathways present in soil microcosms. Pathways highlighted in color were overrepresented (>50%) due to exposure to textile dye mixture (TDM) and textile industrial effluent (TIE). Reprinted with permission from Patil et al. (2020).

present in soil microcosms were −10,980, and in the sample augmented with textile effluent were 45,552. They concluded that even though indigenous microbial diversity has the potential to activate the bioremediation process of the pollutants present in the soil, bioaugmentation can enhance the process and carry it out in a more efficient and enhanced manner (Patil et al., 2020).

Another experiment was conducted by Bao et al. (2017), where oil-contaminated soil was analyzed for a functional study of the metagenome present in the sample responsible for the degradation of xenobiotic aromatic compounds. Soil samples that were collected from a polluted area near an oilfield were subjected to Illumina Sequencing to generate the sequence reads. These reads were taken for taxonomic assignments, where it was revealed that Proteobacteria (44.4%), Bacteroidetes (12.4%), Firmicutes (6.2%), and Actinobacteria (6.1%) were the most abundant phyla, among which Pseudomonadales and Alteromonadales (Proteobacteria), Flavobacteriales (Bacteroidetes), Bacillales (Firmicutes), and Actinomycetales (Actinobacteria) were the key players in xenobiotic aromatic compound degradation. Incidence-based coverage estimator (richness = 247) and Shannon index (diversity = 4.4, evenness = 0.80) were calculated for diversity. Functional interpretation of the genes that were identified was performed, which revealed various degradation pathways related to a broad set of compounds like xylene, toluene, DDT, and bisphenol, among many other compounds. These bacterial communities can play their roles as rich reservoirs for identifying the potential enzymes responsible for the biodegradation of these harmful compounds by employing the key bacterial metagenomes for bioremediation techniques (Bao et al., 2017).

Due to the ever-growing population, the use of xenobiotic compounds has increased, thereby resulting in large amounts of accumulation in the environment. Implementation of new techniques for the removal of such compounds has been employed to eliminate them. One such commonly produced compound is pesticides, which accounts for their low biodegradability. A study conducted by Zhang et al. on the AS samples collected from five full-scale pesticide wastewater treatment plants revealed the bacterial community harboring PDGs (pesticide degrading genes) and biodegradable genes and their functional profiling. Dominant phyla, Proteobacteria (58.02), and among genera *Thauera*, *Methyloversatilis*, *Paracoccus*, and many others were dominant. The relative composition of various genera found in the sample was temperature-dependent and showed variation patterns. In all the samples, PDGs for metamitron were the most abundant (1.284–4.504%), the second most abundant being nicosulfuron (0.501–1.878%), followed by linuron (0.204–1.087%). PDGs including *dhn*, *libA*, *ccdF*, and *hdl-I* were among a few other genes that were found dominant in AS. It was illustrated that the *dhn* gene might be responsible for the transformation of 2,3-dihydro-2,3-dihydroxy metamitron to 2,3-dihydroxy metamitron. The potential hosts of these genes were shown to be predominantly *Dechloromonas*, *Pseudomonas*, and *Thauera*. It was therefore concluded that the bacterial population was dependent on the pesticidal characteristics and variations in temperature due to seasonal change, and that further studies are required to study these potential genes at the metatranscriptome level to enhance the process of bioremediation of these harmful toxicant pesticidal residues present in WWTPs (Fang et al., 2018).

Volatile organic compounds (VOCs) are a class of compounds that are commonly used in paint, pharmaceutical, and other industries for their manufacture. They supposedly have carcinogenic properties, and their contamination in soil and water is associated with a high risk of cancer in humans. A study was conducted by Liang et al. (2020) on the removal of VOCs emitted from a TDWT plant using a combination of a Spray Tower (ST) and a biofilter. The VOCs were identified using GS-MS, an electronic nose (for odorous VOCs), and proton transfer reaction time-of-flight MS (PTR-TOF-MS, for low volatility compounds), which detected the presence of around 50 VOCs (concentration ranging between 1.26 and 2.79 mg/m^3) after a period of 90-day treatment of the effluent. The highest level of VOCs present included nitrogen- and oxygen-containing compounds (85.1%), aliphatic hydrocarbons (85.1%), aromatic hydrocarbons (6.3%), and halogenated hydrocarbons (1.5%). After a treatment period of 90 days, the levels of VOCs decreased significantly: NAOCC (1,630 to 547 µg/m^3, removal efficiency – 66.7%), AIHs (149 to 48 µg/m^3, 67.9%), AH (131 to 116 µg/m^3, 11.7%), and HHs (43.9 to 21.0 µg/m^3, 52.1%). The structure of the bacterial community present in the biofilter was identified, which gave the following results on the phyla level: Proteobacteria (80.5%), which indicated that it played a major role in the breakdown of VOCs. Removal of NAOCCs was attributed to phyla Proteobacteria, whereas at the genus level, a correlation of *Thiobacillus*, *Rhizobiales*, *Pseudarthrobacter*, Methylobacteriaceae, *Novosphingobium*, Hydrogenophilaceae, and Planctomycetaceae with AIHs and HHs was observed. The study concluded that biofilters can be an effective bioremediation strategy for the removal of VOCs, and the overall performance of the ST-biofilter is good as it can provide effective growth conditions for the survival of the bacterial diversity that can be employed in the degradation of VOCs (Liang et al., 2020). Another study was conducted by Watahiki et al. (2019), at a bioaugmentation site near a chemical company contaminated with chlorinated ethylenes (a type of VOC) like *cis*-dichloroethylene (cDCE), trichloroethylene (TCE), and vinyl chloride (VC), which are widely used in industries for processes like paint removers, vapor-degreasing, and metal cleaning. These compounds can accumulate and persist for decades, hence are a serious contaminant. The site was assessed for its ecological impact by using a multi-omics approach. The augmentation site, which was surrounded by wells, and *Rhodococcus jostii* strain RHA1, responsible for the degradation of chlorinated compounds like PCB and TCE, was cultured and injected into the bioaugmentation site, and the samples were collected on days 1, 6, 19, and 54 after injection. The DNA and RNA were extracted from the samples, and metagenome sequencing was performed for taxonomic and functional assessment. The α-diversity of the population was analyzed using MG-RAST. The sequencing results showed that the total number of 16S rRNA genes eventually decreased from day 0 (1.5×10^7 copies/mL) to day 54 (7.6×10^6 copies/mL). Comparative analysis revealed an increase in the proportion of *Rhodococcus* (including RHA1) – 0–76.5%, *Bacillus* – 8.5%, and others including *Mycobac-*

terium, Geobacillus, and *Paenibacillus*. There was also an increase in e-Proteobacteria and g-Proteobacteria, even though they were not injected into the site. The dominant gene expressed in RHA1 immediately after infection was biphenyl-degrading enzymes. It was noteworthy that these enzymes were activated in the presence of the substrate triclosan, which indicated that perhaps the aromatic compounds and alkane oxygenases might be responsible for the primary degradation of TCE and cDCE. The multiomics approach provided beneficial information on the bacterial consortium and predicted the pathways responsible for breaking down the contaminants present in bioremediation sites (Watahiki et al., 2019).

A study was conducted by Malematja et al. (2019) on the petrochemical wastes produced from the fuels generated by the Fischer-Tropsch (F-T) process to explore the microbial consortium responsible for the bioremediation of F-T wastewater containing volatile fatty acids (VFAs), used for chemical treatment, which results in the presence of increased dissolved components in the wastewaters. Here, specifically, an APH mixture was utilized for the study. The sample was supplemented with the APH mixture to enrich it with a microbial consortium that degrades APH. The enriched sample was incubated for 10 days. GC analysis of the treated samples showed a decline in APH concentrations in the samples: acetone (82.5%), propionic acid (85%), and hexanoic acid (75.8%) (here COD was used for measuring removal efficiencies). The Illumina Miseq platform revealed the dominance of Proteobacteria (55–92%), Actinobacteria (5–33%), and Firmicutes (0.08–9%) in samples after 10 days of incubation. Furthermore, an increase in pH from 6.5 to ~ 9.3 also validated the removal of VFAs from the sample. It was noteworthy that the bacterial consortium was present maximally in the sample from day one, which indicated that these bacteria thrive best at high concentrations and abundance of the substrate. *Rhodococcus* sp. produce rhamnolipid biosurfactants, which can help in the emulsification of the VFAs, eventually increasing the substrate availability and providing a nutrient-rich medium for other bacterial members. The study has demonstrated the role of the bacterial consortium in degrading harmful VFAs present in petrochemical wastewater (Malematja et al., 2020).

Perfluorinated compounds (PFCs) are another class of pollutants that are persistent in soil and bioaccumulate as a result of industrial and anthropogenic activities, synthesized chemicals, and are industrially used as surfactants, aqueous fire-fighting foams, and so on. Cai et al. (2020) studied how bacterial communities respond in the presence of PFCs and the effect on the soil microbial community under PFOS and PFOA stress by exploring the function of the bacterial community. Four samples containing PFOA (1,000 mg/kg), PFOS (3,000 mg/kg), PFOS (1,000 mg/kg), and a control were prepared, incubated for 10 days, and taken for sequencing. The diversity of the bacterial consortium was measured, which displayed low diversity and high soil bacterial richness, indicating that bacterial richness could increase under PFAS stress in soil. The most dominant phyla, accounting for >95.8%, included Proteobacteria (*Pseudomonas*, Betaproteobacteria-dominant), Acidobacteria, Actinobacteria, Chloroflexi, Firmicutes, and Gemmatimonadetes. These play a significant role in the degradation of these toxic substances through processes like defluorinating and shortening of carbon chains of PFCs. KEGG analysis revealed a higher abundance of carbon metabolism and glycolysis than the control sample. Through carbohydrate-active enzyme databases, enzymes identified (in decreasing order) were GTs, carbohydrate esterases, GHs, carbohydrate-binding modules, and polysaccharide lyases, especially in PFOS-treated samples. Fisher's exact test on class taxonomy revealed the highest gene counts of GTs and GHs (responsible for enzymatic breakdown to polymeric substrates). The studies are consistent with other related studies as well and suggest that the dominant bacterial species and their dominant pathways could be a promising tool for the bioremediation of PFCs (Cai et al., 2020).

Aquaculture, one of the growing food industries that provides 20% of the animal proteins consumed by humans, has a negative impact on the environment, as their culturing in the water bodies produces large amounts of excreta rich in phosphorus and nitrogen, contaminating the ponds and resulting in infections and species stress. To eliminate the contamination in these water bodies, metagenomic studies are conducted on the biofilms on the effluents to analyze the microbial diversity responsible for the biodegradation of the harmful compounds present in these aquacultures due to effluent discharge and to employ them as a bioremediation strategy to achieve sustainable aquaculture practices. A study was conducted by Mendoza et al. (2021) on whiteleg shrimp culture (*Litopenaeus vannamei*) effluents where metagenomic analysis was done on the biofilms present on these effluents to analyze the microbial diversity responsible for the nitrification process, and their implication toward a sustainable white leg shrimp culture. The effluents were collected, and the native microbiota was captured; the DNA that was extracted was amplified. Metagenomic sequencing of these DNA fragments identified 920 bacterial species belonging to 198 families, with several of them involved in the nitrification process including genera like *Nitrosococcus*, *Nitrosomonas*, and *Nitrospinae* (overall 12 genera) with sequences of genes including nitrous oxide reductase (*nosZ*) and nitrite reductase (*nirS*) that are involved in the reduction of ammonium to nitrate and nitrite to nitrate. This microbiota can be seeded and used as a natural agent in the bioremediation process to reduce the nitrogen load to tolerable levels and improve the quality of water (Mendoza et al., 2021).

Antibiotics have been used as therapeutics by humans for decades to combat diseases. However, due to excessive pollution, their presence has increased and they enter the environment through WWTPs. This has led to an increase in antibiotic resistance (AR) in bacteria, which is a serious issue for humans and animals. A study was conducted on the River Yamuna by Salam et al., which is rich in toxic organic and heavy metal load at two different times (June and November) where they analyzed the role of the bacterial consortium in the degradation of these toxicants. Taxonomic analysis using QIIME found OTUs to be 250,904 and 167,020 in June and November, respectively, and a high value of Shannon indices revealed higher phylogenetic diversity in the June sample. Both samples contained Proteobacteria (78% and 80%, respectively) at the most abundant levels in phylogenetic studies. The June sample also had

Bacteroides (14%) and Actinobacteria (3%). Genera *Aeromonas* and *Acinetobacter*, which harbor ARGs (antibiotic resistance genes), were found in the top 10 OTUs of both samples. Functional annotation revealed genes for penicillin binding, peptidoglycan synthesis inhibition, the AmpG-AmpR-AmpC pathway, the Opp-BlaI-BlaZ pathway, and RND efflux pumps for β-lactam efflux. The CARD database identified 662 (538 genes) types of ARGs. The most abundant genes were the rpoB gene (9–10%, resistance to rifampicin), followed by rpoC (~7%, resistance to daptomycin), the gyrA gene (~5%, resistant to fluoroquinolones), and the adeJ gene (1–4%, encoding multiple efflux AdeJ protein), which was present in *Acinetobacter*. There were 40 ARG categories found, out of which that the most abundant was ARG mutant/variant (20.2%), efflux pump providing AR (13.4%, suggesting resistance was acquired by bacteria through efflux mechanisms), and the fluoroquinolone resistance gene (11.7%). Various ARGs were found that were resistant to rifampin, β-lactams, macrolides, chloramphenicol, phenicols, tetracycline, aminocoumarin, and so on. The results imply the application of these resistomes for the bioremediation of antibiotics in the environment. The novel identified enzymes have huge potential to reduce the contamination of persistent antibiotics present in the environment (Mittal et al., 2019).

12.3 Conclusion

The study of bacterial consortia to understand and analyze the functioning of the microbiome, both individually and as a community, to biodegrade toxic persistent pollutants present in the ecosystem can be assisted with the help of metagenomic profiling to classify the bacteria for a deeper understanding of breakdown mechanisms. The use of microbe-assisted technologies for decontaminating industrial wastewaters is a feasible technique for remediating industrial wastewaters loaded with heavy toxicants that cannot be degraded naturally. Microbial studies with the help of NGS technologies are an emerging technology that can aid the process of bioremediation to improve upon the methods and develop new strategies. Various metabolic pathways that are involved in the breakdown of these compounds and the enzymes can be studied efficiently by taxonomic and functional annotation of the samples. The identified novel enzymes have the potential to eliminate contaminants that are persistently present in the environment.

Abbreviations

AAS	Atomic absorption spectroscopy
Eh	Redox potential
RDA	Redundancy analysis
OM	Organic matter
PCR-DGGE	Polymerase chain reaction denaturing gradient gel electrophoresis
ARGs	Antibiotic resistance genes
PDGs	Pesticide degrading genes
KEGG	Kyoto Encyclopedia of Genes and Genomes
AR	Antibiotic resistance
APH	Acetone, propionic acid, and hexanoic acid
PFOS	Perfluorooctane sulphonate
PFOA	Perfluorooctanoic acid
AS	Activated sludge
SM	Soil mixture
ASME	Augmented soil mixture
SME	Soil with textile effluent
ASM	Augmented soil mixture
ASMD	Augmented soil mixture with dye
DDT	Dichlorodiphenyltrichloroethane
PAH	Polyaromatic hydrocarbons
PAE	Phthalates
BaP	Benzo[a]pyrene
COD	Chemical oxygen demand
BOD	Biological oxygen demand
ASP	Activated sludge process
DAF	Dissolved air flotation
GC-MS	Gas chromatography-mass spectrometry
OSPW	Oil sand process water
HRT	Hydraulic retention time
TTC	2,3,5 Triphenyl tetrazolium chloride
PTR-TOF-MS	Proton transfer reaction time-of-flight mass spectrometry
CETP	Common effluent treatment plant
ETP	Effluent treatment

References

Akinpelu, E. A., Ntwampe, S. K. O., Fosso-Kankeu, E., Nchu, F., & Angadam, J. O. (2021). Performance of microbial community dominated by *Bacillus* spp. in acid mine drainage remediation systems: A focus on the high removal efficiency of SO_4^{2-}-, Al^{3+}, Cd^{2+}, Cu^{2+}, Mn^{2+}, Pb^{2+}, and Sr^{2+}. *Heliyon*, Elsevier Ltd. *7*(6), e07241. 10.1016/j.heliyon.2021.e07241

Bao, Y. J., Xu, Z., Li, Y., Yao, Z., Sun, J., & Song, H. (2017). High-throughput metagenomic analysis of petroleum-contaminated soil microbiome reveals the versatility in xenobiotic aromatics metabolism. *Journal of Environmental Sciences (China)*, Elsevier B.V. *56*, 25–35. 10.1016/j.jes.2016.08.022

Barak, H., Brenner, A., Sivan, A., & Kushmaro, A. (2020). Temporal distribution of microbial community in an industrial wastewater treatment system following crash and during recovery periods. *Chemosphere*, Elsevier Ltd. *258*, 127271. 10.1016/j.chemosphere.2020.127271

Bhakta, J. N., Lahiri, S., Bhuiyna, F. A., Rokunuzzaaman, M., Ohonishi, K., Iwasaki, K., & Jana, B. B. (2018). Profiling of heavy metal(loid)-resistant bacterial community structure by metagenomic-DNA fingerprinting using PCR–DGGE for monitoring and bioremediation of contaminated environment. *Energy, Ecology and Environment*, Joint Center on Global Change and Earth System Science of the University of Maryland and Beijing Normal University *3*(2), 102–109. 10.1007/s40974-017-0079-2

Cai, Y., Chen, H., Yuan, R., Wang, F., Chen, Z., & Zhou, B. (2020). Metagenomic analysis of soil microbial community under PFOA and PFOS stress. *Environmental Research*, Elsevier Inc. *188*, 109838. 10.1016/j.envres.2020.109838

Chaudhari, A. U., Paul, D., Dhotre, D., & Kodam, K. M. (2017). Effective biotransformation and detoxification of anthraquinone dye reactive blue 4 by using aerobic bacterial granules. *Water Research*, Elsevier B.V. *122*, 603–613. 10.1016/j.watres.2017.06.005

Chaudhary, K., Saraswat, P. K., & Khan, S. (2019). Improvement in fluoride remediation technology using GIS based mapping of fluoride contaminated groundwater and microbe assisted phytoremediation. *Ecotoxicology and Environmental Safety*, Elsevier Inc. *168*(June 2018), 164–176. 10.1016/j.ecoenv.2018.10.007

Fang, H., Zhang, H., Han, L., Mei, J., Ge, Q., Long, Z., & Yu, Y. (2018). Exploring bacterial communities and biodegradation genes in activated sludge from pesticide wastewater treatment plants via metagenomic analysis. *Environmental Pollution*, Elsevier Ltd. *243*, 1206–1216. 10.1016/j.envpol.2018.09.080

Guo, G., Liu, C., Hao, J., Tian, F., Ding, K., Zhang, C., Yang, F., Liu, T., Xu, J., & Guan, Z. (2021). Development and characterization of a halo-thermophilic bacterial consortium for decolorization of azo dye. *Chemosphere*, Elsevier Ltd. *272*, 129916. 10.1016/j.chemosphere.2021.129916

Han, X., Wang, F., Zhang, D., Feng, T., & Zhang, L. (2021). Nitrate-assisted biodegradation of polycyclic aromatic hydrocarbons (PAHs) in the water-level-fluctuation zone of the Three Gorges Reservoir, China: Insights from in situ microbial interaction analyses and a microcosmic experiment. *Environmental Pollution*, Elsevier. *268*, 115693. https://doi.org/10.1016/j.envpol.2020.115693

Kachienga, L., Jitendra, K., & Momba, M. (2018). Metagenomic profiling for assessing microbial diversity and microbial adaptation to degradation of hydrocarbons in two South African petroleum-contaminated water aquifers /631/80 /631/326 /704/172 /13 /13/89 /82 /82/58 /82/75 /82/80 /82/83 / 82/16 article. *Scientific Reports*, Springer US. *8*(1), 1–6. 10.1038/s41598-018-25961-0

Kim, E., Yulisa, A., Kim, S., & Hwang, S. (2020). Monitoring microbial community structure and variations in a full-scale petroleum refinery wastewater treatment plant. *Bioresource Technology*, Elsevier. *306* (March), 123178. 10.1016/j.biortech.2020.123178

Krishnamoorthy, R., Choudhury, A. R., Jose, P. A., Suganya, K., Senthilkumar, M., Prabhakaran, J., Gopal, N. O., et al. (2021). Long-term exposure to azo dyes from textile wastewater causes the abundance of Saccharibacteria population. *Applied Sciences (Switzerland)*, *11*(1), 1–8. 10.3390/app11010379

Krishnaswamy, V. G., Aishwarya, S., & Kathawala, T. M. (2020). Extrication of the microbial interactions of activated sludge used in the textile effluent treatment of anaerobic reactor through metagenomic profiling. *Current Microbiology*, Springer US. *77*(9), 2496–2509. 10.1007/s00284-020-02020-4

Kumar, D., Patel, Z., Pandit, P., Pandit, R., Patel, A., Joshi, M., & Joshi, C. (2021). Textile industry wastewaters from Jetpur, Gujarat, India, are dominated by Shewanellaceae, Bacteroidaceae, and Pseudomonadaceae harboring genes encoding catalytic enzymes for textile dye degradation. *Frontiers in Environmental Science*, *9*(September), 1–15. 10.3389/fenvs.2021.720707

Kumar, V., & Chandra, R. (2020). Metagenomics analysis of rhizospheric bacterial communities of *Saccharum arundinaceum* growing on organometallic sludge of sugarcane molasses-based distillery. *3 Biotech*, Springer International Publishing. *10*(7), 1–18. 10.1007/s13205-020-02310-5

Lee, H., Lee, D. W., Kwon, S. L., Heo, Y. M., Jang, S., Kwon, B. O., Khim, J. S., Kim, G. H., & Kim, J. J. (2019). Importance of functional diversity in assessing the recovery of the microbial community after the Hebei Spirit oil spill in Korea. *Environment International*, Elsevier. *128*(May), 89–94. 10.1016/j.envint.2019.04.039

Liang, Z., Wang, J., Zhang, Y., Han, C., Ma, S., Chen, J., Li, G., & An, T. (2020). Removal of volatile organic compounds (VOCs) emitted from a textile dyeing wastewater treatment plant and the attenuation of respiratory health risks using a pilot-scale biofilter. *Journal of Cleaner Production*, Elsevier Ltd. *253*, 120019. 10.1016/j.jclepro.2020.120019

Liu, Y., Huang, Y. H., Lü, H., Li, H., Li, Y. W., Mo, C. H., & Cai, Q. Y. (2021). Persistent contamination of polycyclic aromatic hydrocarbons (PAHs) and phthalates linked to the shift of microbial function in urban river sediments. *Journal of Hazardous Materials*, Elsevier B.V. *414*(February), 125416. 10.1016/j.jhazmat.2021.125416

Lukhele, T., Nyoni, H., Mamba, B. B., & Msagati, T. A. M. (2021). Unraveling bacterial diversity in oil refinery effluents. *Archives of Microbiology*, Springer Berlin Heidelberg. *203*(3), 1231–1240. 10.1007/s00203-020-02062-z

Mahajan, M., Manek, D., Vora, N., Kothari, R. K., Mootapally, C., & Nathani, N. M. (2021). Fungi with high ability to crunch multiple polycyclic aromatic hydrocarbons (PAHs) from the pelagic sediments of Gulfs of Gujarat. *Marine Pollution Bulletin*, Elsevier Ltd. *167*(March), 112293. 10.1016/j.marpolbul.2021.112293

Malematja, T. P., Ijoma, G. N., Selvarajan, R., & Matambo, T. (2020). Revealing the bacterial community profiles during the degradation of acetone, propionic and hexanoic acids-components of wastewater from the Fischer-Tropsch process. *International Microbiology*, *23*(2), 313–324. 10.1007/s10123-019-00106-z

Mendoza, O., Pretell, K., Ramírez, B., Sandoval, J., Caballero, J. L., & Dorado, G. (2021). Metagenomic analyses of biofilms on whiteleg shrimp (*Litopenaeus vannamei*) effluents: Implications for worldwide aquaculture bioremediation and environmental sustainability in the current trend of climate change and global warming–state of the art and exp. *Reviews in Fisheries Science and Aquaculture*, Taylor & Francis. *29*(4), 431–444. 10.1080/23308249.2020.1820720

Mishra, S., Chen, S., Saratale, G. D., Saratale, R. G., Ferreira, L. F. R., Bilal, M., & Bharagava, R. N. (2021). Reduction of hexavalent chromium by *Microbacterium paraoxydans* isolated from tannery wastewater and characterization of its reduced products. *Journal of Water Process Engineering*, Elsevier Ltd. *39* (October), 101748. 10.1016/j.jwpe.2020.101748

Mittal, P., Pk, V. P., Dhakan, D. B., Kumar, S., & Sharma, V. K. (2019). Metagenome of a polluted river reveals a reservoir of metabolic and antibiotic resistance genes. *Environmental Microbiomes*, *14*(1), 1–12. 10.1186/s40793-019-0345-3

Muangchinda, C., Rungsihiranrut, A., Prombutara, P., Soonglerdsongpha, S., & Pinyakong, O. (2018). 16S metagenomic analysis reveals adaptability of a mixed-PAH-degrading consortium isolated from crude oil-contaminated seawater to changing environmental conditions. *Journal of Hazardous Materials*, Elsevier. *357*(May), 119–127. 10.1016/j.jhazmat.2018.05.062

Muangchinda, C., Yamazoe, A., Polrit, D., Thoetkiattikul, H., Mhuantong, W., Champreda, V., & Pinyakong, O. (2017). Biodegradation of high concentrations of mixed polycyclic aromatic hydrocarbons by indigenous bacteria from a river sediment: A microcosm study and bacterial community analysis. *Environmental Science and Pollution Research*, *24*(5), 4591–4602. 10.1007/s11356-016-8185-9

Pandit, P. R., Kumar, R., Kumar, D., Patel, Z., Pandya, L., Kumar, M., & Joshi, C. (2021). Deciphering the black box of microbial community of common effluent treatment plant through integrated metagenomics: Tackling industrial effluent. *Journal of Environmental Management*, Elsevier Ltd. *289*(October 2020), 112448. 10.1016/j.jenvman.2021.112448

Patil, S. M., Suryavanshi, M. V., Chandanshive, V. V., Kurade, M. B., Govindwar, S. P., & Jeon, B. H. (2020). Regeneration of textile wastewater deteriorated microbial diversity of soil microcosm through bioaugmentation. *Chemical Engineering Journal*, Elsevier. *380*(August 2019), 122533. 10.1016/j.cej.2019.122533

Prakash, A. A., Rajasekar, A., Sarankumar, R. K., AlSalhi, M. S., Devanesan, S., Aljaafreh, M. J., Govarthanan, M., & Sayed, S. R. M. (2021). Metagenomic analysis of microbial community and its role in bioelectrokinetic remediation of tannery contaminated soil. *Journal of Hazardous Materials*, Elsevier B.V. *412*(January), 125133. 10.1016/j.jhazmat.2021.125133

Puthusseri, R. M., Nair, H. P., Johny, T. K., & Bhat, S. G. (2021). Insights into the response of mangrove sediment microbiomes to heavy metal pollution: Ecological risk assessment and metagenomics perspectives. *Journal of Environmental Management*, Elsevier Ltd. *298*(August), 113492. 10.1016/j.jenvman.2021.113492

Rathour, R., Jain, K., Madamwar, D., & Desai, C. (2021). Performance and biofilm-associated bacterial community dynamics of an upflow fixed-film microaerophilic-aerobic bioreactor system treating raw textile effluent. *Journal of Cleaner Production*, *295*. 10.1016/j.jclepro.2021.126380

Sánchez-Reyes, A., Bretón-Deval, L., Mangelson, H., Salinas-Peralta, I., & Sanchez-Flores, A. (2022). Hi-C deconvolution of a textile dye–related microbiome reveals novel taxonomic landscapes and links phenotypic potential to individual genomes. *International Microbiology*, Springer International Publishing. *25*(1), 99–110. 10.1007/s10123-021-00189-7

Santhanarajan, A. E., Rhee, C. Y., Sul, W. J., Yoo, K., Seong, H. J., Kim, H. G., & Koh, S. C. (2022). Transcriptomic analysis of degradative pathways for azo dye acid blue 113 in *Sphingomonas melonis* B-2 from the dye wastewater treatment process. *Microorganisms*, *10*(2), 1–17. 10.3390/microorganisms10020438

Selvarajan, R., Sibanda, T., Venkatachalam, S., Kamika, I., & Nel, W. A. J. (2018). Industrial wastewaters harbor a unique diversity of bacterial communities revealed by high-throughput amplicon analysis. *Annals of Microbiology*, *68*(7), 445–458. 10.1007/s13213-018-1349-8

Sharma, P., Tripathi, S., & Chandra, R. (2021). Metagenomic analysis for profiling of microbial communities and tolerance in metal-polluted pulp and paper industry wastewater. *Bioresource Technology*, Elsevier Ltd. *324*(January), 124681. 10.1016/j.biortech.2021.124681

Tripathi, S., Purchase, D., Al-Rashed, S., & Chandra, R. (2022). Microbial community dynamics and their relationships with organic and metal pollutants of sugarcane molasses-based distillery wastewater sludge. *Environmental Pollution*, Elsevier Ltd. *292*(PA), 118267. 10.1016/j.envpol.2021.118267

Tripathi, S., Sharma, P., & Chandra, R. (2021). Degradation of organometallic pollutants of distillery wastewater by autochthonous bacterial community in biostimulation and bioaugmentation process. *Bioresource Technology*, Elsevier Ltd. *338*(July), 125518. 10.1016/j.biortech.2021.125518

Verma, S. K., & Sharma, P. C. (2020). NGS-based characterization of microbial diversity and functional profiling of solid tannery waste metagenomes. *Genomics*, Elsevier. *112*(4), 2903–2913. 10.1016/j.ygeno.2020.04.002

Watahiki, S., Kimura, N., Yamazoe, A., Miura, T., Sekiguchi, Y., Noda, N., Matsukura, S., et al. (2019). Ecological impact assessment of a bioaugmentation site on remediation of chlorinated ethylenes by multi-omics analysis. *Journal of General and Applied Microbiology*, *65*(5), 225–233. 10.2323/jgam.2018.10.003

Wu, H., Li, Y., Zhang, W., Wang, C., Wang, P., Niu, L., Du, J., & Gao, Y. (2019). Bacterial community composition and function shift with the aggravation of water quality in a heavily polluted river. *Journal of Environmental Management*, *237*(June 2018), 433–441. 10.1016/j.jenvman.2019.02.101

Wu, H., Wang, M., Zhu, S., Xie, J., Preis, S., Li, F., & Wei, C. (2019). Structure and function of microbial community associated with phenol co-substrate in degradation of benzo[a]pyrene in coking wastewater. *Chemosphere*, Elsevier Ltd. *228*, 128–138. 10.1016/j.chemosphere.2019.04.117

Xu, W., Zhang, Y., Cao, H., Sheng, Y., Li, H., Li, Y., Zhao, H., & Gui, X. (2018). Metagenomic insights into the microbiota profiles and bioaugmentation mechanism of organics removal in coal gasification wastewater in an anaerobic/anoxic/oxic system by methanol. *Bioresource Technology*, *264*, 106–115. 10.1016/j.biortech.2018.05.064

Yan, C., Wang, F., Geng, H., Liu, H., Pu, S., Tian, Z., Chen, H., Zhou, B., Yuan, R., & Yao, J. (2020). Integrating high-throughput sequencing and metagenome analysis to reveal the characteristic and resistance mechanism of microbial community in metal contaminated sediments. *Science of the Total Environment*, Elsevier B.V. *707*, 136116. 10.1016/j.scitotenv.2019.136116

Zhang, J., Zhu, C., Guan, R., Xiong, Z., Zhang, W., Shi, J., Sheng, Y., et al. (2017). Microbial profiles of a drinking water resource based on different 16S rRNA V regions during a heavy cyanobacterial bloom in Lake Taihu, China. *Environmental Science and Pollution Research*, *24*(14), 12796–12808. 10.1007/s11356-017-8693-2

Zhang, L., Zhang, Y., & Gamal El-Din, M. (2019). Integrated mild ozonation with biofiltration can effectively enhance the removal of naphthenic acids from hydrocarbon-contaminated water. *Science of the Total Environment*, Elsevier B.V. *678*, 197–206. 10.1016/j.scitotenv.2019.04.302

Zheng, M., Han, H., Shi, J., Zhang, Z., Ma, W., & Xu, C. (2021). Metagenomic analysis of aromatic ring-cleavage mechanism in nano-Fe_3O_4-activated coke enhanced bio-system for coal pyrolysis wastewater treatment. *Journal of Hazardous Materials*, Elsevier B.V. *414*(February), 125387. 10.1016/j.jhazmat.2021.125387

Index

https://doi.org/10.1515/9783110793635-013

Also of interest

Extremophiles.
A Paradox of Nature with Biotechnological Implications
Shah, Dey (Eds.), 2022
ISBN 978-3-11-078837-2, e-ISBN 978-3-11-078848-8

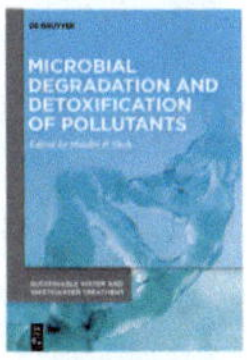

Microbial Degradation and Detoxification of Pollutants.
Shah (Ed.), 2023
Life in Extreme Environments Volume 6
ISBN 978-3-11-074327-2, e-ISBN 978-3-11-074362-3

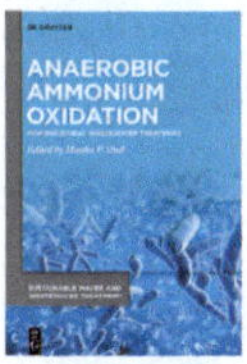

Anaerobic Ammonium Oxidation.
For Industrial Wastewater Treatment
Shah (Ed.), 2023
ISBN 978-3-11-077992-9, e-ISBN 978-3-11-078009-3

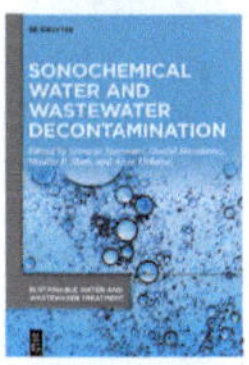

Sonochemical Water and Wastewater Decontamination.
Merouani, Hamdaoui, Shah, Dehane (Eds.), 2024
ISBN 978-3-11-113791-9, e-ISBN 978-3-11-113794-0

Smart Wastewater Technologies.
For Industrial Wastewater Treatment
Shah (Ed.), 2025
ISBN 978-3-11-077993-6, e-ISBN 978-3-11-078005-5

About the series

Sustainable Water and Wastewater Treatment discusses the recent advances in water and wastewater treatment research and processes covering bio-remediation, bio-degradation, molecular approaches, electro-biochemical, and sonochemical strategies. It also evaluates the possible applications of these corrective strategies to remove toxic pollutants from the environment.

This series describes the limitations and challenges of wastewater treatments. In addition, it covers various advanced and innovative technologies to remove toxic and hazardous pollutants present in wastewater in a sustainable manner. Moreover, it considers the application of mathematical modeling and different emerging molecular tools to wastewater treatment. Particular attention is given to water reuse and recovery of added-value products from wastewater since it would contribute to a circular and bio-based economy.

Editor-in-Chief
Maulin P. Shah, Enviro Technology Ltd., India

Please send any book proposals to Maulin P. Shah.